Techniques for Downstream Process for Biologic Drugs and Vaccines

Developments in Applied Microbiology and Biotechnology

Techniques for Downstream Process for Biologic Drugs and Vaccines

Basanta Kumara Behera

Advanced Centre for Biotechnology, M.D. University, Rohtak, Haryana, India

Director, Department of Biotechnology, MITS Gwalior, Gwalior, Madhya Pradesh India

Head, Centre for Bioenergy Technology, Amity University, Noida, Uttar Pradesh, India

ELSEVIER

ACADEMIC PRESS

An imprint of Elsevier

Academic Press is an imprint of Elsevier
125 London Wall, London EC2Y 5AS, United Kingdom
525 B Street, Suite 1650, San Diego, CA 92101, United States
50 Hampshire Street, 5th Floor, Cambridge, MA 02139, United States
The Boulevard, Langford Lane, Kidlington, Oxford OX5 1GB, United Kingdom

ISBN 978-0-443-19157-2

For information on all Academic Press publications
visit our website at https://www.elsevier.com/books-and-journals

Publisher: Stacy Masucci
Acquisitions Editor: Kattie Washington
Editorial Project Manager: Billie Jean Fernandez
Production Project Manager: Sajana Devasi P K
Cover Designer: Christian Bilbow

Typeset by STRAIVE, India

Contents

About the author

Basanta Kumara Behera was a professor of biotechnology at three distinguished Indian universities, where he was regularly associated with teaching and research for postgraduate-level courses on topics related to basic biotechnology, industrial biotechnology, biomass process technology development, biopharmaceuticals, microbial process development, drug designing, bioenergy management, and biomass processing technology since 1978. His career included a number of academic, government, and industrial positions ranging from assistant professor to professor and technical adviser in nationally and internationally reputed industries. In 2009, he joined an MNC company as a full-time adviser for specialty chemicals production and drug design through microbial process technology. Professor Behera is associated with nationally and internationally reputed companies as a technical adviser for biopharmaceutical product manufacturing under cGMP norms. He has the credit of authoring several books published by CRC Press and Elsevier Inc. (Academic Press), United States; Springer Nature Verlag, Germany; CABI, United Kingdom.

CHAPTER 1

Preharvesting conditions and planning for downstream process

Abbreviations

ATMP	advanced therapy medical product
BCG	Bacillus Calmette-Guerin
CapEX	capital expenditures
CCFS	cell culture fermentation sciences
CDER	Centre for Drug Evaluation and Research
CHO	Chinese Hamster Ovary
CMO	Contract Manufacturing Organization
CQA	critical quality attribute
DFDD	dosage form design and development
GMP	good manufacturing practice
HTA	health technology assessment
LCS	logistics and clinical supply
PAC	phenylacetylcarbinol
PPS	purification process sciences
QbD	quality by design
QRM	quality risk management
TPP	target product profile
TPQP	target product quality profile
WHO	World Health Organization

1.1 Introduction

The process upstream ends with fermented broth containing the target product and other unwanted materials. The method of processing harvested broth for the isolation and purification of therapeutic bio-molecule is known as downstream process (Fig. 1.1). The therapeutic recombinant proteins or other biomolecules of interest are from biological sources, such as animal cells, insects, plant cells, and mammalian cells [1,2].

It is an essential process in the manufacturing of biopharmaceuticals, such as antibiotic, hormones (e.g., insulin and human growth hormone), antibodies (e.g., infliximab and abciximab), and vaccines; antibodies and enzymes used in diagnostics; industrial enzymes; natural fragrance; and flavor compounds.

The various sequential stages for downstream process include removal of insoluble and unwanted material, purification of crude extract, and polishing of the product (Fig. 1.1).

Techniques for Downstream process for Biologic Drugs and Vaccines. https://doi.org/10.1016/B978-0-443-19157-2.00002-9

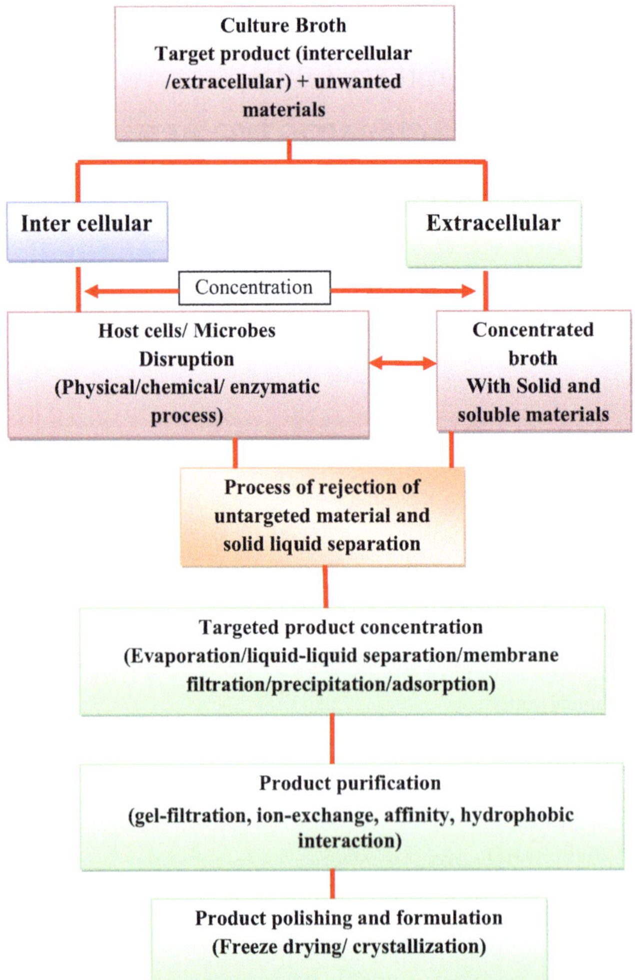

FIG. 1.1

An outline of important steps involved in downstream process.

The downstream process cost of fermented broth mainly depends on the preharvesting state of the broth generated at the end of the process. The freshly obtained fermented broth is, generally, in dilute condition containing the product of interest along with many unwanted materials like complex mixture of organic substances and residual unused nutrients, west by products, and foam (Fig. 1.1).

Downstream process is a highly expensive process, which plays a crucial role in final price fixation. In downstream process production of target biologic is expressed in term of titer value. A biopharmaceutical product can be produced batch wise or in a continuous process. It has been experienced that a shift to continue process from batch process not only saves time but also reduces the production cost. Following are the detailed sequential process for biopharmaceuticals (Fig. 1.2).

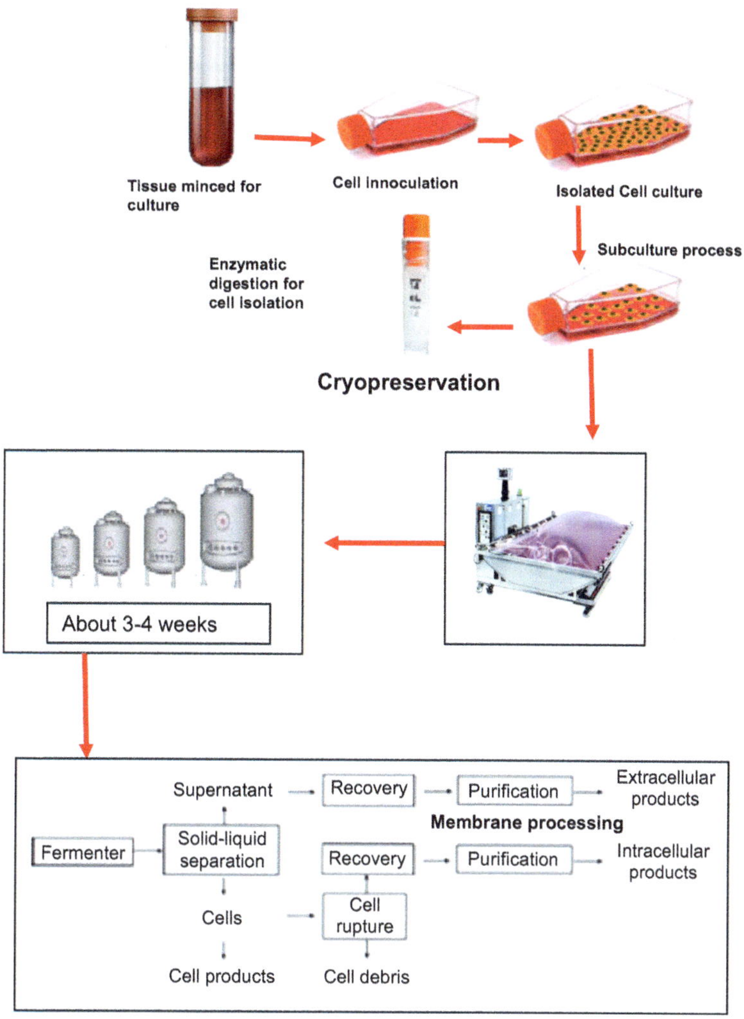

FIG. 1.2

Schematic presentation of downstream process to obtain biopharmaceutical.

1.2 History of biopharmaceutical development

Commonly, low molecular weight drugs are known as generic drug (after the expiry of patent), derived by chemical or partially biochemical methods. Because the active chemical substance is the same, the medical profile of generic is equivalent in performance [3].

The small size drugs are easy to deliver into the body, either as tablets or liquids. Acceptability of small drugs in the body is easy and rapid. Biologics are large-molecule biopharmaceuticals composed of more than 1300 amino acids and other molecules. Due to the high molecular weight, biologic drugs

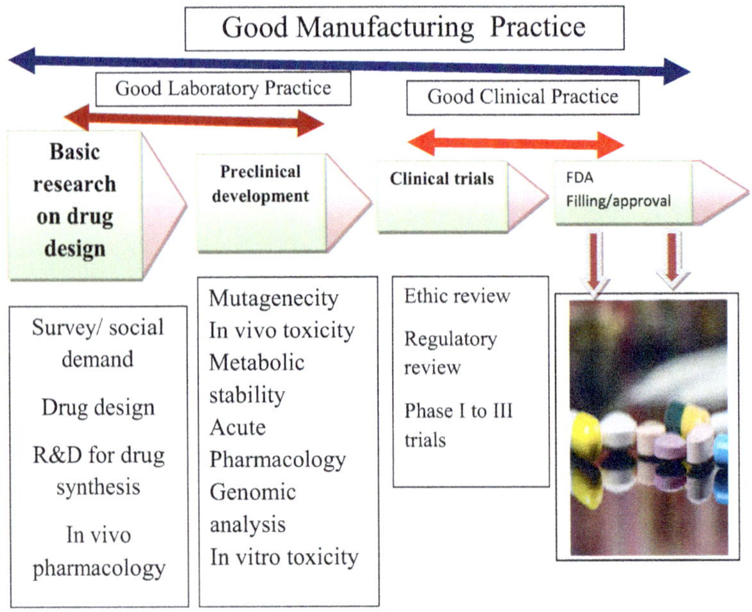

FIG. 1.3

Biopharmaceutical drug development processes.

are directly delivered into blood stream through intravenous injection or muscular injection. The overall process for biologic drugs development and production is complex, time consuming, and expensive (Fig. 1.3).

The large-molecule products include vaccines, blood, blood components, therapeutic gene, and stem cells. In 1978, the first recombinant DNA human insulin was prepared by David Goedded and his colleagues by utilizing and combining the insulin A- and B-chains expressed in *Escherichia coli*. Thereafter, Genentech and Lilly signed an agreement to commercialize rDNA insulin. In 1982, the first insulin utilizing rDNA technology, Humulin R (rapid) and N (NPH, intermediate-acting) were marketed. In due time, with the development of tools and technique in recombinant DNA technology, more and more recombinant proteins and peptides were discovered in both research laboratory and biopharmaceutical and biotechnology industries. Biologic drugs development has been a challenging task with highly complex steps for isolation, separation, and purification of product of interest. Time taken for development and marketing of biopharmaceutical drugs is about 10–12 years of intense research and development. The larger size therapeutic biomolecules such as nucleic acids [4,5], amino acids (peptides and proteins), therapeutic gene, small RNA (SiRNA), and vaccines are very promising strategies [5], but with through clinical protocols [6,7]. Proteins and peptides represent major classes of biopharmaceuticals [8].

Major kinds of biopharmaceuticals include:

- Blood factors (Factor VII and Factor IX)
- Thrombolytic agents (tissue plasminogen activator)

- Hormones (insulin, glucagon, growth hormones, gonadotropins)
- Hematopoietic growth factors (Erythropoietin, colony-stimulating factors)
- Interleukin-based products (interleukin-2)
- Vaccines (Hepatitis B surface antigen)
- Monoclonal antibodies
- Additional products (tumor necrosis factor, therapeutic enzymes)
- Interferon (interferon's -α,-β,-γ)

1.3 Biologic, biosimilar, and biobetter

Various terminologies have been used for the expression of biologics. Some regulatory agencies use the terms *biological medicinal products* or *therapeutic biological products* to express larger size biomolecules, such as protein or nucleic acid-based drugs differentiating them from products like blood, blood components, or vaccines, which are commonly isolated directly from a biological source. The advanced *therapy medical products* (ATMPs) is being used by the European Medicines Agency to express biopharmaceutical products.

Biopharmaceutical drugs are available in the market under the categories biologic, biobetter, and biosimilar (Fig. 1.4).

A biopharmaceutical, also known as biological product or biologic, is any pharmaceutical drug product manufactured from materials extracted from or semisynthesized from biological sources. It is interesting to note that by the introduction of a specific gene protein with specific amino acids sequence can be produced in different types of genetically modified host cells. But, the same protein produced by different manufacture processing is having different characteristics. So, in order to have differentiation in biopharmaceuticals the basic therapeutic protein is referred as biologic drug or reference medicine, where the following ones are denominated biosimilar. Biosimilar are different from biologic drug due to posttranslational modifications (glycosylation, phosphorylation) and different manufacturing process. The term biobetter refers to therapeutic macromolecules of the next generation having more efficient pharmacokinetic potential and structural stability [9–11].

The state of the art of separation of biopharmaceutical of interest is the most challenging process. The culture medium (broth) obtained after upstream process contains many impurities, besides the targeted therapeutic proteins. In the beginning of downstream process, the unwanted biomolecules and other extracellular products are separated. The overall process for downstream is consisting of multiple steps in order to remove host cells-related impurities (e.g., host cell protein and nucleic acids); process-related impurities (e.g., buffer, leached ligands, and antifoam); and product-related impurities (e.g., aggregate, clipped, and species) [12–14].

1.4 Status of biological product in harvested broth

The target biological product in fermented broth may be intercellular or extracellular. The recovery of a soluble product (MW range 500–2500 Da), such as an antibiotic, organic acid, or animal vaccine from fermentation broth takes several processing steps. Hence, it is desirable to apply strategies for isolation

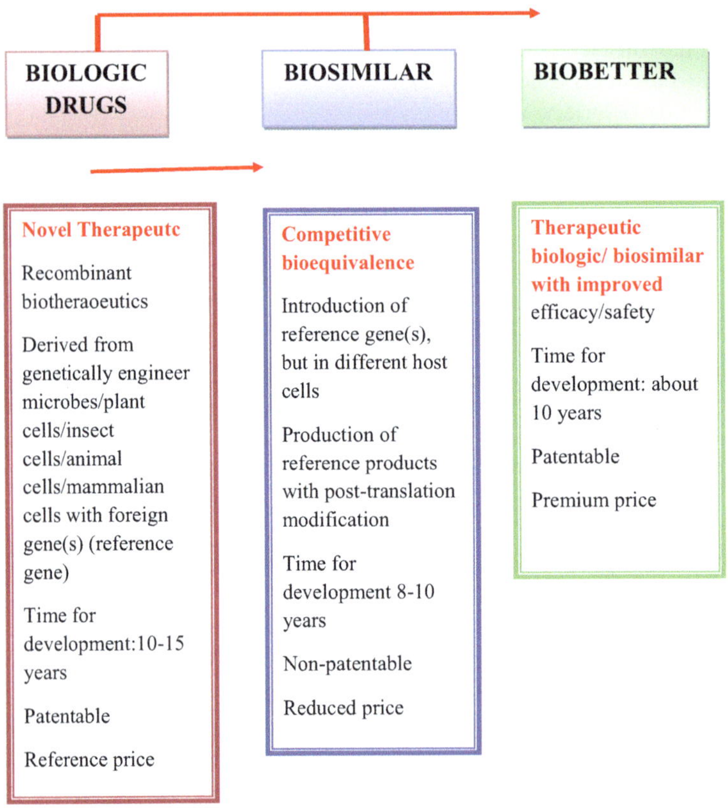

FIG. 1.4

Different between biologic, biosimilar, and biobetter.

of target biologic in maximum quantity, immediately with minimum time and less process cost. The recovery cost of targeted biologic is a variable factor ranging from as low as less than 15% to as high as 70% of the total manufacturing cost [15–21].

The extraction and purification of biologics like monoclonal antibody, recombinant therapeutic proteins can account for 80%–90% of the total processing costs [22].

1.4.1 Composition of harvested broth

The composition of fermented broth is mainly dependent on the type of medium and the nature of biocatalyst used for fermentation process.

A variety of media exists, but invariably contains a carbohydrate source, a nitrogen source, water, salts, and micronutrients. For example, in the production of wine, the medium is grape; in the production of bioethanol is an inexpensive carbon source; for the culture of animal cells serum is the most common medium, but other medium containing plasma, lymph, and amniotic fluid are also used. Generally, carbon sources are typically sugars or other carbohydrates. But in case of substrate transformations the source of carbon differs. For example, for vinegar production the source of carbon is alcohol or something else altogether. For large scale fermentation like ethanol production, inexpensive sources of carbohydrates, such as molasses, corn steep liquor [23], sugarcane juice, or sugar beet juice are used to minimize costs. In some of the specific fermentation process, purified glucose, sucrose, glycerol or other sugars, which reduce variation and help to ensure the purity of the final product are preferred as better carbon source.

Biological catalysts used in fermentation process also need nitrogen for synthesis of proteins, nucleic acids, and other cellular components [24]. On the basis of capability of microorganisms used in fermentation process, nitrogen may be provided as bulk proteins, such as soy meal; as predigested polypeptides such as peptone or tryptone; or as ammonia or nitrogen source. Cost is also an important factor in the choice of a nitrogen source.

The fermented broth also contains growth factors and trace nutrients to support the microorganisms, incapable of producing all the vitamins they require. Yeast extract is a common source of micronutrients and vitamins for fermentation media. Although the unrefined carbon and nitrogen sources contain trace elements such as iron, zinc, copper, manganese, molybdenum, and cobalt, but may to be added when purified carbon and nitrogen sources are used. Often, gas is formed in the fermentation process or sometimes gas is to be bubbled. This process tends to form a layer of foam, since fermentation broth typically contains a variety of foam-reinforcing proteins, peptides, or starches. So, antifoaming agent is used to prevent such foam forming process.

Production of biomass or microbial cells is sometimes the targeted product of fermentation. Examples include single cell protein, bakers yeast, lactobacillus, *E. coli*, and others.

A typical fermentation is associated with the followings steps [2]:

- Production of biomass (viable cellular material like yeasts)
- Production of extracellular metabolites
- Production of intracellular metabolites
- Transformation of substrate into product

These types are not necessarily disjoined from each other, but provide a framework for understanding the differences in approach. The biological catalysts used in fermentation are in general microorganisms, particularly bacteria, algae, and fungi, such as yeasts and molds. Cell cutlers from plants and animals, such as Chinese Hamster Ovary (CHO) and insect cells are also used for biologic drug production. So, broth obtained after fermentation process contains the deadly biomass of a specific biocatalysts. In addition, it contains target product, by-products, residual nutrient, cell debris, etc. The target product is available as intercellular or extracellular material but in small quantities.

1.4.2 Extracellular metabolite(s)

Metabolites that are produced during the growth phase of the organism are called primary metabolites. A common example is ethanol or lactic acid, produced during glycolysis (Fig. 1.5).

Citric acid is produced by some strains of *Aspergillus niger* as part of the citric acid cycle to acidify their environment and prevent competitors from taking over. Glutamate is produced by some *Micrococcus* species, and some Corynebacterium species produce lysine, threonine, tryptophan, and other amino acids.

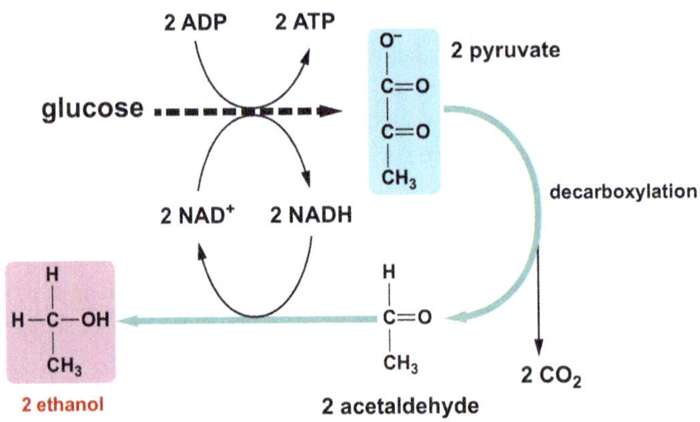

FIG. 1.5

Glycolysis process for the production of ethanol.

1.4.3 Secondary metabolites

Secondary metabolites are those produced during the stationary phase, called secondary metabolites (penicillin, cyclosporine A, gibberellins, and lovastatin). Mostly, monoclonal antibodies are produced during the early growth phase (3 days); however, most of the antibodies are produced during a later growth phase (3–10 days).

Typically, secondary metabolites are not produced in the presence of glucose or other carbon sources, which would encourage growth, and like primary metabolites, they are released into the surrounding medium without rupture of the cell membrane. Some cultures will not produce secondary metabolites in the presence of phosphate [25].

Earlier, *E. coli* was used for commercial level of production of biologics, but gradually industrialists preferred to use eukaryotic cells like CHO cells and insect cells for biopharmaceutical products [26,27].

1.4.4 Intracellular components

Variety of therapeutic proteins either in the form of enzymes (catalase, amylase, protease, pectinase, and cellulose) or therapeutic proteins (recombinant proteins, such as insulin, hepatitis B, vaccine, interferon, granulocyte colony-stimulating factors, and streptokinase) are commonly produced as intracellular product during fermentation process [26,28].

In order to obtain the intracellular product of interest, the cells are ruptured (lysed) at the end of fermentation, and the environment must be manipulated to maximize the amount of the product. In case of the target product is a therapeutic protein, it is to be separated from all other cellular proteins in the lysate to be purified.

1.4.5 Transformation of substrate

The traditional substrate for fermentation industries is sugar (cane or beet) or sugar-based byproducts, such as molasses. The process of substrate transformation involves the transformation of a specific compound into another, such as in case of phenyl acetylcarbinol, and steroid biotransformation, or the transformation of a raw material into a finished product. For example phenylacetylcarbinol (PAC or L-PAC) is an organic compound, which is known as a precursor in the synthesis of pharmaceuticals, such as ephedrine and pseudoephedrine.

A variety of strain of yeast such as *Candida utilis, Torulaspora delbrueckii*, or *Saccharomyces cerevisiae* are used as biocatalyst to produce -PAC by using benzaldehyde and dextrose as substrate, and the enzyme pyruvate decarboxylase. The yeast is left to grow for a period of time, after which the benzaldehyde is introduced into the broth. The yeast then ferments the benzaldehyde into -PAC. The majority of L-PAC is generated in pharmaceutical plants, as an intermediate precursor or in the production of pseudoephedrine.

1.5 Status of target bioproduct in the fermented broth

Currently, fermentation is widely used for the production of biologically derived biomolecules of wide variety of products ranging from therapeutic proteins, enzymes, antibiotics, antibodies, and other biomolecules. These products are produced at industrial-scale fermentation. Through fermentation process small molecule, steroids, and cytotoxic are also derived. Originally, the term "fermentation" was used to describe an anaerobic process to convert starch grain into alcohol, a process still used in first-generation biofuels. This process is pretty old and was used by ancient civilization thousands of years ago, as the origin of biotechnology. But, in due time the concept of fermentation changed (Table 1.1).

The economic success in fermentation technology could possible due to two significant achievements. The first was the development of large-scale sterile fermentation technology for the production of penicillin during the Second World War. The second was the development of genetic engineering tools for horizontal gene transfer, which has evolved from transferring a single gene (insulin) into a bacterium (*E. coli*) to transferring multiple genes.

Currently, biologic drugs are gaining tremendous demand in international market. These therapeutic biomolecules are present in small amount and highly sensitive to temperature and pH of surrounding microenvironment of fermented broth.

Table 1.1 Historical overview of fermentation over the last 350 years.

Period	Technology	Product
1670–1918	Empirical fermentation	Ethyl alcohol, acetic ABE
1918–1943	Deep vat (5 m+) fermentation	Vitamin C, ephedrine, organic acids
1943–1978	Large scale sterile fermentation technology	Antibiotics and steroids, amino acids, vitamins, industrial enzymes
1978–2012	Horizontal gene transfer, mammalian cell culture, high cell density culture	Recombinant proteins, monoclonal antibodies, many other different products
2012	Synthetic biology, bioinformatics	Personalized medicine, chemical manufacturing GMO plants widely accepted, artificial photosynthesis

1.6 Strategic development of biopharmaceutical production planning

With the approval of first recombinant human insulin in 1982, industrialization of biopharmaceutical gained a tremendous boost. In the year 2000, there were about 84 biopharmaceuticals in the global market. This figure increased to about 250, at the end of 2014 [29]. The revenue for biopharmaceutical companies within the United States, Canada, and Australia increased by 24% of earlier market, in the end of 2014 [30].

Since the past decade, capacity planning for multiple biopharmaceutical therapeutics has been in an accelerated state of development due to large network of manufacturing facilities, including contract manufacturers. With the introduction of production planning through a continuous process has reduced the risk of contamination and to some extend the production cost of biologic drugs. FDA's Centre for Drug Evaluation and Research (CDER) supports the pharmaceutical industry at every step of the process. The CDER provides scientific and regulatory advice needed to bring new therapies to the market. Each year, CDER approves a wide range of new drugs and biological products. Below is a listing of new biologics for the year 2020 (excluding vaccines, allergenic products, blood) given in the Table 1.2.

Table 1.2 Novel drug approvals for 2020.

No.	Drug name	Active ingredient	Approval date	FDA-approved use on approval date[a]
53	Gemtesa	Vibegron	12/23/2020	To treat overactive bladder
52	Ebanga	Ansuvimab-zykl	12/21/2020	To treat ebola
51	Orgovyx	Relugolix	12/18/2020	To treat advanced prostate cancer
50	Margenza	Margetuximab (anti-HER2 mAb)	12/16/2020	To treat HER2+ breast cancer
49	Klisyri	Tirbanibulin	12/14/2020	To treat actinic Keratosis of the face or scalp
48	Orladeyo	Berotralstat	12/3/2020	To treat patients with hereditary angioedema
47	Gallium 68 PSMA-11	Gallium 68 PSMA-11	12/1/2020	For detection and localization of prostate cancer
46	Danyelza	Naxitamab-gqgk	11/25/2020	To treat high-risk refractory or relapsed neuroblastoma
45	Imcivree	Setmelanotide	11/25/2020	To treat obesity and the control of hunger associated with pro-opiomelanocortin deficiency, a rare disorder that causes severe obesity that begins at an early age
44	Oxlumo	Lumasiran	11/23/2020	To treat hyperoxaluria type 1
43	Zokinvy	Lonafarnib	11/20/2020	To treat rare conditions related to premature aging
42	Veklury	Remdesivir	10/22/2020	To treat COVID-19
41	Inmazeb	Atoltivimab, maftivimab, and odesivimab-ebgn	10/14/2020	To treat ebola virus
40	Gavreto	Pralsetinib	9/4/2020	To treat nonsmall lung cancer

Table 1.2 Novel drug approvals for 2020—cont'd

No.	Drug name	Active ingredient	Approval date	FDA-approved use on approval date
39	Detectnet	Copper Cu 64 dotatate injection	9/3/2020	To help detect certain types of neuroendocrine tumors
38	Sogroya	Somapacitan-beco	8/28/2020	Growth hormone
37	Winlevi	Clascoterone	8/26/2020	To treat acne
36	Enspryng	Satralizumab-mwge	8/14/2020	To treat neuromyelitis optica spectrum disorder
35	Viltepso	Viltolarsen	8/12/2020	To treat Duchenne muscular dystrophy
34	Olinvyk	Oliceridine	8/7/2020	To manage acute pain in certain adults
33	Evrysdi	Risdiplam	8/7/2020	To treat spinal muscular atrophy
32	Lampit	Nifurtimox	8/6/2020	To treat Chagas disease in certain pediatric patients younger than age 18
31	Blenrep	Belantamab mafodotin-blmf	8/5/2020	To treat multiple myeloma
30	Monjuvi	Tafasitamab-cxix	7/31/2020	To treat relapsed or refractory diffuse large B-cell lymphoma
29	Xeglyze	Abametapir	7/24/2020	To treat head lice
28	Inqovi	Decitabine and cedazuridine	7/7/2020	To treat adult patients with myelodysplastic syndromes
27	Rukobia	Fostemsavir	7/2/2020	To treat HIV
26	Byfavo	Remimazolam	7/2/2020	For sedation
25	Dojolvi	Triheptanoin	6/30/2020	To treat molecularly long-chain fatty acid oxidation disorders
24	Zepzelca	Lurbinectedin	6/15/2020	To treat metastatic small cell lung cancer
23	Uplizna	Inebilizumab-cdon	6/11/2020	To treat neuromyelitis optica spectrum disorder
22	Tauvid	Flortaucipir F18	5/28/2020	Diagnostic agent for patients with Alzheimer's disease
21	Artesunate	Artesunate	5/26/2020	To treat severe malaria
20	Cerianna	Fluoroestradiol F18	5/20/2020	Diagnostic imaging agent for certain patients with breast cancer
19	Qinlock	Ripretinib	5/15/2020	To treat advanced gastrointestinal-stromal tumors
18	Retevmo	Selpercatinib	5/8/2020	To treat lung and thyroid cancers
17	Tabrecta	Capmatinib	5/6/2020	To treat patients with nonsmall cell lung cancer
16	Ongentys	Opicapone	4/24/2020	To treat patients with Parkinson's disease experiencing "off" episodes
15	Trodelvy	Sacituzumab govitecan-hziy	4/22/2020	To treat adult patients with metastatic triple-negative breast cancer who received at least two prior therapies for metastatic disease
14	Pemazyre	Pemigatinib	4/17/2020	To treat certain patients with cholangiocarcinoma, a rare form of cancer that forms in bile ducts
13	Tukysa	Tucatinib	4/17/2020	To treat advanced unresectable or metastatic HER2-positive breast cancer

Continued

Table 1.2 Novel drug approvals for 2020—cont'd

No.	Drug name	Active ingredient	Approval date	FDA-approved use on approval date
12	Koselugo	Selumetinib	4/10/2020	To treat neurofibromatosis type 1, a genetic disorder of the nervous system causing tumors to grow on nerves
11	Zeposia	Ozanimod	3/25/2020	To treat relapsing forms of multiple sclerosis
10	Isturisa	Osilodrostat	3/6/2020	To treat adults with Cushing's disease who either cannot undergo pituitary gland surgery or have undergone the surgery but still have the disease
9	Sarclisa	Isatuximab	3/2/2020	To treat multiple myloma
8	Nurtec ODT	Rimegepant	2/27/2020	To treat migraine
7	Barhemsys	Amisulpride	2/26/2020	To help prevent nausea and vomiting after surgery
6	Vyepti	Eptinezumab-jjmr	2/21/2020	For the preventive treatment of migraine in adults
5	Nexletol	Bempedoic acid	2/21/2020	To treat adults with heterozygous familial hypercholesterolemia or established atherosclerotic cardiovascular disease who require additional lowering of LDL-C
4	Pizensy	Lactitol	2/12/2020	To treat chronic idiopathic constipation (CIC) in adults
3	Tazverik	Tazemetostat	1/23/2020	To treat epithelioid sarcoma
2	Tepezza	Teprotumumab-trbw	1/21/2020	To treat thyroid eye disease
1	Ayvakit	Avapritinib	1/9/2020	To treat adults with unresectable or metastatic gastrointestinal stromal tumor (GIST)

[a]*Used in USA.*

The main target lying behind for developing biopharmaceutical company is to manufacture well-design biologic drugs, which can be body friendly of patients for restoring health and transforming the lives to a normal stage with minimum side effect. In order to develop quality biologic drugs, a biopharmaceutical company has to follow guide lines (Good Manufacturing Practice, GMP) as laid down by the World Health Organization (WHO) [1]. GMP for biologic drugs was first published by the WHO in 1992. This provides risk-based approaches for manufacturing biologic drugs. Brief information on WHO guidelines for biologic drugs is explained in the Table 1.3. The WHO guidelines are applicable to National regulatory authorities (NRAs) of any country, globally. This is to maintain harmony in quality products for human use that are risk-free.

The biopharmaceuticals are derived from living resources. So, these are prone to higher temperatures and variables pH. So, it is necessary to maintain maximum care during a manufacturing process and product marketing for end users. The World Health Organization (WHO) has framed suitable protocols for quality control from safety and security point of views:

- Selection of suitable host cell for the target biologic production.
- Maintenance in consistency for the production of biologics, and their preservation in cell bank.

Table 1.3 The WHO guidelines for manufacturing biologic drugs.

No.	Nature of assignment development/ manufacturing	Purpose
1	Biological product development and manufacturing	World Health Organization; 1992: Annex 1 (WHO Technical Report Series, No. 822; http://www.who.int/biologicals/publications/trs/areas/vaccines/gmp/WHO_TRS_822_A1.pdf?ua=1, accessed 8 November 2015)
2	Pharmaceutical products	2013: Annex 2 (WHO Technical Report Series, No. 986; http://www.who.int/medicines/areas/quality_safety/quality_assurance/TRS986annex2.pdf?ua=1, accessed 8 November 2015)
3	Manufacturing practices for sterile pharmaceutical products	2011: Annex 6 (WHO Technical Report Series, No. 961; http://www.who.int/medicines/areas/quality_safety/quality_assurance/GMPSterilePharmaceuticalProducts TRS961Annex6.pdf?ua=1, accessed 8 November 2015)
4	Active pharmaceutical ingredients	2010: Annex 2 (WHO Technical Report Series, No. 957; http://www.who.int/medicines/areas/quality_safety/quality_assurance/GMPActivePharmaceuticalIngredientsTRS957Annex2.pdf?ua=1, accessed 8 November 2015)
5	Good manufacturing practice for medicinal products	Part I, Annex 2. Manufacture of biological medicinal substances and products for human use. Pharmaceutical Inspection Convention and Pharmaceutical Inspection Co-operation Scheme (PIC/S); 1 March 2014 (http://www.fda.gov.ph/attachments/article/224762/pe-009-11-gmp-guide-xannexes.pdf, accessed 8 November 2015)
6	Good manufacturing practice for medicinal products	Part II. Pharmaceutical Inspection Convention and Pharmaceutical Inspection Co-operation Scheme (PIC/S); 1 March 2014 (http://www.medsafe.govt.nz/regulatory/Guideline/PE_009-8_GMP_Guide%20_Part_II_Basic_Requirements_for_API.pdf, accessed 8 November 2015)
7	EU guidelines for good manufacturing practice for medicinal products s for human and veterinary use	Annex 2: Manufacture of biological active substances and medicinal products for human use. Brussels: European Commission; 2013
8	EU guidelines for good manufacturing practice for medicinal products s for human and veterinary use	Part I. Chapter 6: Quality control. Brussels: European Commission; 2005
9	Current good manufacturing practice for finished pharmaceuticals	Code of Federal Regulations Title 21, Vol. 4, revised 1 April 2014. Silver Spring, MD: United States Food and Drug Administration; 2014 (http://www.accessdata.fda.gov/scripts/cdrh/cfdocs/cfcfr/CFRSearch.cfm?CFRPart=211&showFR=1, accessed 4 July 2015)
10	WHO good practices for pharmaceutical quality control laboratories	In: WHO Expert Committee on Specifications for Pharmaceutical Preparations: forty-fourth report. Geneva: World Health Organization; 2010: Annex 1 (WHO Technical Report Series, No. 957

Continued

Table 1.3 The WHO guidelines for manufacturing biologic drugs—cont'd

No.	Nature of assignment development/ manufacturing	Purpose
11	Good manufacturing practice for drugs (2010 revision)	Beijing: China Food and Drug Administration; 2011 (http://eng.sfda.gov.cn/WS03/CL0768/65113.html, accessed 8 November 2015)
12	WHO guidelines on good manufacturing practices for blood establishments	In: WHO Expert Committee on Specifications for Pharmaceutical Preparations: forty-fifth report. Geneva: World Health Organization; 2011: Annex 4 (WHO Technical Report Series, No. 961; http://www.who.int/bloodproducts/publications/GMP_Bloodestablishments.pdf?ua=1, accessed 2 February 2016)
13	WHO good distribution practices for pharmaceutical products	In: WHO Expert Committee on Specifications for Pharmaceutical Preparations: forty-fourth report. Geneva: World Health Organization; 2010: Annex 5 (WHO Technical Report Series, No. 957; http://www.who.int/medicines/areas/quality_safety/quality_assurance/GoodDistributionPracticesTRS957Annex5.pdf?ua=1, accessed 2 February 2016)
14	WHO guidelines on quality risk management	In: WHO Expert Committee on Specifications for Pharmaceutical Preparations: forty-seventh report. Geneva: World Health Organization; 2013: Annex 2 (WHO Technical Report Series, No. 981; http://www.who.int/medicines/areas/quality_safety/quality_assurance/Annex2TRS-981.pdf?ua=1, accessed 2 February 2016)

World Health Organization, Annex 1, WHO Technical Report Series, No. 822, 1992. http://www.who.int/biologicals/publications/trs/areas/vaccines/gmp/WHO_TRS_822_A1.pdf?ua=1 (Accessed 8 November 2015).

- Propagation of host cells with consistency in expression of foreign gene(s).
- Extraction, purification, and crystallization of target therapeutic through standard protocol approved by regulatory authority.
- Recombinant DNA/protein engineering development for target biologic.
- Hybridoma technique development.
- Packing, logistic system management, and supply chain management.
- Disposal of rejected biological drugs.

1.6.1 Planning strategies

Biotechnology or biopharmaceutical companies with growing portfolios of commercial biologic drugs confront the challenge of selecting suitable host cells, formulating medium- and long-term production plans for several drugs while minimizing cost. The risk factor for successful production of biologic drugs mainly depends on correct capacity of planning, acquisition of land and building, choosing whether to build in-house facilities, or outsource manufacturing to a contract manufacturing organization (CMO) [1,2,31,32]. So, developing a comprehensive conceptual manufacturing planning strategy needs proper assessment of the cost, risk, and time trade-offs of each option [33,34].

So, the basic approach for planning a biotechnology company is to have deep survey on the necessity of specific therapeutic molecule, its assessment for public acceptability, and suitable resources needed to achieve the targeted objectives. For biopharmaceutical product development, the blueprint is designed in phase wise by ensuring success in the completion of project to its final form.

The overall planning process for developing blueprint of a biopharmaceutical company includes the followings:

- Strategic plans for the market survey on a target biologic drug in contest with public demand.
- Pilot level R&D work with research institute or in house laboratory on drug design and development.
- Planning for developing infrastructure for manufacturing process, including all types of civil constructions.
- Contract basis manufacturing of turnkey project, for long-term basis.

Program planning involves the meticulous analysis of the current status of the subjects carrying the main theme of the project to be undertaken. Besides this, the process of planning also includes a clear objective; timely action plan, and strategies for timely outcome of successful results.

Mostly, work planning is immediately followed by program planning. Work planning is the short-term planning involved in arrangement of adequate budget, preparation of Capital expenditures (CapEX); fixation of responsibility; indicators for monitoring and measuring progress and results. The success of a company mainly depends on how meticulously the CapEX is design to acquire, upgrade, and maintain physical assets, such as properties, buildings, an industrial plant, technology, and equipment. An example CapEX preparation of a biologic drug is given in the Table 1.4.

Table 1.4 An example of CapEx (capital expenditure) preparation for a typical biological product (thiocholchicoside TCS) given before planning for manufacturing.

Capex workings for bio TCS		**Dated**	**03rd July 2013**	**Rev-01**
Prepared by	**XXX[a]**	**Released for review with Mr XYZi**		
Reviewed by	**YYY[a]**	**Basis: TCS 120 kg/year**		
S. no.	**Description**	**Rs. in Lakhs**		**Remarks**
1	**Equipment**			
	Process equipment	Up stream	204.5	Detailed costing as **Annexure 1**
		Down stream	108.50	
	Utility equipment		96.50	Detailed costing as **Annexure 2**
	QC equipment		77.00	Detailed costing as **Annexure 3**
A	**Total cost**		**486.50**	

Continued

Table 1.4 An example of CapEx (capital expenditure) preparation for a typical biological product (thiocholchicoside TCS) given before planning for manufacturing—cont'd

Capex workings for bio TCS		**Dated**	**03rd July 2013**	**Rev-01**
Prepared by	**XXX**	**Released for review with Mr XYZi**		
Reviewed by	**YYY**	**Basis: TCS 120 kg/year**		
S. no.	**Description**	**Rs. in Lakhs**		**Remarks**
2	**Piping and valve cost**			
	Process	10% of A	48.65	Most of the upstream process equipment piping are part of equipment and factored in equipment costing
3	**Utility**	20%	14.00	
B	**Total-Eqpt. and piping**		**549.15**	
4	**Electrical cost**		54.50	Detailed costing as **Annexure 4**
5	**Instrumentation cost**		50.00	Lump sum amount
C	**Total material supply cost**		**653.65**	
6	Packing and forwarding charges	1% of B	5.49	
7	Excise duty	–	–	
8	VAT	–	–	
D	**Total supply cost including P&F**		**659.14**	
9	Transportation cost	2%	13.00	
10	Civil/structural cost		50.05	Detailed costing as **Annexure 5**
11	Effluent treatment plant		50.00	Lump sum amount
12	Fire protection system		23.00	Detailed costing as **Annexure 6**
13	Electrical installation cost		05.00	
14	Insulation and painting cost		15.00	
15	HVAC systems		Not applicable	Detailed costing as **Annexure 7**
E	**Total cost of supply and installation**		**815.14**	
16	Engineering fees	1% OF E	8.15	
17	Contingency	10% of E	81.46	

Table 1.4 An example of CapEx (capital expenditure) preparation for a typical biological product (thiocholchicoside TCS) given before planning for manufacturing—cont'd

Capex workings for bio TCS		Dated	03rd July 2013	Rev-01
Prepared by	**XXX**	**Released for review with Mr XYZi**		
Reviewed by	**YYY**	**Basis: TCS 120 kg/year**		
S. no.	**Description**	**Rs. in Lakhs**		**Remarks**
F	**Total cost with contingency**		**904.75**	
18 19	Insurance **Furniture**	1% of D	6.59 25.00	On supply and installation Detailed costing as **Annexure 8** (±15%)
G	**Total project cost with insurance**		**936.34**	

Assumptions

1. The above costing is only an estimate without any detailed engineering
2. Equipment cost is based on the budgetary quote obtained end of 2013
3. Equipment sizing are done considering120 kg TCS/Annum
4. Equipment identification and Costing is done based on understanding of the process in Shake flask level and 30 L fermenter level

Process equipments specification and cost—Annexure 1

S. no.	Equipment description Process	Capacity W.V (L)	G.V (L)	Specification	Unit cost Rs. Lakhs
1.0	Media preparation and fermentation				
	Germ plasma			Refrigerator (−40°C), (−10°C)	05
	Incubator room			Incubator (360 L)—3 nos (Temp—28°C, 33°C, and 37°C res.)	05
	Shaker and laminar flow area			Double deck shaker (150 flasks holding capacity, pilot shaker(300 flask holding capacity), Incubator shaker (300 flask holding capacity)	10
	Bio safety cabinet room			Bio safety cabinet—1 nos (horizontal)	5
	Autoclave room			Autoclave-2 nos (capacity-150 L)	3
1.1	Media preparation vessel jacketed with agitator	300	500	Semiautomated system, Top driven agitator with marine impeller Single mechanical seal Suitable for CIP and SIP Handhole for powder addition	10

Continued

Table 1.4 An example of CapEx (capital expenditure) preparation for a typical biological product (thiocholchicoside TCS) given before planning for manufacturing—cont'd

Process equipments specification and cost—Annexure 1					
	Equipment description	Capacity			Unit cost
S. no.	Process	W·V (L)	G.V (L)	Specification	Rs. Lakhs
				Jacket: Plant steam supply, chilled water supply and return with automated valves, sensors: temperature, level	
1.3	Seed vessel-I, skid mounted, bottom mounted agitator	5	10	Semiautomated system Top driven agitator with two set rushton turbine impellers Single mechanical seal Suitable for SIP Jacket: plant steam supply, chilled water supply and return with automated valves Automated temp control loop with heat exchanger and pump in the jacket Automated pH control	28
1.7	Production fermenter, skid mounted (3–4 VVM)—2 nos	1000	1500	Semiautomated system Bottom driven agitator with three set rushton turbine impellers Double mechanical seal with thermosiphon Suitable for CIP/SIP Jacket: plant steam supply, chilled water supply, and return with automated valves Automated temp control loop with heat exchanger, pump in the jacket Automated pH control Air flow control, flow meter and control valve, automated pressure control Sensors: temperature, pH, DO	75
1.9	Sterile filtration unit (pre and final 0.2 M) for glucose media and nutrient solution	100 LPH	4 nos	0.45 and 0.2 μm	12
1.10	Metering pump for glucose feed, plunger/ diaphragm type, 0 to 100 LPH or by air pressure	100 LPH	4 nos		4

Table 1.4 An example of CapEx (capital expenditure) preparation for a typical biological product (thiocholchicoside TCS) given before planning for manufacturing—cont'd

Process equipments specification and cost—Annexure 1					
	Equipment description	**Capacity**			**Unit cost**
S. no.	**Process**	**W·V (L)**	**G.V (L)**	**Specification**	**Rs. Lakhs**
1.11	Alkali vessel, suitable for SIP, with sterile air filter	50	75	Semiautomated system Top driven agitator with marine impeller Single mechanical seal Suitable for CIP and SIP Handhole for powder addition Jacket: plant steam supply, chilled water supply and return with automated valves Sensors: temperature, level, pH	2.5
1.12	Acid dosing vessel, Nalgen bottle with peristaltic pump	50	75	Semiautomated system, top driven agitator with marine impeller Single mechanical seal Suitable for CIP and SIP Handhole for powder addition Jacket: plant steam supply, chilled water supply and return with automated valves Sensors: temperature, level, pH	2.5
1.13	Antifoam vessel, suitable for SIP, with sterile air filter	50	75	Semiautomated system Top driven agitator with marine impeller Single mechanical seal Suitable for CIP and SIP Handhole for powder addition Jacket: plant steam supply, chilled water supply and return with automated valves Sensors: temperature, level	2.5
1.14	Harvested broth vessel with agitator	600	750	Semiautomated system Top driven agitator with marine impeller Single mechanical seal Suitable for CIP and SIP Hand hole for powder addition Jacket: plant steam supply, chilled water supply and return with automated valves Sensors: temperature, level, pH	7

Continued

Table 1.4 An example of CapEx (capital expenditure) preparation for a typical biological product (thiocholchicoside TCS) given before planning for manufacturing—cont'd

Process equipments specification and cost—Annexure 1					
	Equipment description	**Capacity**			**Unit cost**
S. no.	**Process**	**W·V (L)**	**G.V (L)**	**Specification**	**Rs. Lakhs**
1.15	CIP station	200	550	Fully automated system Single tank with provision for dosing acid and alkali Heat exchanger and Circulation pump for supply and recirculation Sensors: conductivity, temperature	10
1.16	Mobile CIP system, pumps		1 nos	Mobile trolley with centrifugal pump, PUMP with VFD Sensors: conductivity sensor	5
1.17	Water for injection holding tank		1000 L		5
1.18	Glassware for lab process				3
	Sub total—upstream				**204.5**
2	**Down stream process**				
2.1	TFF filter system	1000 LPH			100
2.2	Buffer vessel for TFF system—2 nos	500	650		10
2.3	Solid mass collection tank/biomass kill tank	500	750	Semiautomated, SS316, Ra 0.8, jacketed, anchor type agitator, CIP able Sensors: level and temperature	5
2.4	Permeate collection tank	600	750	Semiautomated, SS316, Ra 0.8, Jacketed, CIP able Sensors: level and temperature	3
2.5	Permeate feed pump to resin column	LPH			01.0
2.6	Water feed tank	1000	1200		05
2.7	Methanol feed tank	1000	1200		2
2.8	XAD resin	100 kg		Amber lite resin XAD 1180	10
2.9	XAD column	150 × 22 cm		4 Glass columns each having 50 kg capacity	10
2.10	Holding tank	500		Intermittent holding tank for Aq layer	1.5
2.11	Aqueous fraction collection tank	600	750	Semiautomated, 304 nonjacketed, CIP able Sensors: level and temperature	2.5

Table 1.4 An example of CapEx (capital expenditure) preparation for a typical biological product (thiocholchicoside TCS) given before planning for manufacturing—cont'd

Process equipments specification and cost—Annexure 1					
	Equipment description	**Capacity**			**Unit cost**
S. no.	**Process**	**W·V (L)**	**G.V (L)**	**Specification**	**Rs. Lakhs**
2.12	Finished material processing system (milling)	LS		Equipment required for finished dried product processing units: IC bins, split valves, blender, milling, filling, and packing (cGMP quality)	5
2.13	Finished material processing system (Shifter)	LS		Equipment required for finished dried product processing units: IC bins, split valves, blender, milling, filling, and packing (cGMP quality)	2
2.14	Finished material processing system (packing)	LS		Equipment required for finished dried product processing units: IC bins, split valves, blender, milling, filling and packing (cGMP quality)	5
2.15	Process pumps			As per requirements (around 15 pumps)	05
	Sub total—down stream				**108.5**
	Downstream process equipment				
1.0	SS holding tank		500	Intermittent holding tank of methanol Ext mixture	2
2.0	SS reactor		500	Reactor for methanol concentration	05
3.0	SS condenser		$10\,m^2$	Condenser for the above	2.5
4.0	Receiver		500	Receiver for the methanol concentration reactor	2
5.0	Glass column		9″ × 3.5 m	For separation of CCS and CCE	3
6.0	Mixing vessel		500	For silica and solvent slurry preparation	3
7.0	Holding tank		500	For dosing tank of solvent (CCS and CCE)	2
8.0	Holding tank		200	Intermittent holding tank of Ext mixture	1.5
9.0	SS reactor		500	For CCS Ext concentration	06
10.0	Condenser		$10\,m^2$	For the above	2.5
11.0	Receiver		500	For the above	2

Continued

Table 1.4 An example of CapEx (capital expenditure) preparation for a typical biological product (thiocholchicoside TCS) given before planning for manufacturing—cont'd

Process equipments specification and cost—Annexure 1					
	Equipment description	Capacity			Unit cost
S. no.	Process	W·V (L)	G.V (L)	Specification	Rs. Lakhs
12.0	Holding tank		100	For the Ext concentrate collection (CCS)	1
13.0	SS crystallizer		500	For CCS crystallization	6
14.0	Condenser		$5\,m^2$	For the above	2
15.0	Receiver		200	For the above	1.5
16.0	Holding tank		50	For charging DNS for crystallization	1
17.0	Agitated Nutsche filter drier			For filtering CCS	10
18.0	Holding tank		100	For collecting CCS ML	1
19.0	Holding tank		2000	For collecting and preparation of Ext (CCS)	4
20.0	RVD		12 trays	For drying the CCS	10
TCE					
21.0	SS reactor		5000	For CCE Ext concentration	05
22.0	Condenser		$10\,m^2$	For the above	2.5
23.0	Receiver		500	For the above	2
24.0	Holding tank		100	For the Ext concentrate collection (CCE)	1
25.0	SS crystallizer		500	For CCE crystallization	6
26.0	Condenser		$5\,m^2$	For the above	2
27.0	Receiver		200	For the above	1.5
28.0	Holding tank		50	Charging ethyl acetate for crystallization	1
29.0	Agitated Nutsche filter drier			For filtering CCE	05
30.0	Holding tank		100	For collecting CCE ML	1
31.0	Holding tank		500	For collecting and preparation of Ext (CCE)	2
32.0	Holding tank		500	Methanol day tank	2
33.0	Holding tank		500	DNS ML collection tank (CCS)	2
34.0	Process pumps		8	For transferring the contents between the process vessels	5.0
35.0	Scrubber				05
					96.5

Table 1.4 An example of CapEx (capital expenditure) preparation for a typical biological product (thiocholchicoside TCS) given before planning for manufacturing—cont'd

Utility equipments specification—Annexure 2							
	Black utility	**Capacity**	**HP**	**Usage %**	**Power consumption (KWH)**	**Power consumption (KWH)-single fermenter**	**Cost in Lakhs**
1	Process load	2 ton	78.15	75	43.96		25
	Plant steam generation (boiler, water tank, fuel tank)						
	Feed water pump		5	75	2.81		
	Furnace oil pump		5	75	2.81		
2	Chilled water generation system, 3x100TR and pumps	100 TR					10
	Chilled water circulation pumps		100	20	15		
	Chilled water circulation pumps		50	100	37.5		
	Holding tanks-primary and secondary						
	Chiller compressor		300	30	67.5		
3	Chilled brine and Pumps	10 TR					05
	Chilled brine compressor		30	50	11.25		
	Holding tanks-primary and secondary (insulated)						
	Chilled brine circulation pumps		20	100	15		
5	Cooling water and pumps	100 TR × 2 nos					05
	Fan for cooling tower		50	100	37.5		
	Circulation pumps for cooling tower		100	100	75		

Continued

Table 1.4 An example of CapEx (capital expenditure) preparation for a typical biological product (thiocholchicoside TCS) given before planning for manufacturing—cont'd

	Utility equipments specification—Annexure 2						
	Black utility	**Capacity**	**HP**	**Usage %**	**Power consumption (KWH)**	**Power consumption (KWH)-single fermenter**	**Cost in Lakhs**
6	Compressed instrument air (generation, filter, dryer, receiver), 7.5 kg/cm^2	100 NCFM	50	70	26.25		10
	Air compressor receiver						
	Compressor for instrument air						
7	Compressed process air (generation, filter)	100 NCFM	50	70	26.25		10
	Compressor for process air						
8	Air dryer for compressed air	500 NCFM					2
9	Purified water system	1 kL per day					5
10	Vacuum pumps	3	20				2
11	Hot water system	10 m^3	10				3.0
	Total electrical load in utility		**868**				
	HVAC systems		80	100	60.0		
	Total cost		948				**77.00**

Instruments for quality control—Annexure 3				
Instrumentation				
S. no.	**Instruments related to strain improvement**		**Specification**	**Amount in Rs**
1	Centrifuge		1500 rpm	05.00
2	Centrifuge		500 rpm	01.00
3	Balance		0–10 kg	00.10
4	Sonicator			03.00
5	Oven		300 L	01.50

Table 1.4 An example of CapEx (capital expenditure) preparation for a typical biological product (thiocholchicoside TCS) given before planning for manufacturing—cont'd

Instruments for quality control—Annexure 3				
Instrumentation				
S. no.	**Instruments related to strain improvement**		**Specification**	**Amount in Rs**
6	Lyophyser			07.00
7	Microscope (stereo) with camera			05.00
8	Double beam spectro photometer			05.00
9	pH meter			0.5
10	Meter balance			0.5
11	HPLC			**10.0**
	Total			**38.60**

Cost estimation for electrical—Annexure 4				
S. no.	**Description**	**Qty**	**Unit cost in Lakhs**	**Approx. cost in lakhs**
1	Main changeover panel	1	10	05.00
2	Subswitch board	1	7.5	05.00
3	MCC panel (process, utility, and HVAC)	3	4	05.00
4	MPDB	1	0.5	0.5
5	MLDB	1	0.5	0.5
6	PDB's	3	0.25	0.75
7	LDB's	3	0.25	0.75
8	Power cables	Lot	05	05
9	Cable trays and MS support	Lot	4	4
10	Termination materials	Lot	3	3
11	Lighting material	Lot	5	5
12	FLP items	Lot	5	5
13	Earthing materials	Lot	5	5
13a	Capacitor panel	1	3	3
14	DG set (600 kva)and Control panel with AMF	1	45	00
15	Statutory requirements	Lot	5	5
16	Miscellaneous	Lot	3	3
17	Transformer yard extension			
	Total electrical cost			**54.5**

Continued

Table 1.4 An example of CapEx (capital expenditure) preparation for a typical biological product (thiocholchicoside TCS) given before planning for manufacturing—cont'd

Fire protection systems—Annexure 6

S. no.	Item	Qty	Remarks/specification	Amount in Rs
1	EHS (environment, health, and safety)			10
2	Fire alarm system and public address system			1
3	Venting and scrubber system			2
4	Drainage system	C		5
5	Fire hydrant system			5
			Total	23

Furniture for bio plant—Annexure 8

S. no.	Furniture for	Specification	Amount in Rs
1.0	Molecular biology and strain improvement	4 Working table, 1 wall rack laboratory stool-8	5
2.0	Analytical laboratory	2-Fume hood, 2-working table, 8-stool, 1-wall rack	5
3.0	Immobilized reactor room	Working table-1, 1-wall rack	0
4.0	Gel bead preparation	Working table-1, 1-wall rack	0
5.0	Instruments room	Working table-04, 1-wall rack	5
6.0	Wash area	Working table(SS)-01, 1-wall rack	2.5
7.0	Office room	Computer, table, chairs, etc.	02.00
8.0	Central control room (DCS panel room) and documentation room	Chairs, tables, etc.	01
		Total	**25.5**

Table 1.4 An example of CapEx (capital expenditure) preparation for a typical biological product (thiocholchicoside TCS) given before planning for manufacturing—cont'd

Layout

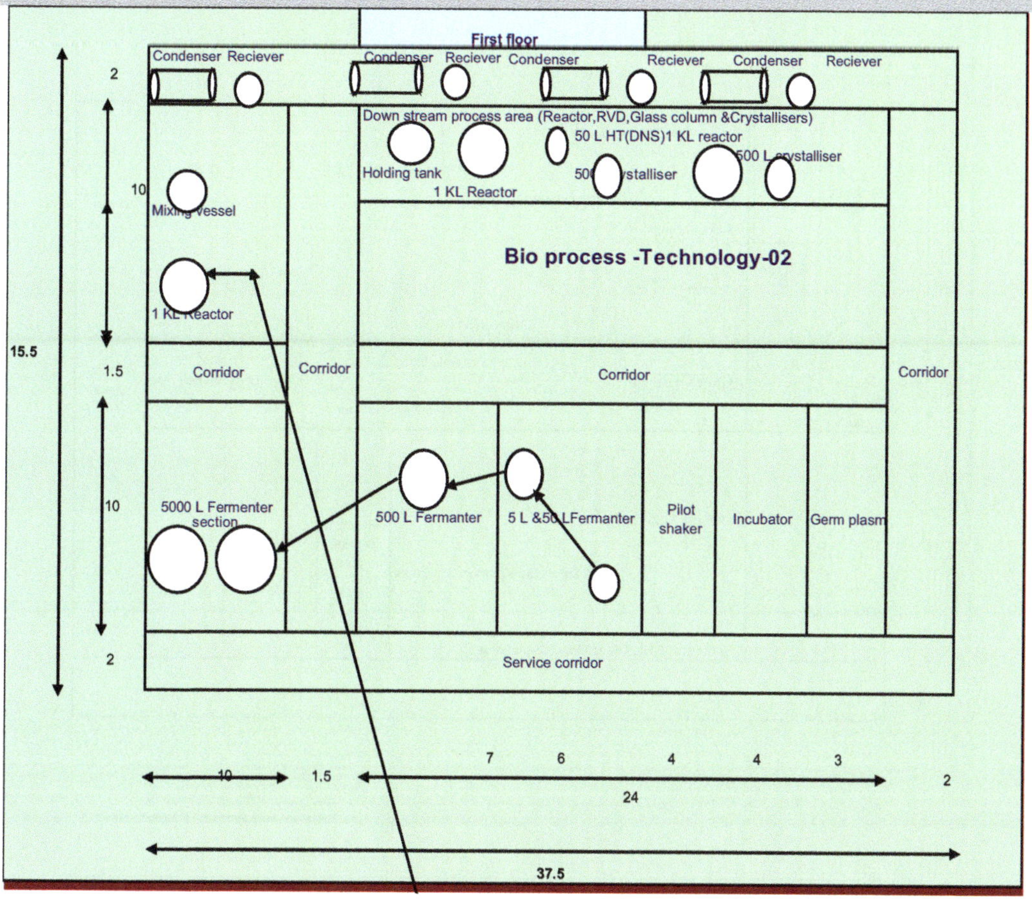

Continued

Table 1.4 An example of CapEx (capital expenditure) preparation for a typical biological product (thiocholchicoside TCS) given before planning for manufacturing—cont'd

Layout

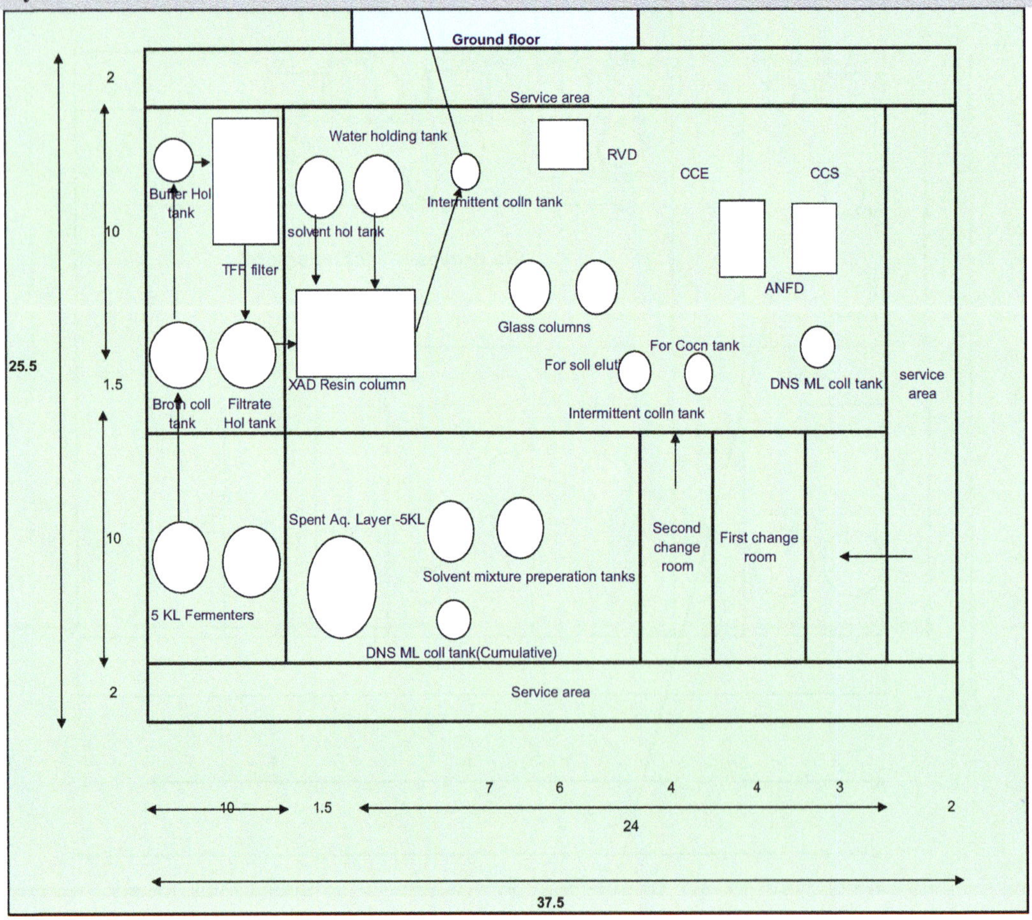

Table 1.4 An example of CapEx (capital expenditure) preparation for a typical biological product (thiocholchicoside TCS) given before planning for manufacturing—cont'd

Layout

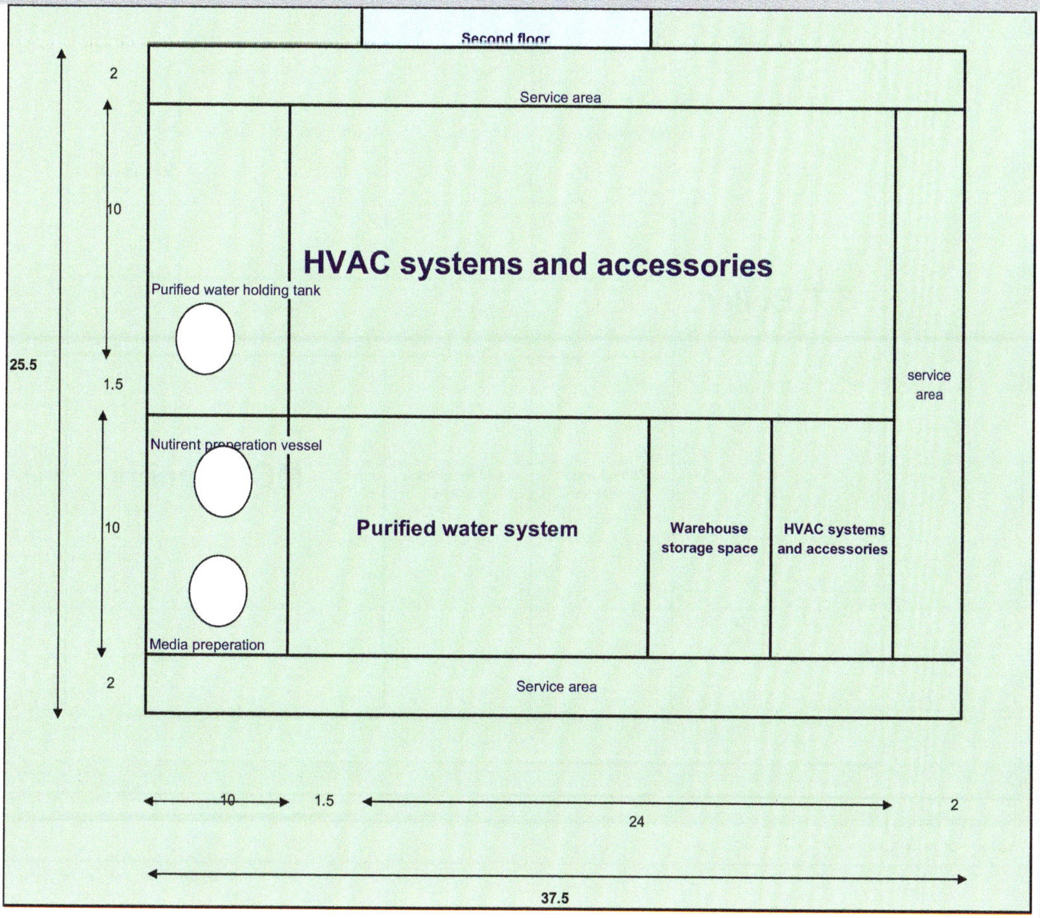

Continued

Table 1.4 An example of CapEx (capital expenditure) preparation for a typical biological product (thiocholchicoside TCS) given before planning for manufacturing—cont'd

Layout

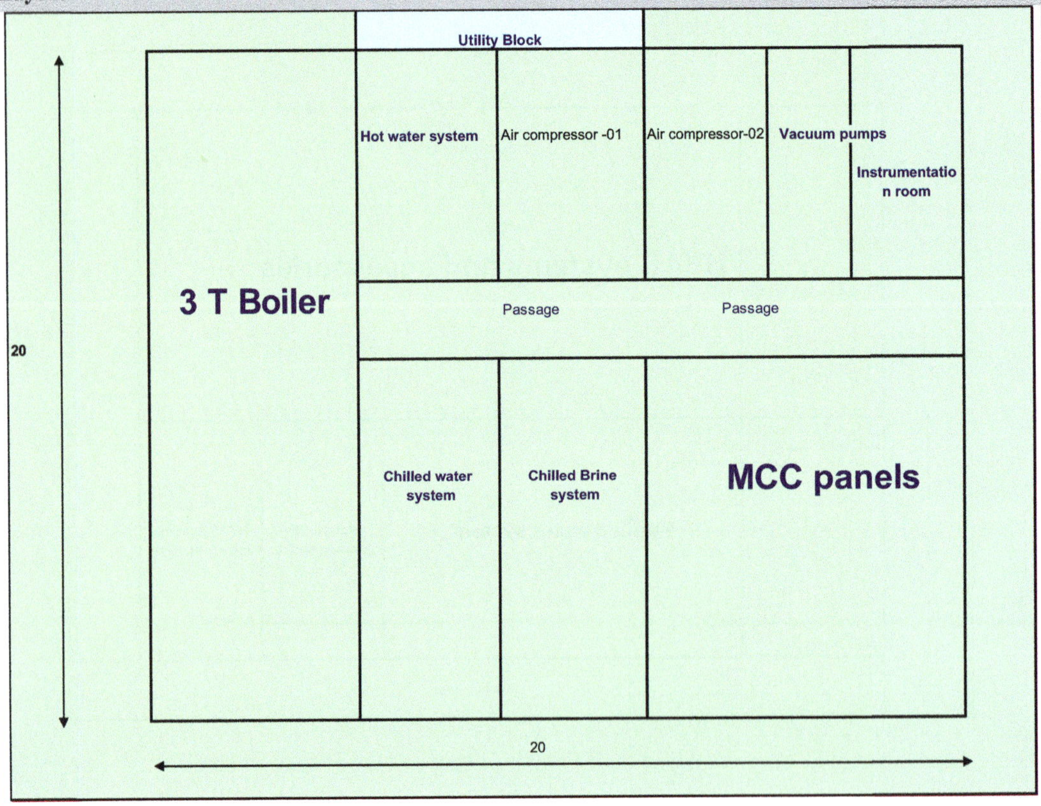

[a]*Name not to be disclosed.*

1.6.2 Planning for successful product development

Design of a new biopharmaceutical needs meticulous planning for successful launch of a quality product. However, due to tough regulatory issues, companies face high stakeholder expectations, numerous market-access hurdles, and stiff competition. These issues can be well-preventable, if the overall manufacturing process follow proper pipelines of guidance design, before the launching a product. The steps in work planning process are:

(i) Through survey of biological product to be launched.
(ii) Conceptual development of highly resourceful task force.
(iii) Availability of noncompetitive livestock.
(iv) Proper interlink between technical wing and administrative system.
(v) Development of innovative integrated manufacturing process.
(vi) Development of sustainable supply chain system.
(vii) Well-planned biological waste management process.

(i) Through survey of biological product and quality assurance

The well-established companies have a market-access team related to the demand of the product to be launched. To have a successful development of product, company must create a cross-functional market-access team comprised of individuals from R7D, sales and marketing, and medical affairs. This team should begin dialogue with payers and health technology assessment (HTA) agencies early and then keep them apprised of the decisions made throughout the R&D process. The team should especially communicate the types of patient populations that are being targeted and the expected benefits. Such exchange of thoughts would be helpful to payers and HTA agencies regarding efficacy and safety data while conducting clinical studies.

The biopharmaceuticals are highly sensitive to the nature of manufacturing process and temperature, due to their biological origin. So, quality by design (QbD) is the most essential aspect while launching a biologic in the global market (Fig. 1.6).

The QbD is not merely a regulatory aspect of manufacturing biopharmaceutical product development and testing but to assure the overall performance of the product in depth on quality assurance point of view and its long-term side effect to the user from safety and security aspects. Besides this, maximum attention and precaution must be taken for product performance profile (target product profile (TPP), target product quality profile (TPQP), and critical quality attribute (CQA).

(ii) Conceptual development of highly resourceful task force

A biopharmaceutical industry should have an open and collaborative culture to bring sustainability to quality production. The task force of a biopharmaceutical company should have the quality of combining the agility of a biotech with the scale and support the expectation of a company for sustainable quality production. In this connection, it is the duty of task force to bring together diverse minds united in breaking new ground to improve patient outcomes. The task force should have the potential to work with a variety of molecular formats across multiple therapeutic areas. The task force should follow adaptive, science-driven approach, and apply cutting-edge technologies to develop industry-leading cell lines, production process, analytical methods, formulations, and drug delivery systems and devices. In addition, the task force should have the skill to develop innovative, high-yielding, and efficient production process for even the

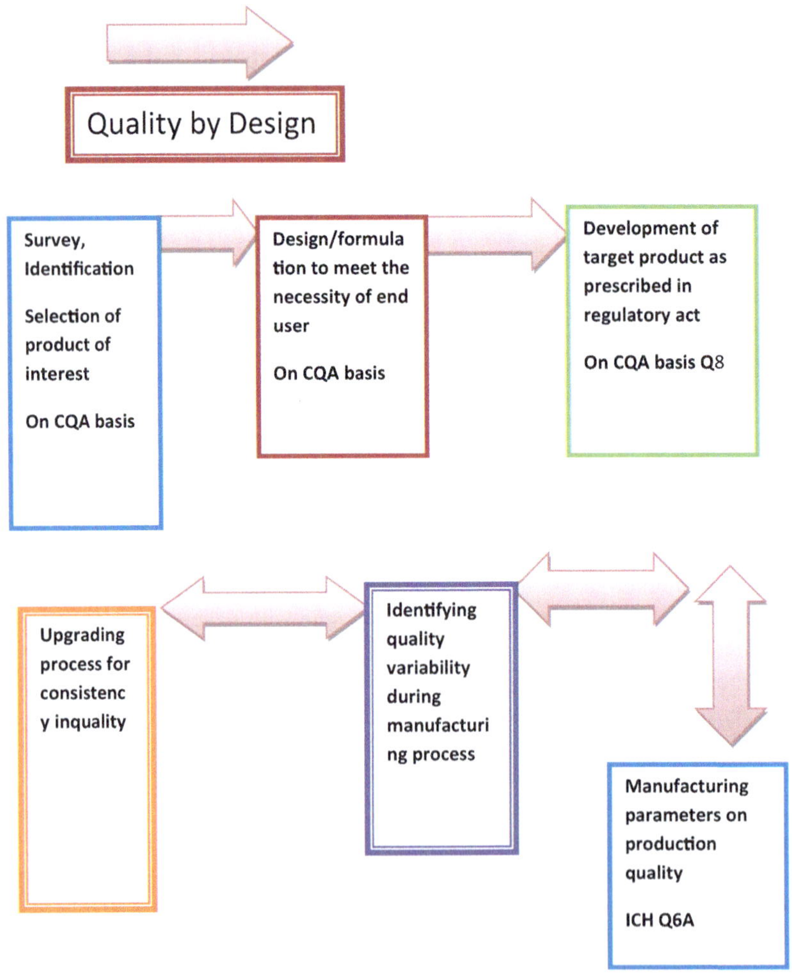

FIG. 1.6

Quality by design.

most challenging and novel molecules. The task force also has the ability to provide clinical products with stable formulations, and developing a range of drug delivery options, in combination with advanced device development.

The responsibility of task force is to be ambitious to push the boundaries of science by creating innovative ways to make and deliver life-changing medicines to patients. Following are some of the important responsibility of different task forces to bring consistency to quality product:

(a) Document preparation and communication

The task force manages the authoring and reviewing of documents that are submitted globally to regulatory agencies to support clinical development from First-in-Man trials through marketing applications. The task force delivers well-drafted, scientifically sound, and regulatory-compliant protocols for all biologics working in the company. The member of the task force should be document specialists who can well assists with formatting and quality control of regulatory submissions in connection with quality biopharmaceutical development.

(b) Task force for data science and modeling

The task force helps scientists and engineers working in technical process development fields. The team also provides support with drug specification setting task interaction with regulatory agencies and postapproval activities.

(c) Task force for biomass processing

Earlier, fermented process for biologic drug production was limited to batch process or continuous process. But with the development of integrated upstream and downstream processes the task force should have sound background of engineering principles to survey, evaluate, and implement transformative solutions and technologies. The task force should have potential rapid delivery of cost-effective cures to patients by implementing state-of-the-art manufacturing and scaling up of bioprocesses through continuous innovation and engineering solutions.

(d) Cell culture fermentation sciences (CCFS)

It is responsible of task force to develop CCFS design, optimizes and validates cell culture and fermentation processes for a wide range of biological products. The team generates optimal cell lines and is also responsible for technology transfer into manufacturing. The task force should have skill for high-throughput scale-down models development to formulate chief media and processes to support production. The team should have depth of knowledge on antibody productivity and incorporates the latest development in automation and analytics into its work.

(e) Purification process sciences (PPS)

The portfolio for PPS is an important responsibility for the task force to optimize the process for a wide range of biological products, including monoclonal antibodies, therapeutic proteins, therapeutic oligonucleotides, viral therapies, and cellular therapies. It is the responsibility of the team to produce high quality and safe biologic drugs for patients.

(f) Dosage form design and development (DFDD)

During clinical trials, it is most important to dosage form design and development of pharmaceutically stable, competitive, patient-centric biopharmaceutical, including formulation, drug delivery devices for clinical and commercial use. So, a biopharmaceutical company should go for highly skilled task force for developing quality drug delivery systems like syringes, autoinjections, on-body injections, and inhalation devices. In addition, the task force team has to be sound knowledge on process engineering and packing assesses dosage form manufacturability, and develop robust packing, fill/finish and device assembly manufacturing process for biologic drugs, which are highly prone to variable temperature and microenvironments.

(g) Flu-MSAT

The Flu Manufacturing Science and Technology team should have updated knowledge on activities linked to live attenuated influenza vaccine (LAIV) that is used every year to vaccinate children. The task force team should have multidisciplinary and experts in virology, upstream and downstream process development, and analytical development.

(h) Analytical sciences group

The team should have a sound technical background on the state-of-the-art physicochemical analytical methods and uses the full range of analytical techniques and imaging analyses to detech, characterize, and quantify product variants ranging from single atomic modifications to visible particles. The team also has work experience of handling monoclonal antibodies, bispecific antibodies, antibody-drug conjugates, and other novel products such as parenteral and oral dosage synthetic peptides, therapeutic viruses, and nucleic acid products. The team should have potential to product and process development from emergence out of basic research, throughout clinical trials, and then supports commercial operations.

(i) Logistics and clinical supply (LCS)

The LCS team is responsible for drug supply chain management system and should assure the end user, the safety and security of biological drugs for end user.

(j) Bioassay group

The team should have potential for working collaboratively, mainly focusing on developing, optimizing and validating functional and impurity methods to support the characterization, GMP lot release, stability, and in-process testing of biotherapeutic products to address unmet medical needs.

(k) Manufacturing sciences

The main responsibility of this team is to bring amazing biologic drugs to patients. The team should have responsibility of supporting and developing fully-integrated process scale-up and biomanufacturing functionality, using a range of technologies to produce the preclinical and early through late stage clinical drug products that advance in pipeline for production process. The team should have potential to actively collaborate with its counterparts in technical process development; provides cGMP manufacturing of cell banks; manage and assure the supply of raw materials; perform nonconformance identification and resolution activities; and provide integrated engineering, technical service, and process automation functions.

(iii) Availability of noncompetitive feed stock

Timely availability of quality and cheap feed stock is of primary importance for a biotechnology or biopharmaceutical industry. The biopharmaceutical products are of microbes, plants, animals, insects, and mammalian cells origin and need different types of feed stocks.

Generally, glucose is the basic energy source for the growth and development of variety of host cells, especially bacteria and fungi) responsible for the production of biologics. Molasses is one of the cheap feedstocks used as the source of carbohydrate and other minerals. Besides this, starch is also used as a source of carbohydrate, by enzymatic hydrolysis. Million-ton of molasses is used for biofuel (ethanol), organic acids (lactic acid, acetic acid, formic acid, oxalic acid, uric acid, malic acid, and amino acid) production through fermentation process. Mostly, *E. coli*, *Corynebacterium glutamicum*, *Pseudomonads*, *Bacilli*, and Baker's yeast are used as host cells for the above purposes.

But, the first generation feedstock is derived from crop plants as energy-containing molecules like sugars, oils, and cellulose with limited availability. So, scientists are in process of utilizing nonfood feedstocks like straw, bagasse, and forest residues as second-generation feedstock. Still, it has been in process to use algal biomass as third-generation feedstock as carbon source. With the improvement in tools and technology for harvesting sunlight, development of photobiology solar cell, and electro fuels are supposed to bring new revolution in carbohydrate feedstock development from solar light, water, and CO_2 available as exhaust gas from industries [35].

Currently, instead of using animal-derived feedstock; people have started using nonanimal-derived origin hydrolysates, such as plant hydrolysates or yeast hydrolysates for recombinant protein and antibiotic production. These sources of hydrolysates is free from any pathogenic contamination (Bacteria or virus), and support cell growth, better as compared to animal origin.

Natural mediums like animal body fluids or medium of tissue extraction like plasma, serum, lymph, and chicken embryo leaching solution are used as feedstock for mammalian cell culture. This natural culture medium contains nutrients, various somatomedins and hormones for supporting growth and development of mammalian cells. These medium are complex in nature and the overall composition varies from batch to batch during the extraction process.

Currently, serum from cattle is widely used for animal cell culture. The main advantages of cattle serum (bovin calf serum, newborn serum, calf serum, and fetal serum) are the availability of adequate resource, easy way of preparation, and long-application time.

(iv) Proper interlink between technical wing and administrative system

A biopharmaceutical company needs well-defined manpower for administration control, manufacturing process, product quality maintenance, etc. In traditional pharmaceutical industries it has been easy to maintain consistency in product quality due to the use of well-defined chemical and physical techniques. But in biologic drug manufacturing process the involvement of biological processes becomes a challenge to maintain consistency in product manufacturing process and development. The biological properties of host cells are inherently variable in features. So, it is obvious to strictly maintain quality risk management (QRM) in biologic drugs production process. So, a consortium of scientists and technicians is needed for bringing sustainability in biologic drug production on quality basis.

Before engaging manpower, it is necessary to ensure the employees are free from any conterminous diseases. The manpower working in the core units should have regular health checkups. Workers with contaminated diseases are threat to biologic drug quality control. Employers associated with the production division should have up-to-date training in disinfection, personal hygiene, and microbial risk management. Nonskilled labors, administrative staff, and technical experts in machineries management employs indirectly associated with the manufacturing process should be kept away from the biologic drug production division. Personnel working in animal maintenance division should have restricted movement in production division in order to have control over cross contamination. Staff assigned to work with *Bacillus Calmette-Guerin* (BCG) or any such products should be restricted for working in any other infections agents.

(v) Development of innovative integrated manufacturing process

The detailed blueprint for manufacturing biologic drugs includes the starting point for suitable host cell selection, isolation, cultivation, and preservation in cell bank, followed by pilot level production of targeted biologic and development of turnkey scale production for commercialization. The entire manufacturing process is organized under the strict guidelines of the Good manufacturing process (GMP).

The overall process for biologic drugs manufacturing is strictly governed by GMP. On the basis of product types, end-user, and purpose of use, the biologic drugs are categorized into followings:

- Monoclonal antibodies
- Recombinant therapeutic proteins
- Growth hormones
- Therapeutic enzymes

The market segments cover biologics are hospitals, specialized clinics, and research and development institutes.

(vi) Development of sustainable supply chain system

Protecting the biopharmaceutical supply chain is a priority for companies ensuring access to life-saving products. The biopharma supply chain involves a complex set of steps that are required to produce a drug, from sourcing and supply of materials, through manufacturing and distribution, to delivery to the end user.

(vii) Well-planned biological waste management process

Handling wastes from biopharmaceutical industries has been a challenging task due to the presence of biological wastes, other than toxic effluent containing a variety of chemicals. The biological wastes from the biopharmaceutical industry include dead host cells, debris from genetically engineered host cells, unwanted organic molecules (proteins, amino acids, lipids, and nucleic acids), and wastes biological liquid (serum-like fluid). Hazardous biological wastes are often more harmful as compared to other chemical pollutants. Before disposal, the wastes of biological origin are to be completely sterilized and processed for further disposal.

References

[1] J.D. Valderrama-Rincon, A.C. Fisher, J.H. Merritt, An engineered eukaryotic protein glycosylation pathway in *Escherichia coli*, Nat. Chem. Biol. 8 (5) (2012) 434–436.

[2] V. Rodriguez, J.A. Asenjo, B.A. Andrews, Design and implementaion of a high yield production system for recombinant expression of peptides, Microb. Cell Factories 13 (2014) 1–10.

[3] Generic Drug Facts, Center for Drug Evaluation and Research, U.S. Food and Drug Administration, 2017. 2017-10-06. Retrieved 2017-11-11.

[4] D.Y. Wangeng, C.C. Wag, K.W. Chy, Therapeutic potentials of gene silencing by RNA inference: principles, challenges, and new strategies, Gene 538 (2) (2014) 217–227.

[5] A. Witrup, J. Lieberman, Knocking down diseases: a progress report on siRNA, therapeutic, Nat. Rev. Genet. 16 (9) (2015) 543–552.

[6] D.J. Crommelin, G. Storm, R. Verrijk, L. De Leed, W. Jiskoot, W.E. Hennik, Shifting paradigms: biopharmaceuticals versus low molecular weight drugs, Int. J. Pharm. 266 (1–2) (2003) 3–16.

[7] D.B.T. Cox, R.J. Plattt, F. Zhang, Therapeutic genome editing: prospects and challenges, Nat. Med. 21 (2) (2015) 121–131.

[8] H.L. Chen, C. Jk, R.B. Tang, Current recommendation for the Japanese encephalitis vaccine, J. Chin. Med. Assoc. 78 (5) (2015) 271–275.

[9] B. Leader, Q.J. Baca, D.E. Golan, Protein therapeutics: a summary and pharmalogical classification, Nat. Rev. Drug Discov. 1 (2008) 21–39.

[10] A. Beck, Biosimilar biobetter and next generation therapeutic antibodies, MAbs 3 (2) (2011) 107–110.

[11] W.R. Strohl, Fusion proteins for half-life extension of biologica as a strategy to make biobetter, BioDrugs 29 (4) (2015) 215–239.
[12] V. Sandeep, J. Parveen, P. Chauhan, Biobetters: the better biologics and their regulatory overview, Int. J. Drug Regul. Aff. 4 (1) (2016) 13020.
[13] A.M. Azevedo, P.A.J. Rosa, I.F. Ferreira, M.R. Aires-Barros, Chromatography-free recovery of biopharmaceuticals through two-phase processing, Trends Biotechnol. 27 (4) (2009) 240–247.
[14] R. As, G. Kapoor, Application of process analytical technology for downstream purification of biotherapeutic, J. Chem. Techynol. Biotechnol. 90 (2) (2015) 228–236.
[15] S. Aiba, A.E. Humphrey, N.F. Millis, Recovery of fermentation products, in: Biochemical Engineering, second ed., 1973, pp. 346–392. New York.
[16] B. Atkinson, I.S. Daoud, Microbial floes and flocculation in fermentation process engineering, Adv. Biochem. Eng. 4 (1976) 41–124.
[17] B. Atkinson, F. Mavituna, Biochemical Engineering and Biotechnology Handbook, second ed., MacMillan, London, 1991.
[18] B. Atkinson, P. Sainter, Development of downstream processing, J. Chem. Technol. Biotechnol. 32 (1982) 100–108.
[19] G.W. Pace, I.H. Smith, Recovery of microbial polysaccharides, in: Abstracts of Communications. Second European Congress of Biotechnology, Eastbourne, Society of Chemical Industry, London, 1981, p. 36.
[20] R.R. Swartz, The use of economic analyses of penicillin G manufacturing costs in establishing priorities for fermentation process improvement, in: Annual Report on Fermentation Processes, 3, 1979, pp. 75–110.
[21] A.J.J. Straathof, The proportion of downstream costs in fermentative production processes (Ed. In Chief), in: M. Moo-Young (Ed.), Comprehensive Biotechnology, second ed., vol. 2, Elsevier, Oxford, 2011, pp. 811–814.
[22] P.M. Doran, Bioprocess Engineering Principles, second ed., Academic Press, Oxford, 2013, pp. 445–578.
[23] R.W. Liggett, H. Koffler, Corn steep liquor in microbiology, Bacteriol. Rev. 12 (4) (1948) 297–311, https://doi.org/10.1128/MMBR.12.4.297-311.1948. PMC 180696 16350125.
[24] P.F. Stanbury, A. Whiitaker, S.J. Hall, Principles of Fermentation Technology, second ed., Butterworth-Heinemann, 1999, ISBN: 978-0750645010.
[25] P.F. Stanbury, Fermentation technology, in: J.M. Walker, R. Rapley (Eds.), Molecular Biology and Biotechnology, Royal Society of Chemistry, 2007, pp. 1–24, ISBN: 978-1-84755-149-8 (Chapter 1).
[26] F.M. Wurm, Production of recombinant protein therapeutics in cultivated mammalian cells, Nat. Biotechnol. 22 (11) (2004) 1393–1398, https://doi.org/10.1038/nbt1026. 15529164. S2CID 20428452.
[27] J.C. Drugmand, Y.J. Schneider, S.N. Agathos, Insect cells as factories for biomanufacturing, Biotechnol. Adv. 30 (5) (2012) 1140–1157, https://doi.org/10.1016/j.biotechadv.2011.09.014. 21983546.
[28] M. De Lourdes, T.M. Polizeli, M. Rai, Fungal Enzymes, CRC Press, 2013, ISBN: 978-1-466-59454-8.
[29] G. Walsh, Biopharmaceutical benchmarks 2014, Nat. Biotechnol. 32 (10) (2014) 992–1000.
[30] Ernst & Young, Beyond Borders—Biotechnology Industry Report 2015, 2015.
[31] T.C. Ransohoff, Considerations impacting the make vs. buy decision, Am Pharm Outsourcing 5 (2) (2004) 52–63.
[32] E.S. Langer, Survey examines trends in the outsourcing relationship, BioPharm Int. 24 (2011) 26–33.
[33] K. Lakhdar, J. Savery, L. Papageorgiou, S.S. Farid, Multiobjective long-term planning of biopharmaceutical manufacturing facilities, Biotechnol. Prog. 23 (2007) 1383–1393.
[34] E. George, N.J. Titchener-Hooker, S.S. Farid, A multi-criteria decision-making framework for the selection of strategies for acquiring biopharmaceutical manufacturing capacity. In: 7th World Congress of Chemical Engineering, Comput. Chem. Eng. 31 (2007) 889–901.
[35] D.E. Cameron, C.J. Bashor, J.J. Collins, A brief history of synthetic biology, Nat. Rev. Microbiol. 12 (2014) 381–390.

CHAPTER

2

Downstream process (harvested broth concentration) phase-I

Abbreviations

BC	basket centrifuge
DC	disc centrifuge
DSP	downstream processing
HBVF	horizontal belt vacuum filters
MAbs	monoclonal antibodies
MCC	multichamber centrifuge
RDVF	rotary drum vacuum filter
SC	scroll centrifuge or decanter
TC	tubular centrifuge
VF	vacuum filtration

2.1 Downstream processes

Downstream processing (DSP) technology is defined as a series of operational processes through which biological materials such as cells, tissue culture fluid, and cell debris are removed from the harvested broth resulting from the overall operational process of upstream process, to get a product of interest in quality and quantity (Fig. 2.1).

2.2 Product recovery

The overall process for obtaining the biological product of interest is based on the modular operations of separation of solids and liquids, and thereafter on the basis of cell treatment, solvent extraction, liquid-liquid phase separation, crystallization and precipitation, adsorption, distillation, chromatography, and membrane filtration. The quality and quantity of the target product in a downstream process are mainly dependent on the stability of the product and its stepwise recovery process. Following are some of the important sequential processes to obtain a polished and formulated product of interest.

The overall DSP includes recovery and purification of biosynthetic products generated during the upstream stage into finished biotechnology products, including whole cells, organic acids, amino acids, solvents, antibiotics, industrial enzymes, antibodies, therapeutic proteins, vaccines, and gums.

Techniques for Downstream process for Biologic Drugs and Vaccines. https://doi.org/10.1016/B978-0-443-19157-2.00003-0

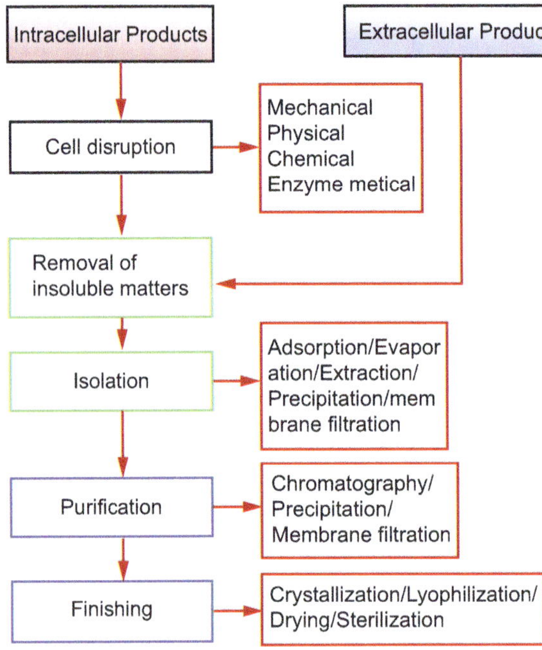

FIG. 2.1

Line diagram of the downstream process to obtain biologic of interest from harvested broth.

The downstream process also includes the sale of those products to other businesses, governments, or private individuals. The type of end user varies depending on the quality of the product which is safe and secure. Regardless of the industry involved, the downstream process has direct contact with customers through the finished product. New methods for process development are being developed. These include the establishment of platform technologies and high-throughput methods of optimization using experimental approaches based on quality by design and the design of experiment [1–3].

Additionally, integration of modeling, simulation of unit operations, and the use of miniplant facilities can be applied in process development (Fig. 2.2).

The complexity of DSP is determined by the required purity of the product at minimum recovery costs. The major factors affecting DSP include microbial morphology, flocculation characteristics, size, and cell wall rigidity. These factors affect filterability, sedimentation, and homogenization efficiency. The presence of fermentation by-products, media impurities, and additives such as antifoaming agents can interfere with DSP steps. Therefore, a holistic approach is required when developing a new industrial purification strategy. For the whole process, both upstream and downstream factors need to be considered; for example, a cheap carbon and energy source containing many impurities could provide initial cost savings but necessitate increased DSP costs [4–7]. Hence, overall cost savings could be achieved with a more expensive but purer product.

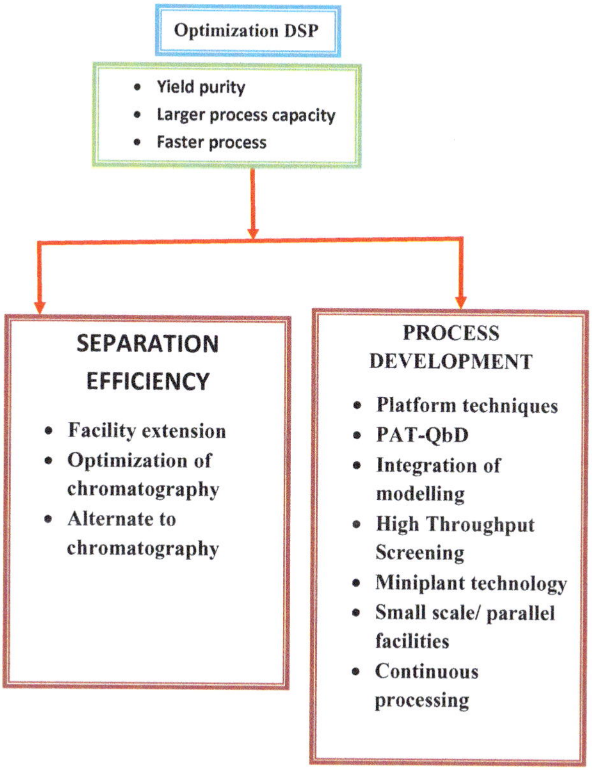

FIG. 2.2

Optimization fields in downstream processing.

2.2.1 Effect of fermentation condition on filtration

As discussed earlier, the water content of the harvested broth is very high with a lot of impurities in terms of cell debris, unused nutrients, by-products, etc., which affect the flow of broth through the filter system. Protein-containing constituents of media, e.g., soybean meal or yeast extracts, impede filtration. An increase in the content of the chief nutrients in the medium causes, as a rule, excessive growth of mycelium and reduces the permeability of the latter. There is a quantitative correlation between the concentrations of medium constituents and the specific resistance of the filter cake. In addition, the deposit of biomass cake over the filter system becomes a limiting factor while processing filtration. The rate of filtration has been found to change substantially, depending on the duration of fermentation. Mycelial proteins and the amount and state of colloids contained in the aqueous phase have been shown to be the principal factors determining the permeability of the cake.

Biomass separation can be achieved by conventional methods such as rotary filter, sedimentation, and centrifugation but such recovery technique must account for the often-labile nature of the by-product(s), the generally low concentration levels, and the complex composition of the media. Membrane filtration is commonly acceptable for industrialists due to its simplicity and procedural protocol. But the disadvantage of this process is to bear operation cost, which is the magnitude of permeate flux [8,9], owing to particle or colloidal suspensions and macromolecular solutions.

So, the appropriate control of both the filtration rate and the rejection (or transmission) of the particles and/or the macromolecules in the filtration processes is of great interest in both industry and academia. Such filtration behaviors are mainly affected by the properties of the filter cake formed by the accumulation of particles and/or macromolecules on the surface of the medium or membrane.

The cake layer's hydraulic resistance depends on the size and shape of microorganisms (such as cell morphology and particle surface properties), which indirectly determine the void fraction in the cake. In cases of cross-flow filtration, rod-shaped particles are orientated in the direction of flow [10,11]. Cake filtration hydraulic resistance then appears to be higher than for cakes of ellipsoidal particles. The compressive nature of a cell cake and particle surface properties also play a role in hydraulic resistance to filtration [12]. A submicron particle also affects the surface interaction on the rate of particle deposition in the cake [13].

Due to the presence of a variety of materials in the fermentation broth, the nature of these interactions is very difficult to assess. It has been reported that for mixtures of protein and yeast cells, protein fouling of the membrane and total filtration hydraulic resistance are reduced in the presence of cells [14–16].

Broth harvesting conditions, including aging, time, pH, and temperature stirring conditions, can influence both the morphology of cells and release of intracellular components. It has also been noticed that the exposure to nutrient-poor broth during cell holding results in changes in filtration hydraulic resistance as a consequence of DNA release [17].

The filtering of microbes mainly depends on the size of the microorganisms, and the MF-Millipore filters (thickness: 150 μm) are chosen accordingly (Table 2.1).

Commonly, 0.2-μ filtrations are frequently referred to as "sterile filtration," reflecting the general belief that all living organisms are excluded from the filtrate. However, it has been observed that a few bacterial species can pass through 0.2-μ filters. Viruses are considerably smaller and will pass through a bacterial filter.

2.2.2 Immediate treatment of the harvested broth

The downstream process covers the isolation, purification, and polishing of the target biological product from the fermented broth obtained at the end of the fermentation process. Biopharmaceuticals are to be isolated and purified to super crystallization form and processed for formulation that would be acceptable as body-friendly therapeutic biomolecules.

Biological products are highly susceptible to microsurroundings while in the fermented broth. As stated earlier, the target biological product may be intercellular or extracellular. Immediately after the fermentation process is overfed, the following steps are to be performed on a priority basis.

Table 2.1 Pore size of MF—millipore filter used for some of the common bacteria used for commercial purposes.

Type of bacteria	Pore size
Bacillus subtilis, *Staphylococcus aureus*, *Klebsiella pneumoniae*, and *Escherichia coli*	0.45-μm 48–96 h
Pseudomonas aeruginosa, *Serratia marcescens*, and *Listeria monocytogenes*	0.3-μm
P. aeruginosa	0.22-μm

2.2.3 Clarification of broth

Clarification is typically the first unit operation in the purification of target biologics such as monoclonal antibodies (MAbs) and other proteins from mammalian cell cultures. This process removes cells and cellular debris from the culture fluid, resulting in a clarified cell culture supernatant that will be suitable for further purification [18].

In the downstream fermentation process, the design of any separation system must consider all stages of pretreatment, biomass concentration, biomass separation, and posttreatment. This encompasses a wide range of equipment and processes, as summarized in Table 2.2.

Pretreatment is used primarily with difficult-to-filter slurries, enabling them to be filtered more easily. It usually involves changing the nature of the suspended biomass by physical means or by adding a

Table 2.2 Components of the solid-liquid separation process.

Pretreatment	Physical	
Chemical	Crystal growth	
Flocculation	Addition of filter aids	
Coagulation	Freezing	
pH adjustment	Aging	
Solid (biomass) concentration		
Thickening	**Clarification**	**Assisted separation**
Gravity	Gravity	
Centrifugation	Centrifugation	
Hydrocyclone		
Delayed cake filters		
High-shear cross-flow		
Low-shear cross-flow		
Microfiltration		
Solid (biomass) separation		
Cake/filtration	**Depth filter**	
Pressure	Granular bed	
Vacuum	Cartridges	
Centrifugal	Precoat	
Gravity	Membrane	
Posttreatment		
Filtrate polishing	**Cake treatment**	**Deliquoring**
Microfiltration		Gravity
Ultrafiltration		Displacement
Ion exchange		Mechanical expression
Reverse osmosis		Thermal drying
Electrodialysis		

Table 2.3 Suitability of cell disruption methods for various ample types.

Technique	Description	Yeast	Plant cells	Mammalian cells	Bacteria
Gentle method					
Osmotic lysis	Suspension of cell in hypotonic solution; cell swell and burst Freezing in liquid nitrogen and subsequent thawing of cell	–	–		–
Freeze thaw process	Freezing in liquid nitrogen and subsequent thawing of cell	–	–		–
Detergent lysis	Suspension of cells in detergent solution to solubilize the cell membrane. This method is usually followed by another disruption method, such as sonication	–	–		–
Enzymatic lysis	Enzymatic digestion in isoosmotic solution. Cellulose and pectinase are used for plant cells, lyticase for yeast cells, and lysozyme for bacterial cells				
Harsher methods					
Sonication	Disruption of cells is carried in cool water to avoid heating damage		–		
French press	Application of shear forces by forcing a cell suspension through a small orifice at high pressure		–		
Grinding	Breaking cells with a mortar and pestle			–	–
Mechanical homogenization	Mechanical homogenization with either a handheld device (e.g., Dounce and Potter-Elvehjem homogenizer), blender, or other motorized devices; this method is best suited for soft, solid tissues	–		–	–
Glass-beads homogenization	Application of gentle abrasion by vortexing cells with glass beads		–		

solid (filter aid) to the suspension to act as a bulking agent to increase the permeability of the cake formed during subsequent filtration. In case of concentrated biomass, part of the liquid may be removed by thickening, using gravity or centrifugation or by hydrocycloning to reduce the liquid volume throughput load on the filter (Tables 2.3 and 2.4).

2.3 Microbial biomass separation

A number of new "assisted separation" techniques (Fig. 2.3) are making their way into the list of technical alternatives. These use a magnetic, electric, or sonic force field (or combination) to provide more effective separation.

Table 2.4 The most common grinding parameters for cell disruption in plastic reaction vials (2 mL).

Sample type	Bead type and quality	Frequency Hz	Grinding time second
Plant tissue	1× metal or ceramic bead (1–6 mm)	60	40–90
1× metal or ceramic bead (1–6 mm)	1× metal or ceramic bead (1–6 mm)	65	60
Stomach, lung, and intestines	1× metal or ceramic bead (1–6 mm)	65	90–120
Liver, brain, and spleen	1× metal or ceramic bead (1–6 mm)	60	45–60
80–90 Skin and fat	2× metal or ceramic bead (1–6 mm)	65	60
Blood vessels (veins, arteries, and capillaries)	1× metal or ceramic bead (1–6 mm)	60	60
Bacteria	25–50 mg glass bead (0.1–0.5 mm)	60	60
Yeast	600 μL glass bead (0.1–0.5 mm)	60	60

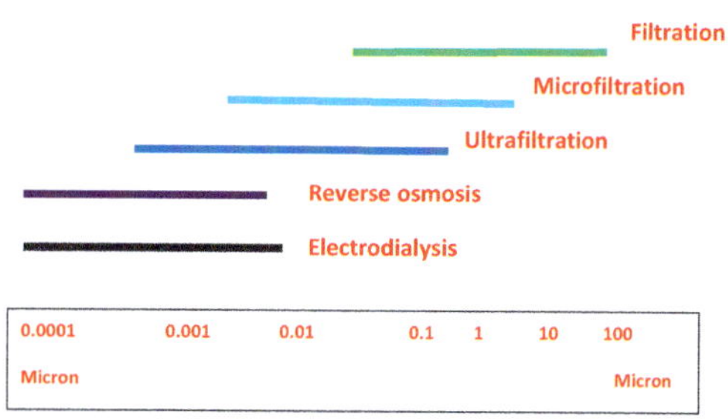

FIG. 2.3

General techniques for the removal of contaminants from liquids relative to the size of the species to be removed.

Separation of solids involves filters, which are classified in many different ways. For the present purses, a division into those in which cakes are formed and those in which the particles are captured in the depth of the medium is adequate. Cake filtration can be further divided into pressure, vacuum, centrifugal, and gravity operations.

Posttreatment processes involve making improvements to the quality of the solid or liquid products. In the case of the filtrate, these operations are often referred to as polishing processes and may involve micro- or ultrafiltration to remove fine substances. Further purification could involve the removal of ionic and macromolecular species by, for example, reverse osmosis, ion exchange, or electrodialysis. The relative position of these separation processes in the spectrum of the size of "particle" to be removed from the liquid is shown in Fig. 2.3. Cake posttreatment processes include washing soluble impurities from the cake voids and removing excess liquid from the voids. Thermal drying is often the final stage of liquid removal.

2.3.1 Principles of separation technique

2.3.1.1 Single-stage process

Single-stage systems can come in many designs, including continuously stirred tank reactors and plug-flow digesters, each with different modes of operation and differences in design and operation. Generally, single-stage systems are simpler than two-stage systems and are cheaper to construct and operate.

Potential limitations of single-stage systems are that conditions within the reactor are not optimal for the various trophic groups of microbes. Although two-stage systems can offer more optimal conditions for methanogenic bacterial populations, it does not mean that single-stage systems are unreliable. Methanogenic bacterial populations can be managed by controlling the feeling rate; thoroughly mixing incoming wastes to avoid peak concentrations of potentially harmful contaminants; co-digesting with other organic wastes to provide essential water content, buffering nutrients, and trace elements; or adding these requirements through the use of chemicals and nutrients.

By means of mathematical models, we can consider a single-stage concentration process as the operation of passing a capture element (in reality, the final volume) through a sample N times greater in volume, until it has passed through the whole sample and captured the target microbes from it. If it takes t seconds for the capture element to pass through its own volume in the sample-specific sweep time), and this remains constant during the capture pas, the time T required for the capture element to reach the end of its pass is as follows:

$$T = t(N - I)$$

2.3.1.2 Two-stage process

The sample can be concentrated in two stages, each of which is less efficient (yields a smaller increase in analytic concentration) than the other. Instead of just one, use n capture elements for the first stage and, after concentrating the sample into these, concentrate these n elements into one in the aforementioned manner. The total time T for concentration is as follows:

$$T = (T_1 + T_2) = t(n + (N/n) - 2)$$

It is minimal when $n = N^{1/2}$, that is, when the overall concentration factor is achieved in two stages of approximate equal effectiveness.

2.3.1.3 Three or more stages

A three-stage process is a two-stage process with a stage added at the front end. If t is the same for all the three processes, the total time needed is as follows:

$$T = t\left[2n + \left(N/n^2\right) - 3\right]$$

It is minimal, when $n = N = N^{1/3}$; that is, three stages of approximately equal effectiveness yield the minimum concentration time. Similarly, p multiplexed processes execute in a minimum time when $n = N^{1/p}$.

The disadvantage of multistage systems is that they are often more complex and therefore more expensive than single-stage systems. The implementation of a digestion system with "support material" is, however, only advised for effluent-type feedstock with low suspended solids; otherwise, digesters are prone to clogging.

Some advantages of multistage systems are as follows:

- Greater biological stability
- Greater ability to cope with fluctuating feedstock volume and quality
- Potentially higher throughputs due to optimal conditions

2.3.2 Filtration

Filtration is a process in which solid particulate matter is removed from a fluid, which can be either liquid or gas, using a porous medium. This process is widely used in the biopharmaceutical industry to separate unwanted materials present in the fermented broth. The efficacy of filtration depends on many factors, such as the size of the organism, the presence of other organisms, viscosity of the medium, and temperature. The filter medium constitutes the separating agent and retains particles according to size while allowing liquid to pass through the filter. In cake filtration, the particles are retained as a cake on the filter medium. The flow through the filter layers is dependent on the area of the filter and the flow resistance provided by the filter medium and cake. If the particles do not penetrate the filter medium, then its flow resistance remains unchanged. However, as the cake layer thickens, it provides increasing resistance. The cake layers, especially in biomass separation, are compressible, and the changing effective pressure difference influences flow through the filter. Filter media can be perforated sintered metal, cloth, synthetic fibers, cellulose, glass wool, ceramics, or synthetic membranes. Several types of filters, such as depth filters, absolute filters, rotary drum vacuum filters (RDVFs), and membrane filters, are commonly used and are described next.

2.3.3 Filter press

A filter press is a tool used in separation, especially in solid-liquid separation using the principle of pressure-driven membrane filtration. History recounts that the first form of a filter press was contained in a British patent from 1853 for obtaining seed oil through the use of numerous pressure cells. The main advantages of the pressure are high filtration velocities and low moisture content in the formed filter cakes, as a result of the high filtration pressures that can be applied. Consequently, filter pressures have long been used for a wider range of applications than any other type of filter. By the middle of the twentieth century, the requirement of mass production in industry had resulted in increasing attention being paid to the automation of filter press operations. A fully automatic horizontal filter press and various other new types were developed in Japan. Filter presses are sometimes called "plate- and-frame filters." This refers to the style of filter element that was most prominent in the mid-1800 until the late 1960s. Although this specialized style of the filter press is still being manufactured, the current manufactured units are more accurately referred to as "recessed chamber" and/or "diaphragram (membrane) filter press" (Fig. 2.4).

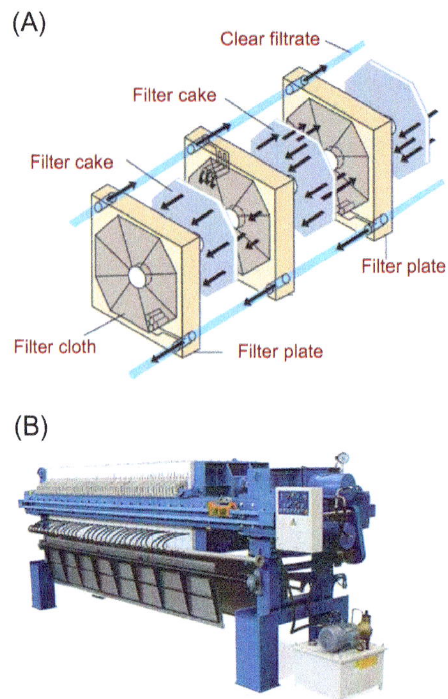

FIG. 2.4

(A) Frame of a filter press showing filter plate, filter cloth, and filter cake; (B) a photo of an industrial type of filter press.

Courtesy of Evoqua Water Technologies.

2.3.3.1 Working principle of a membrane filter press

The basic principle of a filter press is surface filtration driven by pressure. A filter press is built of a sequence of perforated plates alternating with hollow frames (Fig. 2.4A). The plates are covered with a suitable filter medium (cloth) that creates a series of chambers through which the slurry can be forced. Solids are retained in the chambers, and the filtrate discharges into the hollows on the plate surface and drains out.

The slurry deposits solids on the filter cloth as the filtrate leaves the press via discharge ports. The solids continue to deposit and eventually form a cake in the spaces between the filter plates. The pressure behind the slurry (typically 100 psi, but up to 900 psi; 7–60 bar) is provided by a feed pump, which is sometimes a positive displacement or centrifugal pump. With a gravity drain on the filter side of the press, a pressure differential between the feed pressure and the gravity discharge is created across the medium and the filter cake solids as they thicken (Fig. 2.5).

It is the existence of this pressure differential, not just the feed pump pressure, that causes the filtering action to occur. Unwanted solids within the harvested broth flow to the area of cake development with the lowest pressure differential, resulting in a filter cake that builds uniformly over the drain field on either side of the chamber walls.

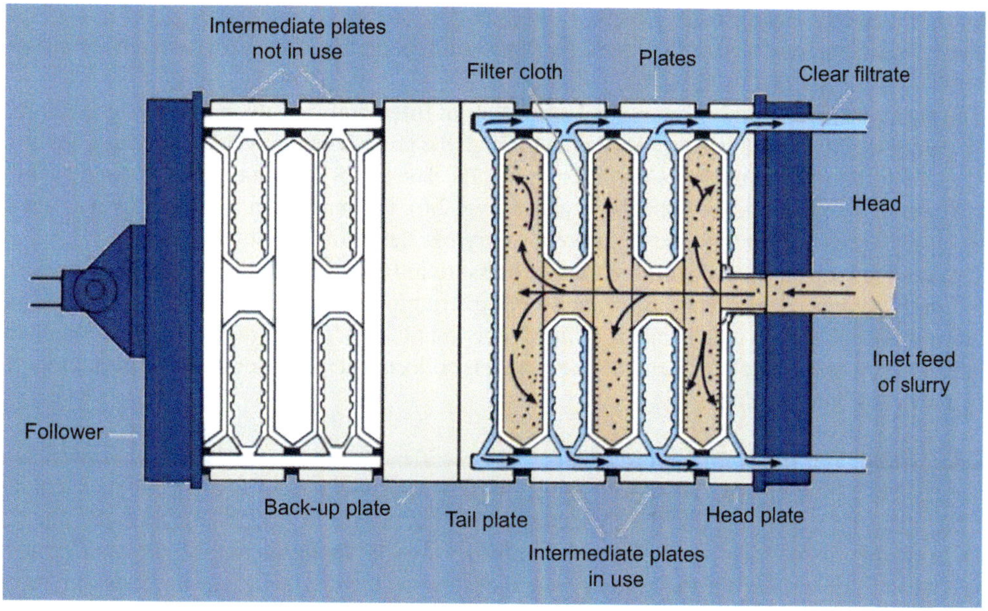

FIG. 2.5

Working principles of a plate filter press.

This fill cycle continues until the filter cakes forming on the chamber walls bridge at the center, completely filling the press with solids. It is at this point that the filtration process becomes complex. Once this is achieved, the hydraulic closer of the press is retracted, the individual filter elements are separated, and the filter cakes are discharged, usually by gravity, to an appropriate receptacle.

The filter cake chambers are formed in either of two ways: by mating two recessed chamber plates or by two flush plates with a cake frame (much like a picture frame) sandwiched in between. The two faces of the filter plate have a drainage surface in the form of ribs, groves, or pips to allow the filtrate to drain behind the cloth to the drainage ports located in each cone of the filter plates. These ports, in turn, are connected to the corner eyes, which carry the filtrate drainage to the fixed end of the filter press. A filter cloth is mounted over each of the two faces of the filter plate. The cloth is joined at the feed eye by an impervious sleeve or tube, known as the barrel neck.

2.3.4 Types of filter press

Filter presses are used across a wide variety of applications, including wastewater, food and beverage, chemicals, mining, and paper industries. Filter presses are predominately used in the process of solid-liquid separation. The three main types of filter presses used in the industry today are (1) the plate-and-frame filter press, (2) the recessed or chamber plate filter press, and (3) the membrane plate filter press. These presses and their major components are described:

2.3.4.1 Plate-and-frame filter press

This is the most common type of filter used for liquid-sold separation (Fig. 2.6) and mainly consists of a filter frame and filter plate (Fig. 2.7).

The plates are metal frames containing large sheets of filter cloth, usually polypropylene. The plates are squeezed together, and the slurry is filtered through the press. The slurry deposits solids on the filter cloth as the filtrate leaves the press via discharge ports. The solids continue to deposit and eventually form a cake in the spaces between the filter plates (Fig. 2.6). For maximum filtrate recovery, air can be blown at high pressure through the filter press to dry out the cake almost completely.

Orifices feed the sludge to be filtered under pressure into the filtration chamber. They are usually placed in the center of the plates, allowing proper distribution of the flow, right pressure, and better drainage of sludge within the chamber. Solid sludge gradually accumulates in the filtration chamber until the final compacted cake is formed. The filtrate is collected at the back of the filtration support and carried away by internal ducts.

2.3.4.2 Recessed plate and filter press

A recessed plate filter press can compact sludge into a dry cake with a density that is dependent on the ultimate compaction pressure and the characteristics of the particular sludge. The recessed plate filter press, or chamber plate filter press as it is also known, is similar in design to the plate-and-frame filter press but has no frame. Each plate serves as both the plate and the frame because the plates are recessed to form the filtrates that are clamped together. The slurry enters through a large central port, filter cake forms in the recessed hollow chamber, and the filtrate passes through the filter medium. As with most filter presses today, recessed plate filter press plates are constructed from polypropylene.

2.3.4.3 Membrane plate filter press

The membrane filter press is a development of the standard recessed plate filter press. The filter construction is similar to that of a recessed plate filter press, but the plates incorporate membranes that are inflated at high pressure to squeeze the filter cake. Normally, the plate pack consists of alternate recessed plates and membrane plates. This design makes it possible to achieve superior results from

FIG. 2.6

Photo of an industrial plate filter press.

FIG. 2.7

Photo of a filter plate covered with filter cloth.

thin cake slurries and can reduce the cycle times of a batch. Membrane plates usually operate at a feed pressure of up to 7 bar and a squeeze pressure of up to 15 bar. Special plates can be designed to accept higher feed and squeeze pressures.

Use of the membrane filtration technique

The pharmaceutical and cosmetic industries typically focus on monitoring their process water for *Pseudomonas* species. The electronics industry monitors for any and all microorganisms because they must keep their process water free from even the smallest organisms. Microbial monitoring in the food and beverage industry typically employs several types of techniques because of the variety of samples that are encountered. Beverage samples can typically be monitored for microorganisms by the MF technique, but when solid samples cannot be liquefied, alternative methods must be used. MF is an effective and acceptable technique for testing fluid samples for microbiological contamination. It involves less preparation than many traditional methods and is one of the few methods that allows the isolation and enumeration of microorganisms. The MF technique also provides information on the presence or absence of microorganisms within 24 h.

2.3.5 Surface filtration (membrane filtration)

The surface filtration process is defined by the retention of particles on the surface of the matter being filtered. Surface filters are made of thin materials such as papers, woven wire, and cloth. In surface filtration, the particles are retained mostly on the surface of the media, forming a layer of material that increases the efficacy or fineness of the particles retained. Generally speaking, this type of filter media is referred to as having "nominal retention"; perhaps initially being 60% to 70% efficient at retaining the targeted particle size and, as the "cake layer" develops, eventually becoming nearly 100% efficient. Nominally rated media is the most common and is less expensive than depth media. The amount of surface area directly correlates to the solid loading capability and related pressure drop (Fig. 2.8).

2.3.6 Ultrafiltration

After fermentation, the liquid broth contains a high density of microorganisms that must be separated from the liquid phase containing the target biologic. This separation is known as fermentation broth clarification and consists of a physical separation of suspended solids from the liquid phase. The first step is the clarification of broth to separate the low molecular weight soluble production from microorganisms and other particulate matters such as cells (in case the target product is extracellular), cell debris husks, colloids, and macromolecules. The two main solutions industrially applied for fermentation broth clarification are filtration and centrifugation. Cross-flow filtration (also known as tangential flow filtration) is used in biotechnology manufacturing. Cross-flow filtration gets its name because the majority of the feed flow travels tangentially across the surface of the filter, rather than into the filter.

Among the different target molecules, some commercially important examples include penicillin and lactic acid recovery from fermentation broth via microfiltration by using microporous membrane filters (MWCO 10,000 to 500,000 Da (Fig. 2.9A–C).

2.3.7 Depth filtration

Depth filter system is commonly used for the clarification of fermented broth before it is subjected to downstream purification. By this filtration process, insoluble particles, soluble materials, and colloidal materials are isolated from the culture broth. Depth filtration is also widely used for the clarification of fermented broth. Cellulosic depth filters can remove significant amounts of DNA, host cell proteins, and virus depending on depth filter media composition and feed-stream pH and conductivity. Generally, depth filters used in biopharmaceutical industry are made of cellulose fibers and filter aids (e.g., diatomaceous earth) bound together by a polymeric resin, which provides the necessary wet strength and imparts a cationic surface characteristic.

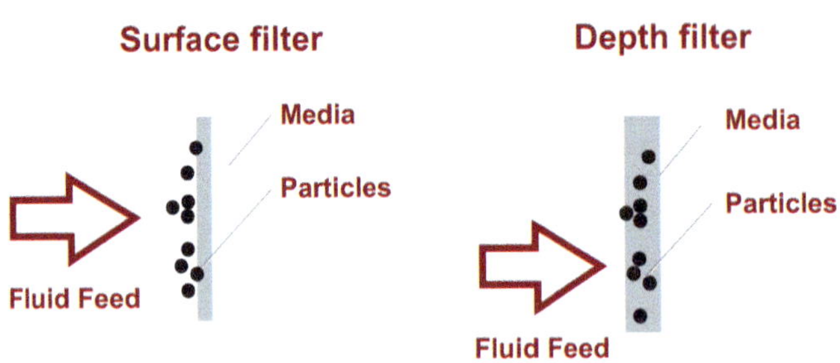

FIG. 2.8

Difference between surface and depth filter.

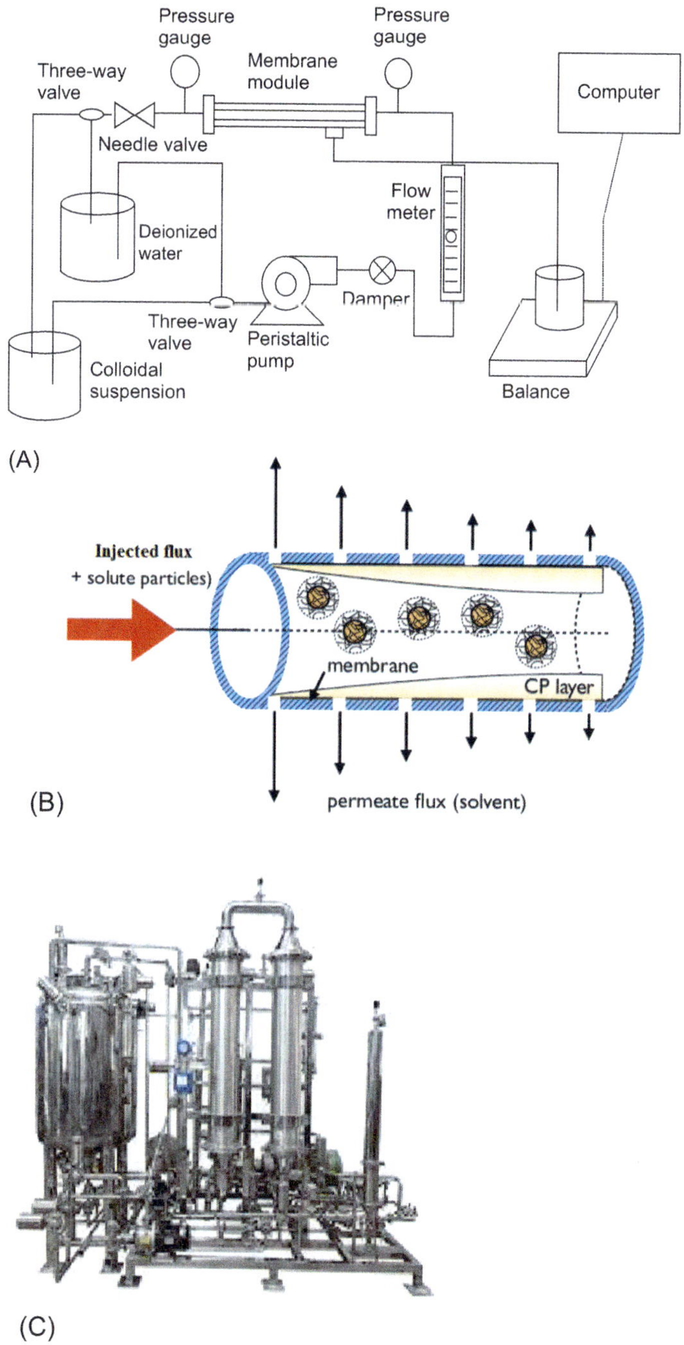

FIG. 2.9

(A) Cross-flow ultrafiltration system; (B) cross-flow ultrafilter line diagram; (C) industrial-level cross-flow filtration system.

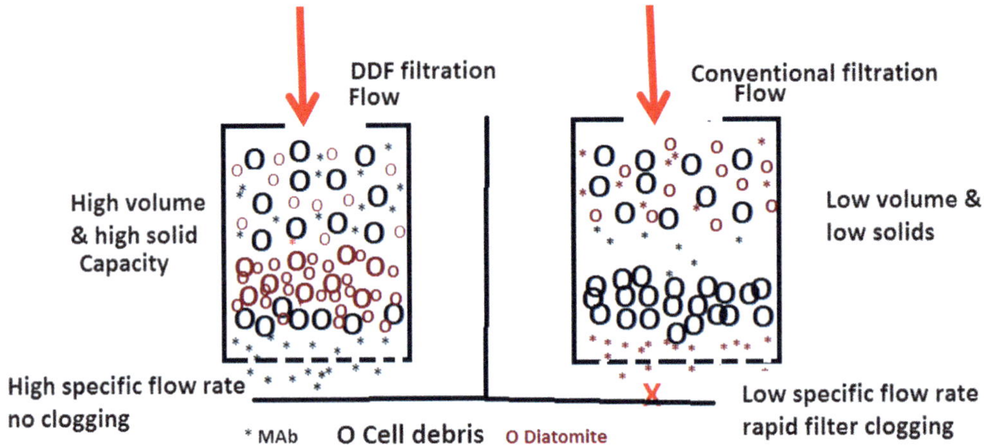

FIG. 2.10

Depth filtration where diatomaceous earth is used as filter aid for the separation of monoclonal antibodies (MAb).

Depth filters typically consist of three components: (i) wet-laid cellulose or polypropylene fibers, (ii) a filter aid such as diatomaceous earth (DE) or perlite, and (iii) a resin binder that typically possesses a positive charge or can be functionalized with common charged legends, such as quaternary amines. Additional depth filter options may incorporate an absolute filter membrane layer available in varying pore sizes as small as o.22 μm. One of the best examples of explaining the efficacy of the diatomaceous earth filtration system is its use in the separation of MAbs from the fermented broth (Fig. 2.10).

The first use of diatomaceous earth (DE) as a filter aid in fractionation of human plasma was reported over five decades ago [19]. Since then, a number of manufacturing processes for the production of intravenous immunoglobulin (IVIG), albumin, and clotting factors have been developed based on that technology [20–22]. Fractionation uses the principles of selective precipitation by pH adjustment, ionic strength, addition of alcohol, and temperature shifts. Precipitation is removed by depth filtration often in combination with diatomaceous earth, which is added as an aid to increase filter throughputs. This method is also widely used for the separation of bacteria and CHO [23].

Following are the few regularly used filter aids:

- Positively charged particles are modified for the separation of negatively charged particles.
- Silica is used for the removal of lipids.
- Carbon is impregnated for specific adsorptive properties.
- Low pyrogenic for pyrogen-sensitive applications.
- The particles are trapped within the matrix, and the fluid passes.

2.3.7.1 Large-scale depth filtration

Large-scale depth filtration systems are common in the bioprocess industry. They use multiround stainless housing with up to 16-in-depth filter modules stacked three or four high to maximum size surface area. Fig. 2.11 shows an industrial-scale depth-filtration harvested system consisting of multiple housings and cartridges for clarifying cell harvest from a 15,000-L bioreactor.

FIG. 2.11

Large-scale depth filter for industrial use.

Before harvest, depth filters are flushed with water or an appropriate buffer to remove loose particulates and extractable from the filter manufacturing process. Once harvesting is completed, the filters are again flushed to recover valuable product held in the housing. Harvest yields as high as 95% are achievable using a postuse flush and ensuring minimum product loss.

Depth filters are widely used in the bioprocess industry for clarifying cell culture broth. Single-use depth-filtration technologies can be used for purposes other than the clarification of cell culture streams. Because the depth-filter media can carry a surface charge, depth filters retain particles smaller than their rated pore size. Studies have shown that, depending on process design, depth filters can also be used in orthogonal virus removal steps, providing up to 4-log viral clearance [24–26].

Other studies have shown that charged depth filters can remove nonenveloped RNA viruses (such as mouse minute virus) at a process capacity of $320\,\mathrm{L\,m^{-2}}$ from a feed containing 2% spiked virus.

2.3.8 Absolute filters

A filter rating indicates the size of the largest particles that can pass through a screen or filter. An absolute filter can cut off 100% (by weight) of solid particles greater than a stated size; the pore size of an absolute filter is the pore size at which any particle of a particular size can be retained with 100% efficiency under strictly defined test conditions. The conditions that should be specified are challenge pressure, detection method, test organism, and concentration methods used for identifying the contaminants.

2.3.9 Vacuum filtration

There are two main types of vacuum filtration (VF): rotary drum filters and horizontal belt filters, both of which operate continuously. Vacuum filters are widely used for clarification of fermentation broths (containing 10%–40% solids by volume with particle sizes of their simplicity of operation and low cost.

2.3.9.1 Rotary drum vacuum filter

The best-known vacuum filters are the RDVFs, which are used for the filtration of filamentous fungi and yeast cells (Fig. 2.12).

They are used to clarify large volumes of liquid, with automatic solid discharge. Tangential flow (cross-flow filtration) is an effective method for separating cells from liquid where high-value product is involved. The parallel motion of fluid to the membrane helps reduce the thickness of the cell layer on the filter surface. RDVFs consist of a rotating drum lined with filter medium that has approximately one third of its outer wall submerged in the slurry. The filtrate is pulled through the wall of the drum by vacuum, toward the center of the drum. The drum rotates such that the cake is dumped into the discharge outlet. That section of the drum wall is then washed before performing filtration again. The process of the RDVF is continuous. Each revolution of the drum consists of cake formation, cake washing (if needed), dewatering or drying, and cake discharge (Fig. 2.13).

Drum filters are designed for all types of liquid-solid separation, including medium concentrate sludge and low-density solids, which do not require efficient washing.

The RDVF, supported by two bearing blocks, is composed of successive cells, which form a horizontal cylinder. The cylinder, in turn, supports the filter cloth, which is partially immersed in the tank. The cells are sealed and interconnected by pipes leading to the distributor, which is divided into several zones, each gathering the various filtrates (mother liquor, washing water, and drying water) coming from the cell suction pipes as they pass during the rotation of the drum. The slurry passes through the trough-shaped tank to be filtered. The tank, which is part of the filter frame, also supports the drum, the cake discharge system, and the slurry agitation device. For cake discharge, several methods can be used such as a string, belt, scraper, or roller.

FIG. 2.12

Front view of the rotary drum vacuum filter.

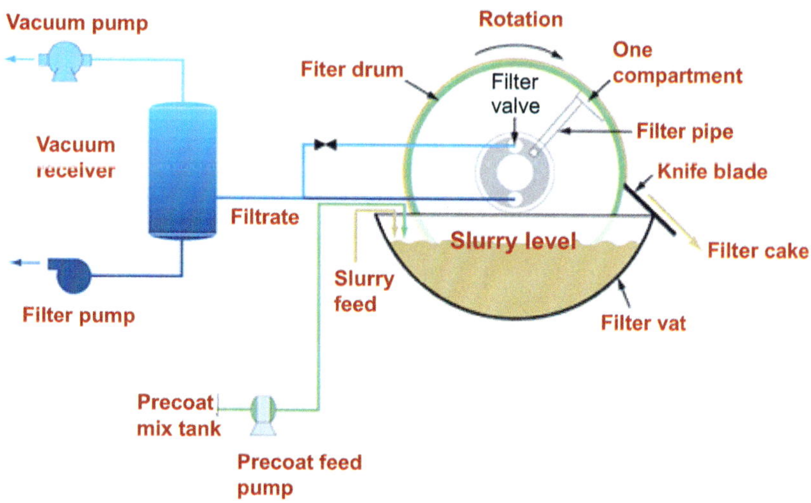

FIG. 2.13

Rotary drum vacuum filter in working condition.

2.3.9.2 Horizontal belt vacuum filters

Horizontal vacuum belts are an ideal solution for applications with very large flow rates, stringent low cake moisture requirements, and the need for highly efficient cake washing as a process step. Horizontal vacuum belt filter (HBVF) presses are used to remove water from liquid wastewater residuals and produce a nonliquid material referred to as "cake." Dewatered residual, or cake, varies in consistency, from that of custard to moist soil. Dewatering serves the following purposes:

- Reduce the vacuum, thus reducing storage and transportation costs
- Eliminates free liquids before landfill disposal
- Reduces fuel requirements if residuals are to be incinerated or died
- Produces a material with sufficient void space and volatile solids for composting when blended with a bulking agent
- Avoids the potential of biosolid pooling and run-off associated with the land application of liquids
- Optimizes subsequent processes such as thermal drying

Horizontal belt filters provide high extraction efficiency, low cake moisture, increased production, and reduced operating costs while achieving maximum filtration area compared with other filter options. Horizontal belt filters maximize cake purity at a minimum specific cost. They are especially suitable for applications requiring low cake moisture and multistage washing. Low energy consumption and a high filtration rate make horizontal belt filters an excellent choice for a wide range of processes (Fig. 2.14).

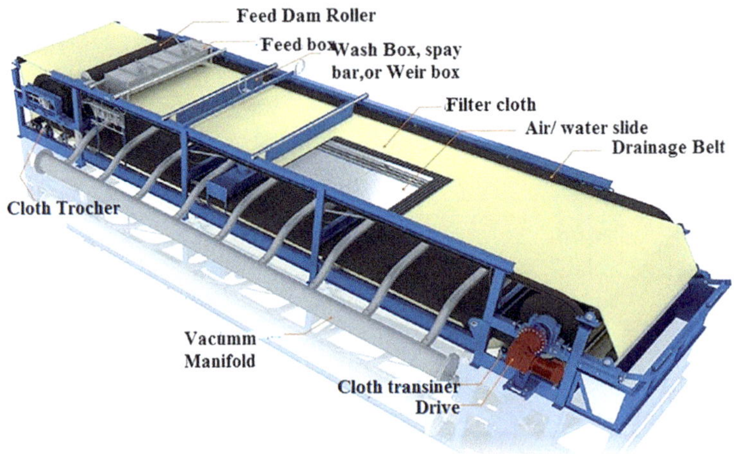

FIG. 2.14

Photograph of the front view of the horizontal vacuum belt filter, showing different functional parts.

2.4 Centrifugation

Before 2000, depth and tangential-flow microfiltration were standard clarification technologies in the biopharmaceutical industry. Process-scale centrifugation was considered to be a significant capital investment, and bioprocessors had limited ability to control the effects of shear on mammalian cells. The following decade saw an industry-wide adoption of centrifugation with a traditional periodic discharge of solids for primary clarification of mammalian cell cultures. That shift to centrifugation was facilitated by the need to scale processes up to handle 10,000 to 15,000-L bioreactors, with increasing harvest cell culture densities and titers. Significant technological advances in centrifugation design and operating conditions minimized shear effects on mammalian cells [27].

The basic principle for centrifugation is based on density differences between the solid product to be separated from the culture broth (i.e., liquid phase). Following are a few common methods used at the industrial level to clarify fermented broth.

2.4.1 Tubular centrifuge

Tubular centrifuges (TCs) comprise a bowl, a specially designed slow acceleration motor, and starter (Fig. 2.15).

The bowl rotates at 15,000 rpm generating a centrifugal force of 16,000 times the gravitational force. The liquid mixture to be separated enters the nozzle of the centrifuge placed at the bottom of the base. The centrifugal force acts on the liquid entering into their specific gravities. The lighter liquid forms the inside layer and the heavier liquid forms the outside (toward wall to bowl) layer. Since the mixture is entering continuously into the bowl, two phases are discharged continuously from two

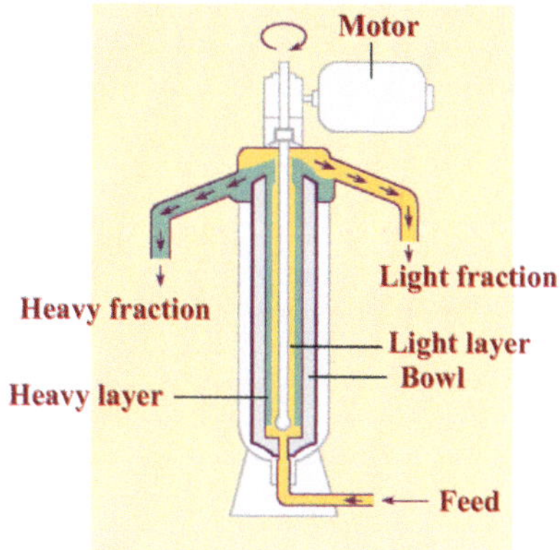

FIG. 2.15

Line diagram of a tubular centrifuge.

separate holes provided on the top portion of the bowl. Solids accumulated inside the bowl can be removed manually after stopping the machine. TCs are commonly used for biocell separation, *E. coli*, plasma/blood fraction, protein precipitation, and yeast cell debris.

2.4.2 Disc centrifuge

Disc centrifuges (DCs) are used for the separation of particle sizes in the range of 3–30 μm. This type of centrifuge can generate forces ranging from 4000 to 14,000 times the gravitational force, thus reducing separation time. These are the most common types of centrifuges to separate microbial biomass in the turnkey scale (Fig. 2.16).

The disk-type centrifuge consists of a vertical stack of thin disks in the shape of cones. The sedimentation process occurs in the radial direction in the space between adjacent cones. The angle of the cones is designed in such a way that upon reaching the inside surface of the cone, heavier material slides down along its surface in a manner that is similar to that of the 37° fixed-angle bottle centrifuge.

Centrifugation is used in a number of bioprocess applications, including clarification of solutions from hormone, vaccine, and enzyme production; bacterial culture harvest; and concentration and separation of blood fractions. Typical centrifuges used for such applications are tubular-bowl designs used in low-volume processes, decanters used when feed-solid concentrations are high (e.g., in sewage water separations), and disc-stack centrifuges.

The latter are commonly used for clarification in the biopharmaceutical industry. Disc-stack centrifuges are fed continuously but discharge solids either periodically or continuously [28]. Continuous solid-discharge disc stack centrifugation has been used in other industries, such as brewing, for the robust removal of solids from process streams [29].

Continuous-discharge centrifugation differs from periodic-discharge centrifugation (Fig. 2.17A and B) in that the solids move upward through channels above the disc stack for continuous discharge of solids streaming out through discharge piping.

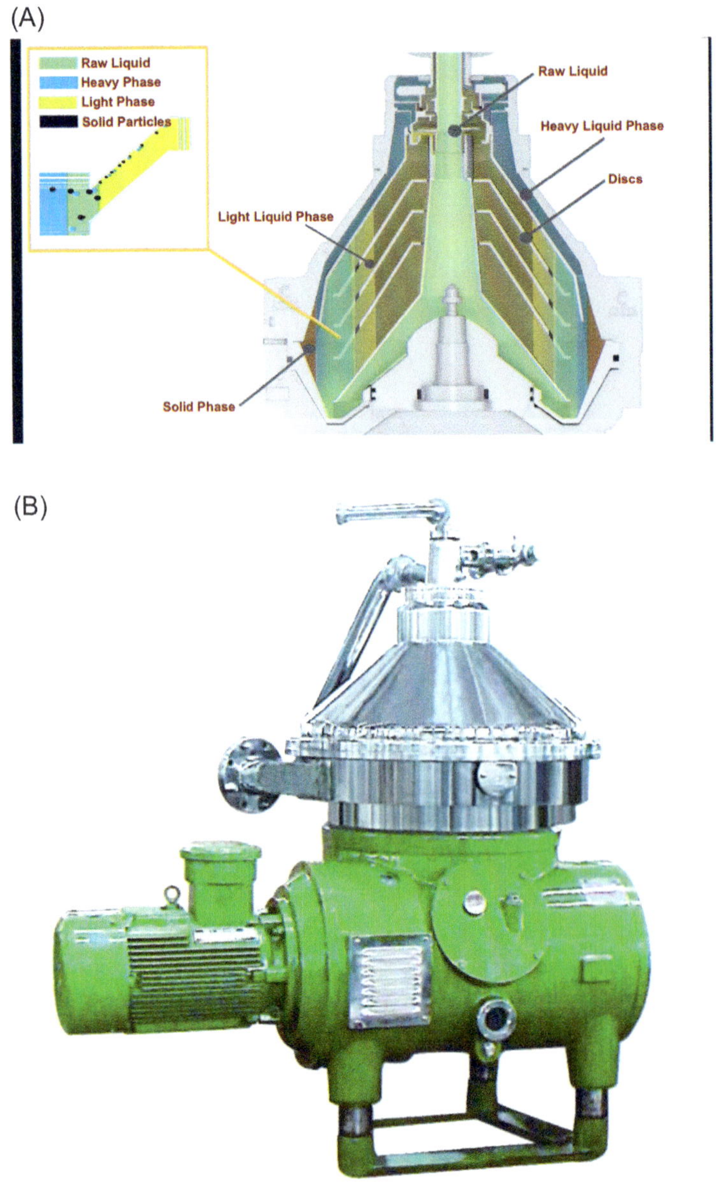

FIG. 2.16

Disc centrifuge, (A) functional line diagram. (B) Photo.

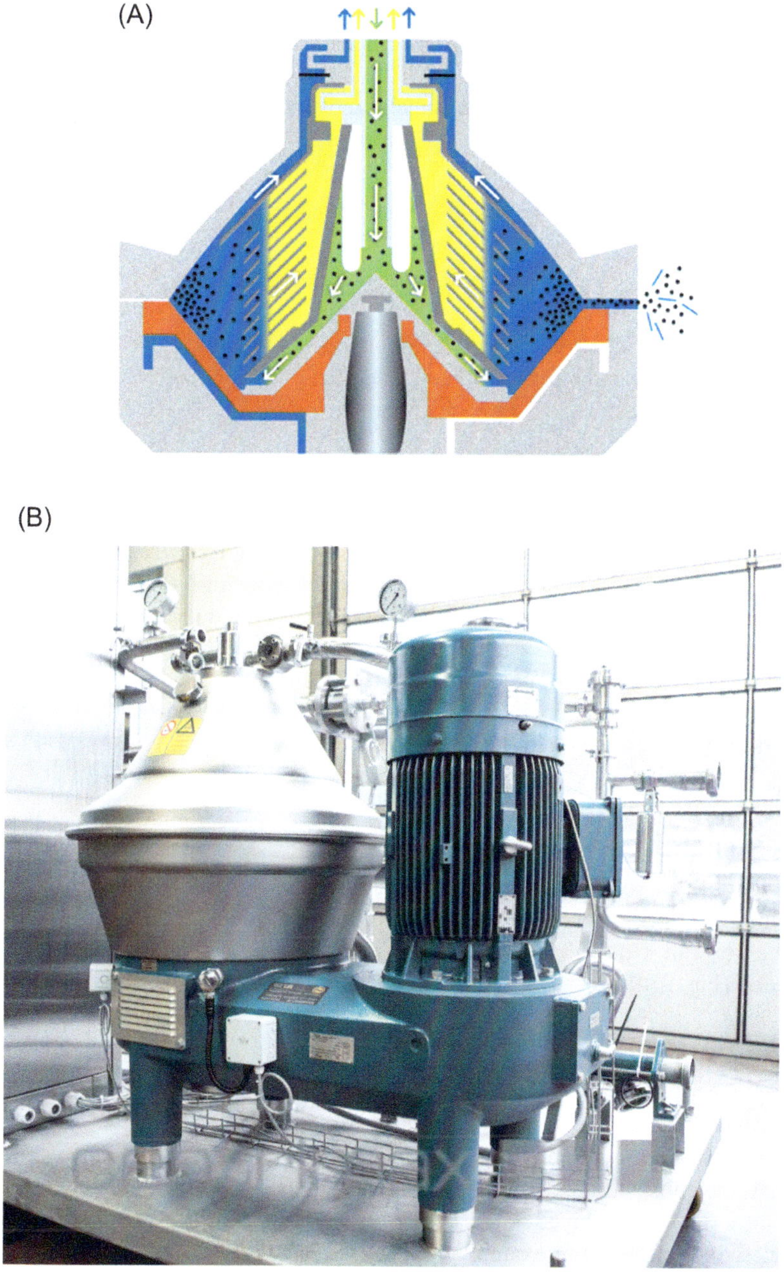

FIG. 2.17

(A) GEA disc-stack centrifuge for solid discharge. (B) Photo.

Continuous solid-discharge disc stack centrifugation has been used in other industries such as brewing for the robust removal of solids from process streams. Continuous-discharge centrifugation differs from periodic-discharge centrifugation (Fig. 2.17A and B) in that rather than solids being discharged periodically, they travel upward through channels above the disc stack for continuous discharge through nozzles into a chamber that has its own centripetal pump (larger than that for centrate collection). It pumps a highly concentrated solid stream out through discharge piping. The percentage of solids in that concentrate stream is relative to its flow rate, which is controlled by a valve.

2.4.3 Multichamber centrifuge

In the multichamber centrifuge (MCC), a series of concentric chambers are mounted within the rotor chamber. The broth enters via the central spindle and then takes a circuitous route through the chambers. Solids collect on the outer faces of each chamber.

2.4.4 Scroll centrifuge or decanter

Scroll centrifuge is also known as a decanter centrifuge or horizontal bowl centrifuge and is commonly used for the clarification of fermented broth to separate biomass in biopharmaceutical industries. A decanter centrifuge is versatile in nature and is used in a variety of industries for solid-liquid separation. The basic operating principle of this system is based on separation via buoyancy. In contrast to a chamber filter press, a solid scroll centrifuge or decanter (SC) can be operated continuously. The high centrifugal force generated during operation is responsible for the separation of finely distributed nonsoluble particles in the culture (Fig. 2.18).

2.4.5 Basket centrifuge

In basket centrifuge (BC), solids/biomass are pulled radically away from the liquid by centrifugal force and collected along the inner wall of the basket. The clarified liquid builds up on the inside and flows up and over the basket wall. On the basis of customer requirement, fully automatic or semiautomatic facility is provided (Fig. 2.19).

2.5 Flotation

It is a simple process of removing biomass from the culture broth. By introducing an inert gas into the culture medium, the cells and other solid particles get absorbed in a gas bubble (Fig. 2.20).

Rising bubbles in the foam layer can be collected and removed. The flotation technique is widely used for a variety of purposes. For example, lipopeptide biosurfactant from *B. subtilis* cultures is removed by continuous bubbling of inert gas for generating foam. The lipopeptides are absorbed in the gas bubble [30,31].

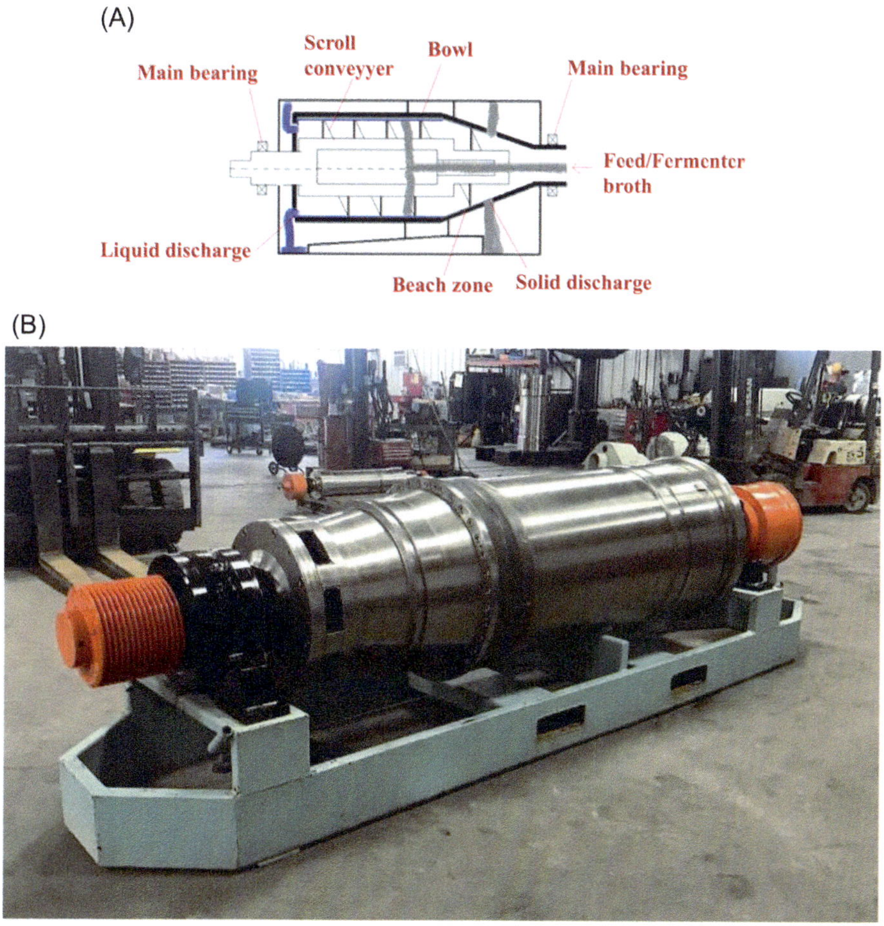

FIG. 2.18

(A) Scroll/conveyyer line diagram. (B) Industrial type photo.

2.6 Flocculation

It is a process in which the suspended hosts are aggregated and separated through the settling process. The process of flocculation depends on the nature of cells and ionic constituents of the medium. The addition of flocculating agents such as inorganic salt, organic polyelectrolytes, and mineral hydrocolloids is often necessary to achieve appropriate flocculation (Figs. 2.21–2.23).

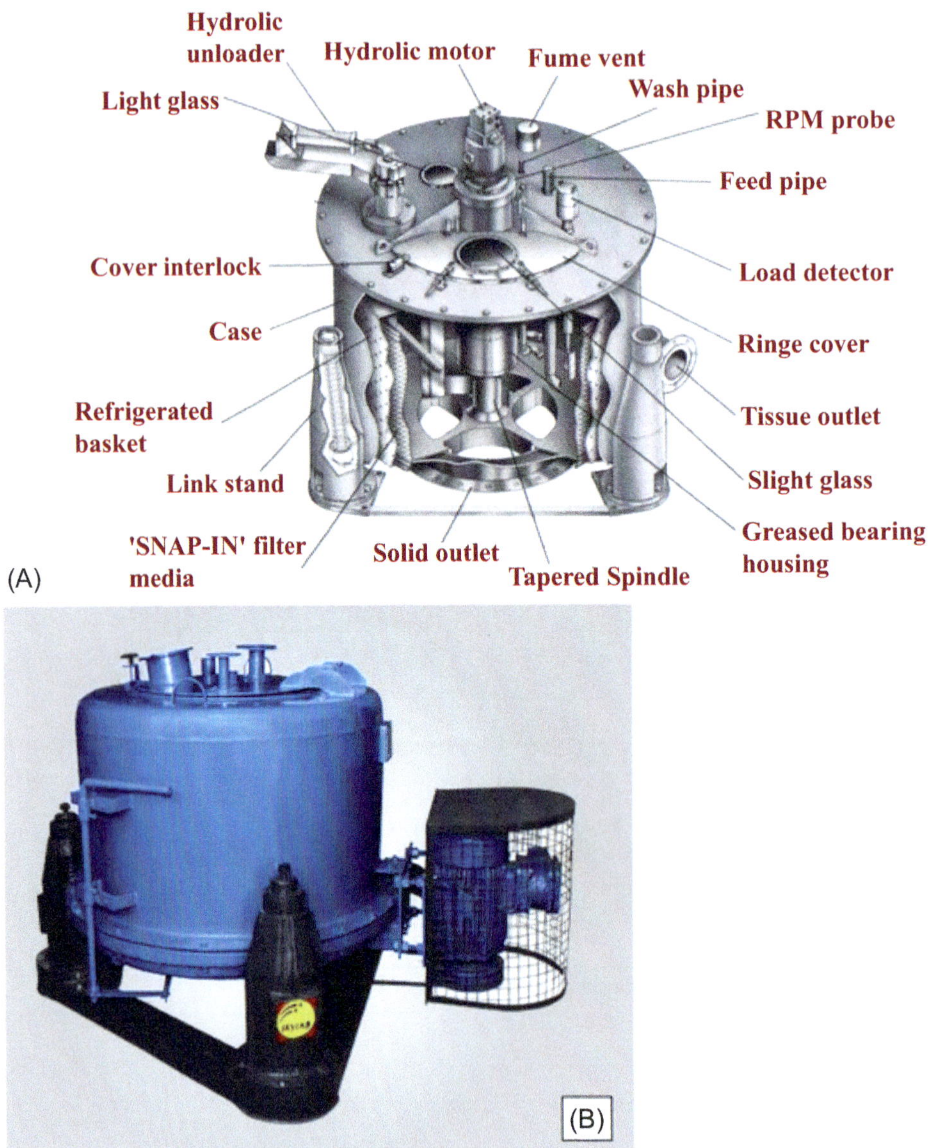

FIG. 2.19

(A) Three-dimensional view of a basket centrifuge; (B) commercial-type basket centrifuge.

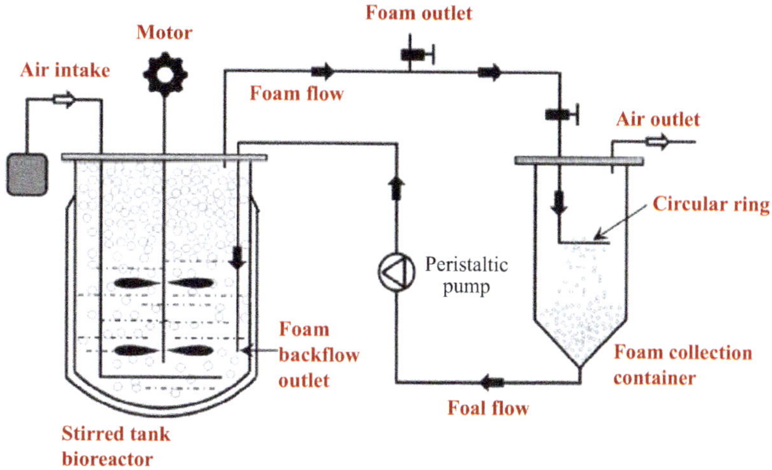

FIG. 2.20

The recovery and purification of a fermentation product through foam formation.

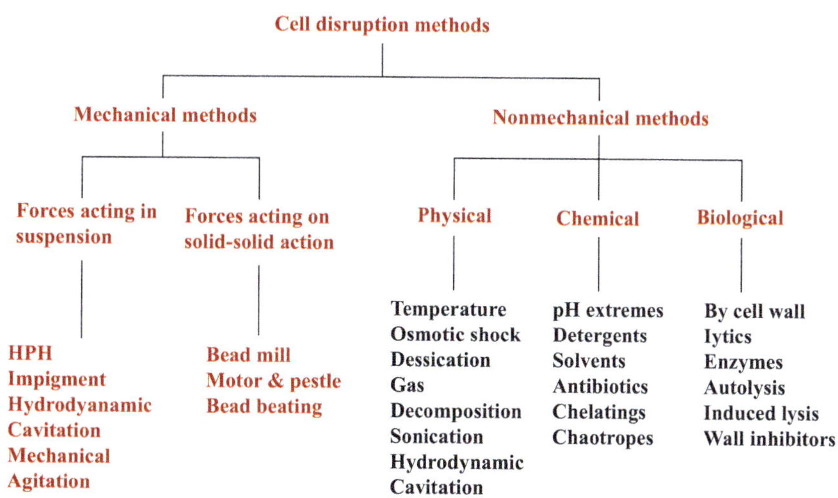

FIG. 2.21

Various methods used for disruption of cells.

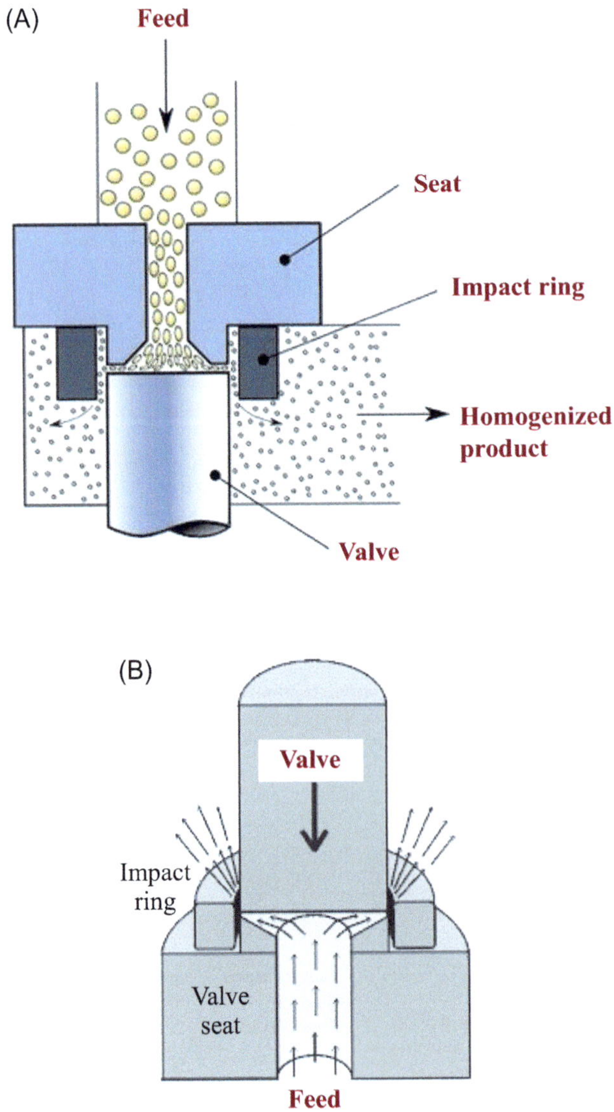

FIG. 2.22

Sterilization and homogenization by means of continuous ultrahigh pressure, also known as UHPH.

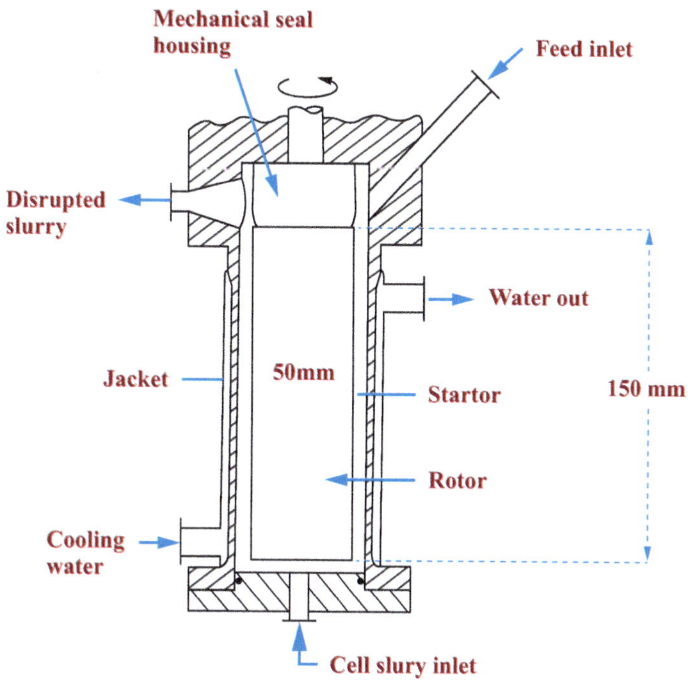

FIG. 2.23

Scheme of a grinding mill for the disruption of cells.

References

[1] A.A. Shukla, J. Thommes, Recent advances in large-scale production of monoclonal antibodies and related proteins, Trends Biotechnol. 28 (2010) 253–261.

[2] R. Bhambure, et al., High-throughput process development for biopharmaceutical drug substances, Trends Biotechnol. 29 (2011) 127–135.

[3] I.J. Del Va, et al., Towards the implementation of quality by design to the production of therapeutic monoclonal antibodies with desired glycosylation patterns, Biotechnol. Prog. 26 (2010) 1505–1527.

[4] J.B. Chon, G. Zarbis-Papastoitsis, Advances in-the production and downstream processing of antibodies, New Biotechnol. 28 (2011) 458–463.

[5] B. Klley, Industrialization of mAb production technology: the bioprocessing industry at a crossroads, mAbs 1 (2009) 443–452.

[6] J. Strub, F. Grote, R. Ditz, Bioprocess design and production technology for the future, in: G. Subramanian (Ed.), Biopharmaceutical Production Technology, first ed., vol. 2, Wiley VCH, Weinhein, 2012.

[7] D. Low, et al., Future of antibody purification, J. Chromatogr. B 848 (1997) 48–63.

[8] A.D. Marschall, P.A. Munro, G. Trägardh, The effect of protein fouling in microfiltration and ultrafiltration on permeate flux, protein retention and selectivity: a literature review, Desalination 91 (1993) 65–108.

[9] G. Belfort, R.H. Davis, A.L. Zydney, The behavior of suspensions and macromolecular solutions in cross flow microfiltration, J. Membr. Sci. 96 (1994) 1–37.

[10] K. Nakanishi, T. Takadoro, R. Matsuno, On the specific hydraulic resistance of cakes of microorganisms, Chem. Eng. Commun. 62 (1987) 187–201.
[11] T. Tanaka, K. Abe, H. Akasawa, H. Yoshida, K. Nakanishi, Filtration characteristics and structure of cake in cross flow filtration of bacterial suspension, J. Fermentation Bioeng. 78 (1994) 455–461.
[12] A.A. McCarthy, D.G. O'Shea, N.T. Murray, P.K. Walsh, G. Foley, Effect of cell morphology on dead end filtration of the dimorphic yeast Kluyveromyces marxiamus var. marxiamus NRRLy2415, Biotechnol. Prog. 14 (1998) 279–285.
[13] P. Bacchin, P. Aimar, V. Sanchez, Model for colloidal fouling of membranes, AICHE J. 53 (1995) 223–228.
[14] N. Arora, R.H. Davis, Yeast cake layers as secondary membrane in dead end microfiltration of bovine serum albumin, J. Membr. Sci. 92 (1994) 247–256.
[15] R.H. Davis, V.T. Kuberkar, Effects of added yeast on protein transmission and flux in cross flow membrane microfiltration, Biotechnol. Prog. 15 (1999) 472–479.
[16] C. Güell, P. Czekaj, R.H. Davis, Microfiltration of protein mixtures and the effects of yeast on membrane fouling, J. Membr. Sci. 155 (1999) 113–122.
[17] Y. Okamoto, K. Ohmori, C.E. Glatz, Harvest time effects on membrane cake hydraulic resistance of Escherichia coli broth, J. Membr. Sci. 190 (2001) 93–106.
[18] H.F. Liu, et al., Recovery and purification process development for monoclonal antibody production, MAbs 2 (5) (2010) 480–499, https://doi.org/10.4161/mabs.2.5.12645.
[19] E.J. Cohn, et al., Preparation and properties of serum and Plasmaproteins, IV: a system for separation into fractions of the protein and lipoprotein components of biological tissue and fluids, J. Am. Chem. Soc. 68 (1946) 459–475.
[20] J. Curling, et al., Production of Plasma Proteins for Therapeutic Use, Wiley-Blackwell, Hoboken, NJ, 2012.
[21] J. More, et al., in: J. Harris (Ed.), Purification of Albumin From Plasma. Blood Separation and Plasma Fractionation, Wiley-Liss, Hoboken, NJ, 1991.
[22] M. Stucki, et al., Investigations of prion and virus safety of a new liquid IVIG product, Biologicals 36 (2008) 239–247.
[23] M. Trexler-Schmidt, et al., Identification and prevention of antibody disulfide bond reduction during cell culture manufacturing, Biotechnol. Bioeng. 106 (3) (2010) 452.
[24] J. Wesner, Reduction of Host-Cell DNA in Mammalian Cell Culture Using Charge Filters at Protein—A Capture for Monoclonal Antibodies, BPI, Korea, 2008. www.Bioprocessintl.com.
[25] J.X. Zhou, U. Gottschalk, Orthogonal Virus Clearance Applications in Monoclonal Antibody Production, Wiley, Boca Raton, FL, 2008, pp. 169–189.
[26] J.X. Zhou, Viral clearance using disposable system in MAb commercial downstream process, Biotechnol. Bioeng. 100 (2008) 488–496.
[27] A. Shukla, J.R. Kandula, Harvest and recovery of monoclonal antibodies from large-scale mammalian cell culture, BioPharm Int. 21 (5) (2008).
[28] A. Pandy, J.A. Teixeira (Eds.), Current Developments in Biotechnology and Bioengineering: Foundations of Biotechnology and Bioengineering, Elsevier, Amsterdam, The Netherlands, 2016.
[29] W.A. Hardwick, Handbook of Brewing, Marcel Dekker, New York, NY, 1995, p. 348.
[30] D.A. Davis, H.C. Lynch, J. Varley, The application of foaming for the recovery of surfactin from B. subtilis ATCC21332 cultures, Enzym. Microb. Technol. 28 (2001) 346–354.
[31] A.J. Rubin, E.A. Cassel, O. Henderson, J.D. Johnson, J.C. Lamb, Microflotation: new low gas-flow rate foam separation technique for bacteria and algae, Biotechnol. Bioeng. 8 (1966) 135–151.

CHAPTER

Biomass concentrations phase II

3

Abbreviations

BLEVE	boiling liquid expanding vapor explosion
BTE	basket type evaporator
CAPEX	capital expenditure
CTFE	centrifugal thin film evaporator
FCE	forced circulation evaporator
FE	flash evaporation
FFE	falling film evaporator
HTE	horizontal tube evaporator
LTVE	low temperature vacuum evaporator
MVR	mechanical vapor recompression
OPEX	operating expenses
RFC	rising falling film concentrator
RFE	rising falling evaporators
RRFC	recirculating version rising falling film concentrator
STVE	short-tube vertical evaporator

3.1 Fermented broth concentrations

The turnkey level of fermented broth obtained after the upstream process contains the target biopharmaceutical of interest, either in an intracellular or extracellular form. To obtain an intracellular target biomolecule, it is necessary to concentrate and separate the biomass through further processing steps. A simple evaporation process is the initial step involved in obtaining biomass in a concentrated form. The detailed method of evaporation technology to be followed is described in this chapter.

3.2 Evaporation

Evaporation refers to a physical process of removal of a solvent from a solution, slurry, or suspension of solid in a liquid. This process can be applied to remove excess water content from fermented broth obtained immediately after the upstream process. Evaporation differs from the other mass transfer operations such as distillation and drying. In distillation, the components of a solution are separated depending upon their distribution between vapor and liquid phases, which is based on the difference

Techniques for Downstream process for Biologic Drugs and Vaccines. https://doi.org/10.1016/B978-0-443-19157-2.00005-4

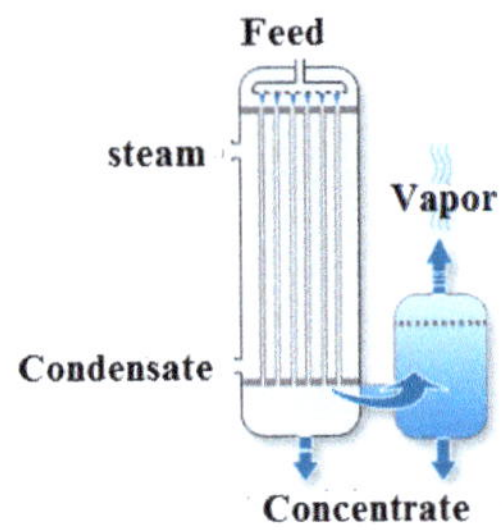

FIG. 3.1

Typical evaporation process of harvested broth to reduce the volume.

of relative volatility of the substance. Removal of moisture from a substance in the presence of a hot gas stream carries away the moisture, leaving a solid residue as the product; this is generally called drying. Evaporation is normally stopped before any damage is done to the target biomolecule present in the fermented broth.

The harvested broth contains about 80% to 90% water. In fermented broth obtained immediately after the end of upstream process, the target product is present in a low concentration (typically 0.1–5 g dm^{-3}). In addition, it also contains intact microorganisms, cell fragments, soluble and insoluble medium components, and other metabolic products. The product may be intracellular, heat labile, and easily broken down by contaminating microorganisms. All these factors become hurdles when safely removing the target product. To ensure good recovery of purified product, it is necessary to have control over the evaporation process to achieve a desirable concentration. The efficacy of an evaporation process depends on the selection of suitable evaporation equipment on the basis of the biocatalyst used and product obtained thereafter (Fig. 3.1). The capacity of the equipment is variable, ranging from a small laboratory scale to an industrial scale. In 1881, Norbert Rillieux developed a multiple-effect pane evaporator for the sugar refining process. He used the steam generated from one pan to heat the sugar juice in the next pan for an energy efficient means of water evaporation.

Evaporation consists of a heat exchanger to boil the solution, with special provision for the separation of liquid and vapor phases. Most industrial evaporators have tubular heating surfaces. The tubes can be horizontal or vertical, long or short, and the liquid can be inside or outside the tubes. There are many ways to classify evaporators on the basis of how the heat exchanger shell and tubes are set, boiling system, etc. A brief description on different types of evaporators follows.

3.2.1 Plate evaporators

Plate evaporator (PE) or thin film plate evaporator is a type of evaporator in which a thin film liquid is passed and flows between plates for the process of evaporation. Plate evaporators are also known as gasketed plate evaporated or frame evaporators (Fig. 3.2).

Falling film plate evaporators can be grouped into single-effect and multiple-effect film plates [1,2].

The basic principle of liquid film in film evaporators is specifically designed to produce a thin film during both the climbing and falling phases [3]. In the case of climbing film evaporators, the feed is introduced at the bottom of the tubes. Due to the evaporation process, the vapor expands, thus causing a thin film of liquid to rise along the tubes. The vapor shear will push the thin film to climb up the wall of the tubes. On the other hand, the feed for the falling film evaporator (FFE) is introduced at the top of the

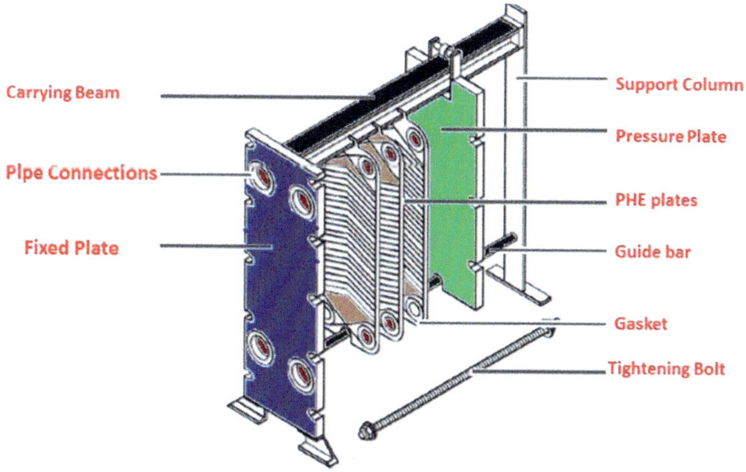

FIG. 3.2

A plate evaporator showing the functional aspects for concentrate harvested broth.

tubes. The liquid flows down the tubes and is evaporated as it descends. The liquid flows down the tubes, driven by the vapor shear stress and gravitation forces.

The thin liquid film in the FFE flows in two ways: concurrent and countercurrent [4]. In the former process, vapor is drawn from the top to the bottom of the tubes, whereas the reverse process occurs countercurrently [4].

Plate evaporators consist of a plate and frame heat exchanger and are mainly used in the food and beverage industry. The major advantages of plate evaporators are they are well adapted to the materials, low headroom is required, and they are easily cleaned and modified on an as-needed basis. The major disadvantage of plate evaporators is the large gasketed area. A leakage can occur if the gasket is not properly selected.

Typical applications of gasketed plate evaporators are stripping applications, removing monomers from polymers, and deodorization. It is typically used for fruit juice, fruit puree, syrup, coffee, and milk.

3.2.2 Falling film evaporator

Commonly, a FFE has a special type of heat exchanger to safely concentrate heat-sensitive components. The FFE is used for a falling heat exchanger (Fig. 3.3A and B).

A FFE is a type of evaporator in which the equipment is placed upside down so that the heat exchanger remains at the top of the equipment. Feed enters the top of the evaporator and then is further distributed to tubes. It consists of a vertical shell and a tube heat exchanger. The solvent is fed into the top of the tube and then flows down the surface of the tube as a thin film. As it heats further, vapor is generated. It increases the heat transfer for the process.

The main advantage of a FFE is its low mean temperature differences between the heating medium and the process stream, typically 3 to 6 K, therefore these devices are ideal for heat recovery in multistage processes [5,6].

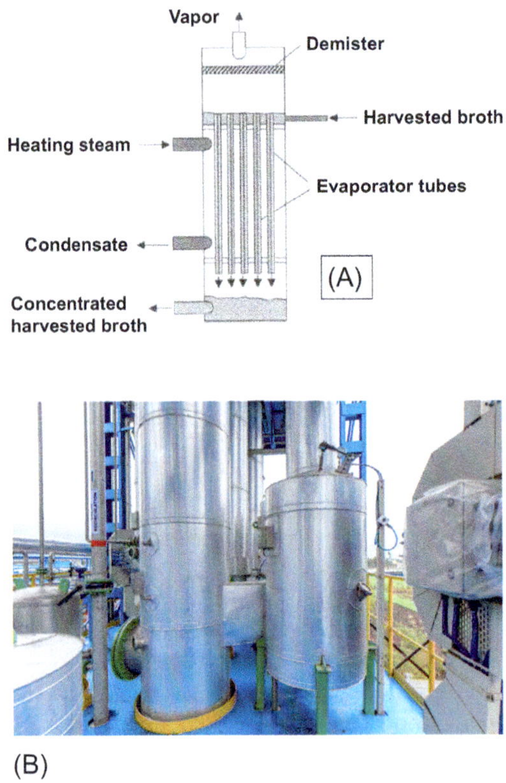

FIG. 3.3

Falling film evaporator principle: (A) line diagram, basic principle; (B) industrial type.

A further advantage of the FFE is the very short residence time of the liquid and the absence of superheating of the same. The residence time inside the tubes is measured in seconds, resulting in safety for heat-sensitive products such as milk, fruit juice, pharmaceuticals, and many others. In addition, FFEs are relatively low cost, have low product hold-up, small space requirement, and good heat transfer. Common disadvantages of FFEs are high headroom requirement, not good for salting and scaling materials, and recirculation is required.

The main applications for FFEs are the concentration of dairy products such as whey, milk protein, skim milk, cream and hydrolyzed milk, sugar solutions, urea, phosphoric acid, and black liquor.

3.2.3 Forced circulation evaporator

The forced circulation evaporator (FCE) is comprised of a heat exchanger, a separator, and circulation pump. The product is sent to the exchanger where it is heated and then sent to the separator. After evaporation, the liquid portion is pumped again to be resent to the exchanger [7,8].

A FCE is usually more costly than a natural circulation evaporator (Fig. 3.4). However, natural circulation evaporators are not suitable for some situations such as:

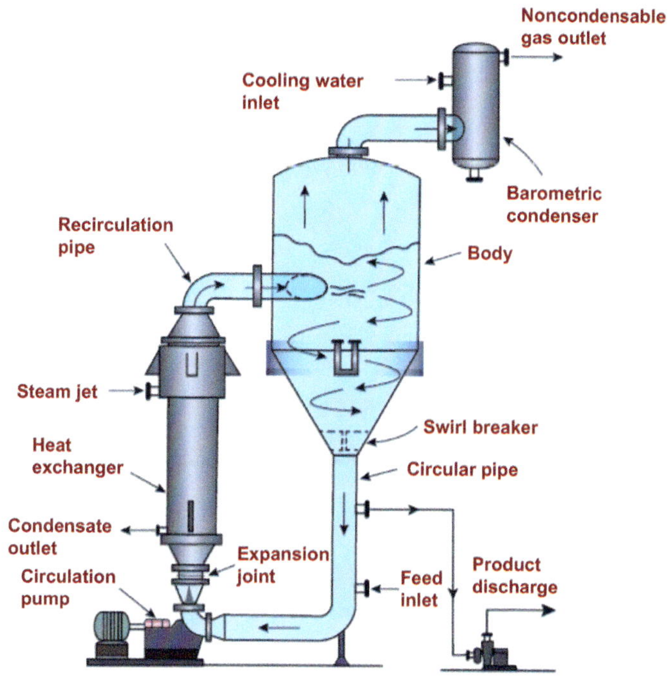

FIG. 3.4

Forced circulation evaporator (vertical tube).

- Highly viscous solutions, because of their low heat transfer coefficient
- Solution containing suspended particles
- Heat-sensitive materials

This type of evaporator is suitable for processing liquors that are susceptible to scaling or crystallization. The liquid is circulated at a high rate through the heat exchanger; boiling is prevented within the unit by a hydrostatic head maintained above the top tube plate. As the liquid enters the separator, where the absolute pressure is slightly less than in the tube bundle, the fluid flashes to form a vapor.

The FCE has many advantages over the other evaporators [9]. In a FCE, the boiled liquid does not enter the system onto the heating surface of a plate-type evaporator, hence minimizing fouling [9]. So, circular evaporators are preferred for liquids with a higher tendency to foul. Minimal fouling also means that the cleaning cycles are not as frequent as with other evaporators such as plate evaporators [10]. These types of evaporators are fairly compact and easy to clean and operate [11].

3.2.4 Centrifugal thin film evaporators (CTFEs)

These evaporate liquid very quickly (in seconds) and are hence suitable for concentrating even heat-labile substrates. In these evaporators, a centrifugal force is used to pass the liquid over heated plates or conical surfaces for instantaneous evaporation (Fig. 3.5).

FIG. 3.5

Centrifugal thin-film evaporator.

3.2.5 Rising falling evaporators (RFE)

The rising falling film evaporator or concentrator (RFC) and the recirculating version (RRFC) work on uniform metered flows entering the bottom of the steam chest and rising through the first pass with rapidly increasing velocity (Fig. 3.6).

After leaving the first pass, the flow of the vapor/liquid mixture is reversed and directed downward through the second pass at extremely high velocity. The mixture then enters the separator where the vapor and liquid are separated. Although not designed as a circulation unit, the RFC provides once-through evaporation of heat-sensitive materials that require minimum retention time. RFCs are particularly adaptable to multiple-effect systems but provide a lesser turndown capability than the RRFC configuration. The characteristics of RFCs are as follows:

- High tube velocity minimizes fouling
- Able to achieve high concentration at low temperatures
- Risk of entrainment is greatly reduced because most of the liquid passes directly out of the steam chest rather than being re-entrained with vapor from the separator
- System may be turned down without loss of efficiency or effectiveness
- Holding time is a matter of seconds (RFC)
- Suitable for evaporation of a wide range of liquid concentrations
- Low headroom requirement

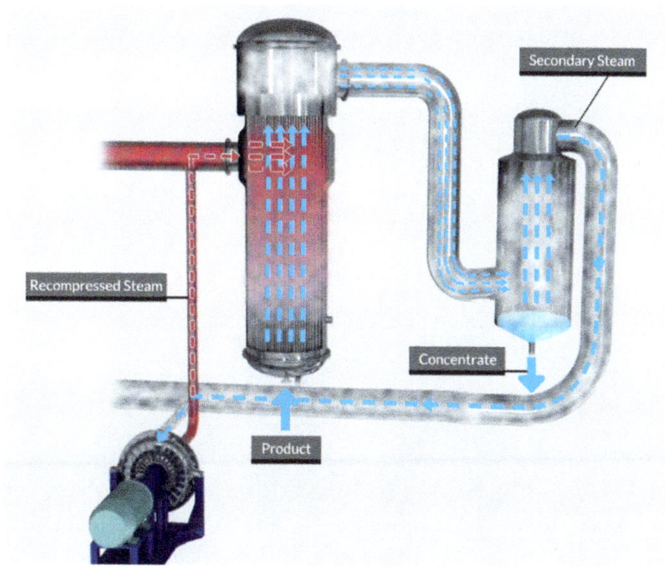

FIG. 3.6

Profile view of a rising film evaporator.

The rising film evaporators are being used for a wide range of purposes like production of polymers, food production, thermal desalination, pharmaceuticals, and solvent recovery process. For fruit juice concentration and processing, the rising film evaporator is used to preserve the natural quality of the juice for long periods of time.

3.2.6 Mechanical vapor recompression

Mechanical vapor recompression (MVR) is a type of evaporator in which the water vapor is compressed, which results in the rise of temperature and pressure (Fig. 3.7) [12].

The rise in temperature increases the temperature difference between the vapor and the fluid. The higher heat transfer results in the heat exchanger. The compressed vapor is again fed back to generate more steam. This is an energy-efficient energy recovery process.

The dairy industry, the brewing industry, sugar industry, pulp industry, and alcohol industry commonly use these types of evaporators.

3.2.7 Low temperature vacuum evaporator

Low temperature vacuum evaporators (LTVEs) can vaporize water at a lower temperature of around 40°C, instead of the normal 100°C, because evaporation takes place in a vacuum (Fig. 3.8A and B) [13].

Vacuum evaporators are useful for recovery of:

- Heat sensitive biomolecules
- Chemicals sensitive to air oxidation, such as cyanide plating baths and stannous tin baths
- Solutions containing volatile components

FIG. 3.7

Industrial-type mechanical vapor recompression (MVR), profile view.

A LTVE mainly consists of an evaporator tank, evaporator heat exchanger, process water circulation pump, refrigerant compressor, condenser heat exchanger, distillate tank, vacuum-producing venture, and distillate circulating pump. The feed water enters the evaporator tank and is circulated by the pump to the evaporator heat exchanger, where the feed water is heated by a high-pressure refrigerant from the refrigerant compressor. The hot water flashes as water vapor into the evaporator tank at the top. This water vapor passes through the mist eliminator and the condenser heat exchanger. The refrigerant from the evaporator heat exchanger passes through the air cooler and expansion valve and becomes cool liquid. The cool refrigerant cools the hot water vapor to a distillate. The distillate is collected in the distillate tank. The distillate is circulated by the distillate circulation pump through a venture, which produces a vacuum in the evaporator tank. The vacuum produced by the venture in the evaporation tank causes the process water to boil at 40°C to 50°C. The concentrated process water from the evaporator tank and the distillate from the distillate tank are automatically drained by the electric control system. All necessary controls and switch gears are part of the system.

3.2.8 Flash evaporation

Flash evaporation (FE) or flashing is an evaporation process caused by a sudden pressure drop sufficiently below the saturation pressure. Due to this sudden drop in pressure, the liquid undergoes a quick phase transition and the sensible heat of the liquid converts into latent heat of evaporation (Fig. 3.9). This process is one of the simplest unit operations. If the throttling valve or device is located at the entry

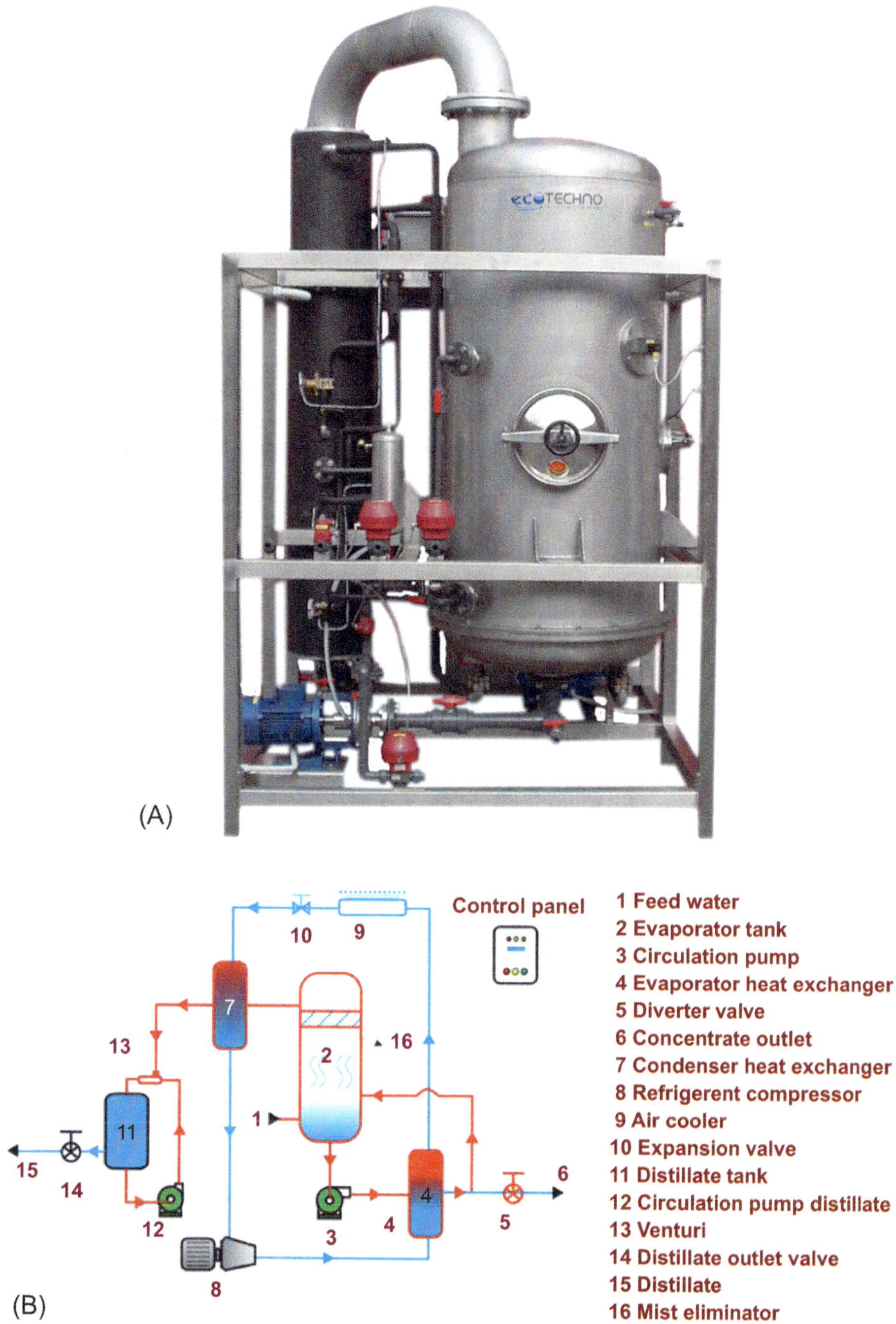

FIG. 3.8

(A) Low temperature vacuum evaporator and (B) working principle.

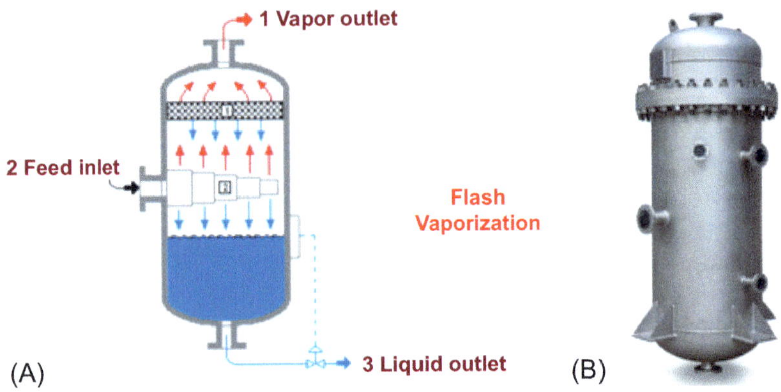

FIG. 3.9

(A) Flash vaporization principle; (B) front profile view (1. De-entrainment mesh pad; 2. Inlet diffuser (distributor); 3. Liquid level control valve).

into a pressure vessel so the FE occurs within the vessel, then the vessel is often referred to as a flash drum [14,15].

In the case of the presence of a single-component liquid like propane or liquid ammonia, a part of the liquid immediately flashes into vapor. Both the vapor and the residual liquid are cooled to the saturation temperature of the liquid at the reduced pressure. This process is also referred to as "auto-refrigeration" and is the basis of most conventional vapor compression refrigeration systems.

In the case of a multicomponent liquid such as a mixture of propane, isobutene, and normal butane, the flashed vapor is richer in the more volatile components than in the remaining liquid.

Uncontrolled FE can result in a boiling liquid expanding vapor explosion (BLEVE).

The flash evaporator is used to concentrate liquid foods such as noodles and make condensed milk, the product of a process that removes water from milk. Similarly, pharmaceutical companies use the evaporator to remove excess moisture from drugs, thus providing stability (Fig. 3.10).

3.2.9 Basket-type evaporator (BTE)

This is a variant of the short-tube evaporator in which the whole tube tank can be easily removed for cleaning. By using vertical rather than horizontal tubes, the natural circulation of heated liquid can be used to give adequate heat transfer. The standard evaporator shown in Fig. 3.11 is an example of this type.

Recirculation of the liquid is through a large downcomer so that the liquid rises through the vertical tubes (of about 5- to 8-cm diameter), boils in the space just above the upper tube plate, and recirculates through the downcomers. The hydrostatic head reduces boiling in the lower tubes, which are covered by the circulating liquid. The length-to-diameter ratio of the tubes is of the order 15:1. The basket evaporator is a variant of the calandria evaporator in which the steam chest is contained in a basket suspended in the lower part of the evaporator, and recirculation occurs through the annular space around the basket.

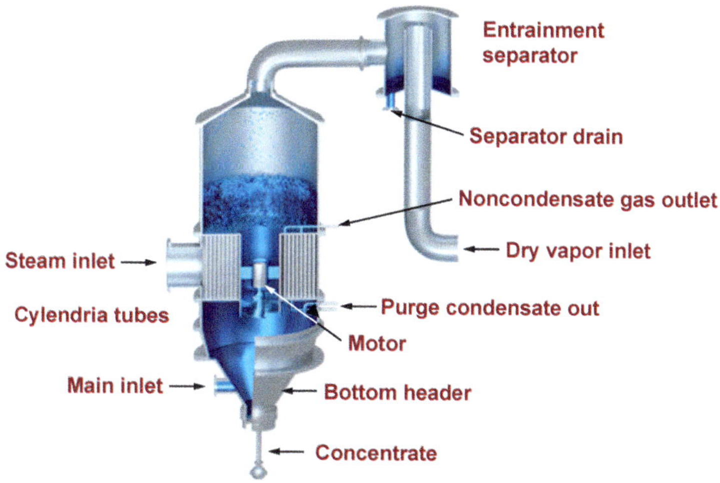

FIG. 3.10

Diagram of a standard basket evaporator.

Courtesy of Thermopedia.

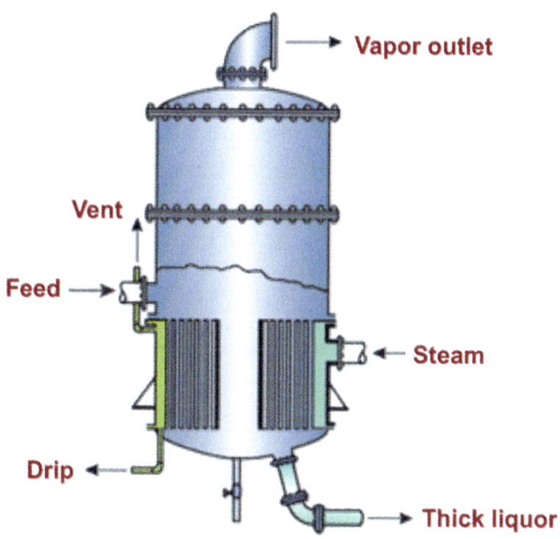

FIG. 3.11

Short-tube calandria-type evaporator.

3.2.10 Short-tube vertical evaporator

Evaporators in which movement of the liquid takes place as a result of a convection current set up by the heating process are called vertical tube evaporators or short-tube evaporators (Fig. 3.11).

A short-tube vertical evaporator (STVE) consists of a short tube bundle (about 1 to 3 m in length) enclosed in a cylindrical shell. This is called the calandria. The feed is introduced above the upper tube sheet, and steam is introduced to the shell or steam chest of the calandria. The solution is heated and partly vaporized in the tubes. The central tube in a calandria is of longer diameter. Typically, the "downcomer" area is taken as 40% to 70% of the total cross-sectional area of the tubes. The circulation rate through the downcomer/downtake is many times the feed rate. The flow area of the downtake is approximately equal to the total tubular flow area.

In standard vertical tube evaporators or short-tube evaporators, the liquid passes through the vertical tubes, and the steam is supplied from outside the tubes. Heat transfer takes place through the tubes, and the liquid inside the tubes is heated. The solvent evaporates and the vapor escapes from the top. The concentrated liquid is collected from the bottom. The main features of vertical tube evaporator construction are as follows:

- Large cylindrical body made of cast iron with a dome-shaped top and bottom.
- Inside the body, the calandria is fitted at the bottom. The calandria consists of a number of vertical tubes of 0.05- to 0.075-m diameter and 1- to 2-m length.
- About 100 such tubes are fitted in a body measuring 2.5 m or more in diameter. Inlet is provided for the steam and feed.
- Outlets are provided for vapor-concentrated products, noncondensed gases, and condensate.

The vertical tube evaporators function as follows:

- Steam is introduced outside the tubes. The condensate passes through the corresponding outlet and no condensed gases escape through the vent.
- The feed is introduced in such a way as to maintain the liquid level slightly above the top of the tubes. The liquid inside the tubes is heated by the steam and begins to boil.
- As the liquid boils, it spouts up through the tube and return through the central downtake. Circulation of hot liquid is set up, which enhances the rate of heat transfer.
- The vapor escapes through the top outlets. Steam is supplied until the required concentration of products is obtained.
- Products are withdrawn from the bottom outlet.

3.2.11 Horizontal tube evaporator

In the case of a horizontal film evaporator (Fig. 3.12), the liquid to be vaporized is distributed evenly on the top of a horizontal tube bundle.

For this purpose, perforated plates and specially designed jet nozzles are applied. A minimum recirculation rate is required to maintain a minimum wetting rate. The liquid characteristics between the tubes are determined by the film's Reynolds number and Kaptiza number. The Kapitza number accounts for the effects of surface tension and viscosity on the turbulence structure near the free surface of the film, whereas the Reynolds number of the liquid film is calculated by the standard definition $Rel = 4 m_0/\mu\text{-}2TTT$, where m_0 is the mass flow rate of liquid phase inlet, kgS^{-1}. The Reynolds number is a dimensionless quantity used to determine the type of flow pattern as laminar or turbulent while flowing through a pipe. It is the ratio of inertial forces to that of viscous forces. The fluid can be encountered between the tubes as droplets, jet, or continuous sheet. As a result of liquid impingement, the heat transfer in horizontal units is generally higher than for vertical evaporators.

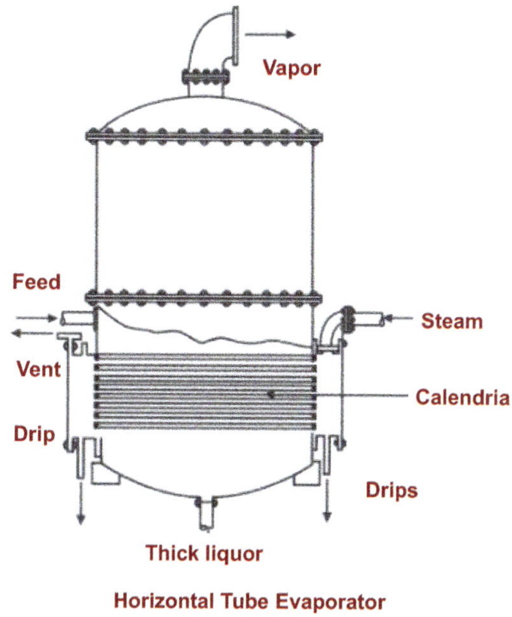

FIG. 3.12

Line diagram of a horizontal tube evaporator.

The horizontal tube evaporator (HTE) is used for heat recovery from gas turbine exhausts and also for recovery from flue gases in refinery and petrochemical furnaces. It is mainly used to remove excess of solvent from the biological suspension in solvent, leaving a concentrated product.

3.3 Methods of feeding of evaporators

When a single evaporator is used for concentration, the vapor coming out of it is condensed and discarded. This type of operation is called single-effect evaporation. When a number of effects are used in series, such that the vapor coming out of one effect is used as a heating medium in the steam chest of the next effect, it is called multiple-effect evaporation (Fig. 3.13).

Single-effect evaporation is simple but fails to effectively utilize the steam, while multiple-effect evaporators evaporate more quantity of water per kilogram of steam consumed in the evaporation process.

This brings about a saving in the steam cost, but at the same time, the cost of material and installation of the evaporator system increases because of the large number of effects involved. There has to be a trade-off between the material cost and the steam cost; the number of effects giving the minimum total cost is the most thermo-economic solution. Although the number of effects in some cases go up to 17, the driving force for heat transfer is the difference in temperature between the steam in the tubes/coils and the product in the pan. The steam temperature is a function of the steam pressure. Water boils at 100°C at 1 atm, but at other pressures, the boiling point changes. At its boiling point, the steam

(A)

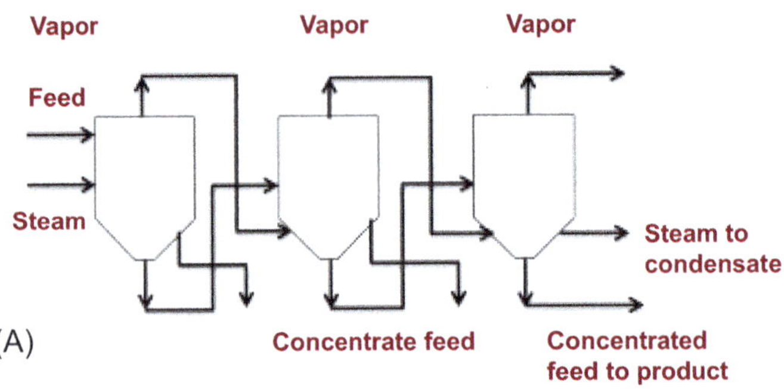

(B)

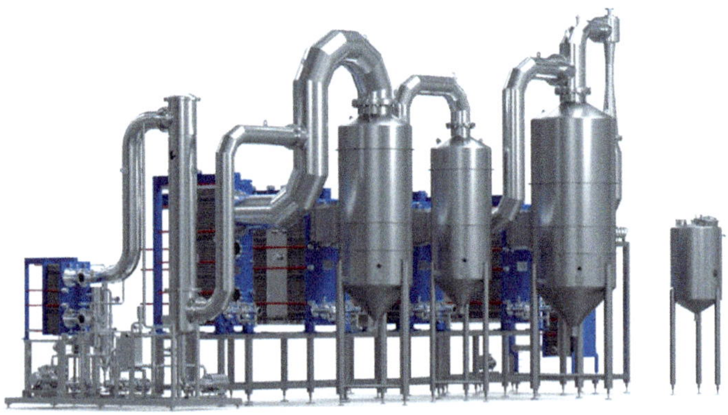

FIG. 3.13

Triple phase evaporator: (A) functional line diagram and (B) photo of triple-phase evaporator used in milk plant.

Courtesy of Alfa Laval.

condenses in the coils and gives up its latent heat. If the steam temperature is too high, burn-on/fouling increases, so there are limits to how high steam temperatures can go. The product is also at its boiling point. The boiling point can be elevated with an increase in solute concentration. This boiling point elevation works on the same principles as freezing point depression, which usually should not exceed 7, because beyond this, the material and installation cost of the evaporator effects offset the saving achieved in the steam cost.

To quicken the process of evaporation and saving time, the vapor from the boiling liquid is condensed, and the concentrated product is withdrawn from the bottom of the evaporator. Although the operation is simple, the device does not use steam efficiently. Typically 1.1 to 1.3 kg of steam is required to evaporate 1 kg of water. When a number of effects are used in series, such that the vapor coming out of one effect is used as a heating medium in the steam chest of the next effect, it is called multiple-effect evaporation. Single-effect evaporation is simple but fails to utilize the steam effectively, while multiple-effect evaporators evaporate more quantity of water per kilogram of steam consumed in the evaporation process The steam consumption per unit mass of water evaporated can be increased by putting more than one evaporator in series such that the vapor from one evaporator is used in the second evaporator for heating. The vapor from the second evaporator is condensed, and the arrangement is called double-effect evaporator. The heat from the vapor generated in the first evaporator is used in the second evaporation. Evaporation of water is nearly doubled in a double-effect evaporation system compared to single-effect per unit mass of steam used. Additional effects can be added in series in the same ways to get a triple-effect evaporator, quadruple-effect evaporator, and so on (Fig. 3.13).

3.3.1 Feed arrangement

The multiple-effect evaporators are classified on the basis of the directions of the heating medium and the feed of the liquid. Thus, the following four types of evaporators are used on the basis of requirement:

(i) Forward feed
(ii) Backward feed
(iii) Mixed feed
(iv) Parallel feed

3.3.1.1 Forward feed

When the thin liquid is pumped into a multiple-effect evaporator and sent in turn through the other effects, this is known as forward feed (Fig. 3.14).

The concentration of the liquid increases from the first effect to the last effect. This pattern of liquid flow is the simplest. It requires a pump for feeding the dilute solution to the first effect; however, it can be carried out without pumps, since the flow is in the direction of decreasing pressure, and control valves in the transfer line are all that is required.

The main advantages are its simplified operational process with less expense; no pump is required for the liquid flow from one effect to the next due to the existence of partial differential pressure; and there is less chance of changes in structure and function of the target product. However, a pump is required to remove the concentrated product from the last effect.

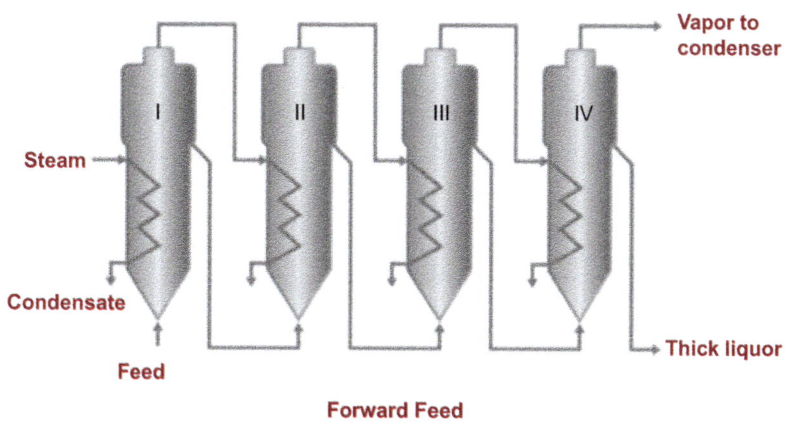

FIG. 3.14

Multiple evaporator (forward feed).

The problems with forward feed are the reduced rate of heat transfer in second and higher effects and the feed should not be below the boiling point because this reduces steam economy by consuming external steam to supply heat. Commonly, the pharmaceutical industry prefers to use this method when the feed is hot or when the final concentrated product might be damaged at high temperatures. Besides this, the pulp and paper industry and sugar factories prefer to use this method.

3.3.1.2 Backward feed

Backward feed is another common method in which dilute liquid is fed to the last effect and then pumped through the successive effects to the first (Fig. 3.15).

This method requires a pump between each pair of effects in addition to the thick liquid pump, since the flow is from low pressure to high pressure. Backward feed often gives a higher capacity than forward feed when the thick liquid is viscous, but it may give a lower economy than forward feed when the feed liquid is cold.

The advantages are at higher temperature of steam, the liquid attains high concentration, lower viscosity, and higher heat transfer rate in first effect.

But the problems with backward feed system are that intereffect pumps are necessary, a higher risk of damage to the viscous product subjected to a higher temperature, and risk of fouling.

3.3.1.3 Mixed feed

In the mixed method, the feed liquid enters at an intermediate effect and flows into the next higher effect until it reaches the last effect of the series. In this process, liquid flows in the forward feed mode. Partly concentrated liquid is then pumped back to the effect before the one to which the fresh feed was introduced for further concentration (Fig. 3.16).

In this operational process, very limited pumping systems are used in backward configuration as flow occurs due to pressure differences whenever applicable. The advantages with the mixed feed

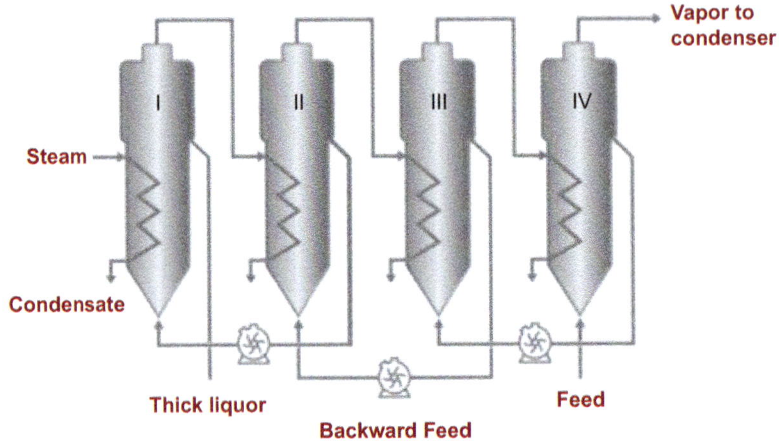

FIG. 3.15

Backward feed in a multiple-effect evaporator, showing functional aspects.

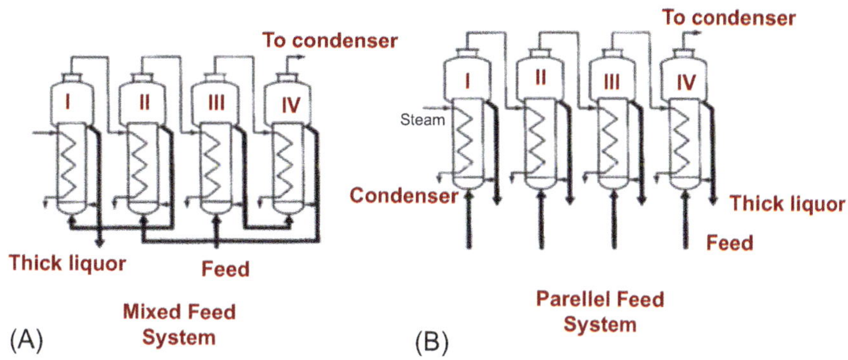

FIG. 3.16

Functional line diagram of (A) mixed feed system, and (B) parallel feed system for multiple evaporators.

process is it has both the simplicity in forward feed operational process and the economy of backward feed; it is used for concentrates of a highly viscous feed. It also carries some disadvantages like more complex piping and instrumentation, and it is more expensive.

3.3.1.4 Parallel feed

In this process, the fresh feed is introduced to each effect, and in this configuration, the product is withdrawn from the same effect in parallel feed operation (Fig. 3.16). In parallel feeding, there is no transfer of liquid from one effect to another. It is used primarily when the feed is saturated and the product is solid containing slurry. This is most common in crystallizing evaporators.

3.3.2 Selection of evaporator

The pharmaceutical industries select an evaporator on the basis of its capacity and economy. Capacity refers to the number of kilograms of water vaporized from all the effects per kilogram of steam used. However, a variety of other factors like feed, solution viscosity, nature of product, solvent undertaken, fouling characteristics, and foaming formation are also important when selecting an evaporator to concentrate a biological product. On the basis of efficacy, it is advisable to select a multiple-effect evaporator rather than a single-effect evaporator system to quicken the process and reduce expenses (Fig. 3.17).

An evaporator's performance is based on three characteristics:

(i) Capacity (kg vaporized/time)
(ii) Economy (kg vaporized/kg steam input)
(iii) Steam Consumption (kg/h) where Consumption = Capacity/Economy

Economy (or steam economy) is the kilograms of water vaporized from all the effects (per kilogram of steam used). For a single-effect evaporator, the steam economy is ca.0.8 (<1), which translates to 0.8 tons of steam needed to evaporate 1 ton of water.

To reduce the expenses for steam evaporation, the multiple-effect design uses the exhaust vapors from the product to heat the downstream evaporation effect and reduce steam consumption.

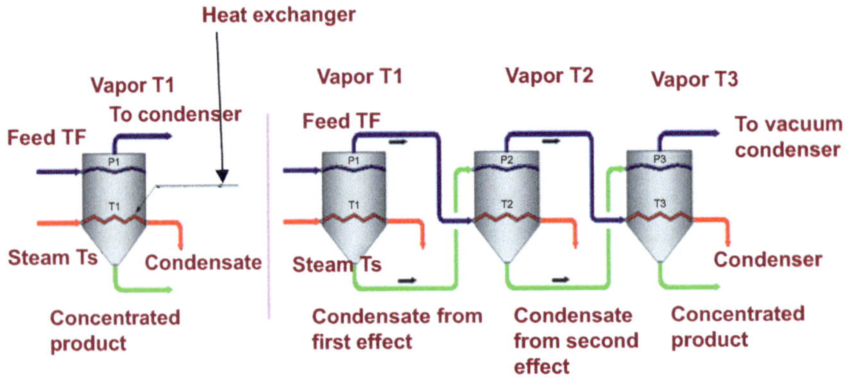

FIG. 3.17

Single- and multiple-effect evaporator configurations.

Evaporators need also pumps, interconnecting pipes, and valves that are required for the transfer of liquid from one effect to another effect, and they increase both the capital expenditure (CAPEX) and operating expenses (OPEX) of the process.

References

[1] D.W. Green, R.H. Perry, Perry's Chemical Engineers' Handbook, eighth ed., McGraw-Hill, New York, 2008. ISBN 0071593136.

[2] R. Billet, Evaporation Technology: Principles, Applications, Economics, VCH, 1989.

[3] J.G. Brennan, A.S. Grandison, Food Processing Handbook, WILEY_VCH Verlag GmbH & Co. KGaA, Germany, 2012.

[4] A. Faghri, Y. Zhang, Transport Phenomena in Multiphase Systems, Elsevier, London, 2006.

[5] K.R. Chun, R.A. Seban, Heat transfer to evaporating liquid films, J. Heat Transfer 93 (197) (1971) 391–396, https://doi.org/10.1115/1.3449836.

[6] A. Alhusseini, K. Tuzla, J. Chen, Falling film evaporation of single component liquids, Int. J. Heat Mass Transf. 41 (12) (1998) 1623–1632, https://doi.org/10.1016/S0017-9310(97)00308-6.

[7] M. Maer, Coulson and Richardson. Chemical Engineering, fifth ed., vol. 2, Butterworth-Heinemann, 2002, pp. 771–822.

[8] R. Perry, D. Green, Perry's Chemical Engineering Handbook, sect. 11–107, 2023, p. 1141.

[9] Evaporation Technology, GEA Wiegand GmbH Process Engineering Division, 2012. Date Reviewed: 9/10/2013.

[10] Forced Circulation Evaporators, Gemina, Machine Point Food technologies, 2012. Date reviewed: 9/10/2013.

[11] S. Ray, Introduction, Advantages and Disadvantages, and Uses Analysis of Forced Circulation Evaporator, Wenzhou Tecellent Machinery Co., Ltd, 13 November 2012.

[12] https://www.energy.gov/sites/prod/files/2014/05/f16/steam11_waste_steam.pdf. (bare URL PDF).

[13] G. Gutiérrez, Á. Cambiella, J.M. Benito, C. Pazos, J. Coca, The effect of additives on the treatment of oil-in-water emulsions by vacuum evaporation, J. Hazard. Mater. 144 (3) (2007) 649–654, https://doi.org/10.1016/j.jhazmat.2007.01.090. 17321675.

[14] S.M. Walas, Chemical Process Equipment: Selection and Design, Butterworth-Heinemann, 1988. ISBN 0-409-90131-8.

[15] Gas Processing Suppliers Association (GPSA), Engineering Data Book, tenth ed., vol. 1, Gas Processing Suppliers Association, Tulsa, Oklahoma, 1987.

CHAPTER

Release of intracellular materials phase-III

4

Abbreviations

CP	chemical permeabilization
DMSO	dimethyl sulfoxide
EDTA	ethylenediaminetetraacetic acid
HOH	high-pressure homogenization
HSH	high-shear homogenizer
HSR	high-shear reactor
HHP	high hydrostatic pressures
OIV	Organization of Vine and Wine
OS	osmotic shock
PEF	pulsed electric fields
RSM	rotor-stator ball mill
SDS	sodium dodecyl sulphate
US	ultrasounds

4.1 Why disrupt the cell?

Cell disruption is crucial for the release of intracellular components (Fig. 4.1) and further processing to obtain a target product of therapeutic value. In this process, the outer boundary or cell membrane is broken down or to release intracellular materials. The fluid containing the contents of lysed cells is called lysate. The cell disruption method that is used must be effective in releasing these cellular compounds reliably, efficiently, and effectively.

4.2 Nature of cell boundaries

Cells are two types: eukaryotic (such as plant cells/mammalian cells) and prokaryotic (such as bacteria). The rigidity of the cell boundary to various mechanical and nonmechanical forces mainly depends on the structural makeup of the boundary layer of prokaryotic and eukaryotic organisms (Fig. 4.2A–C).

Techniques for Downstream process for Biologic Drugs and Vaccines. https://doi.org/10.1016/B978-0-443-19157-2.00004-2

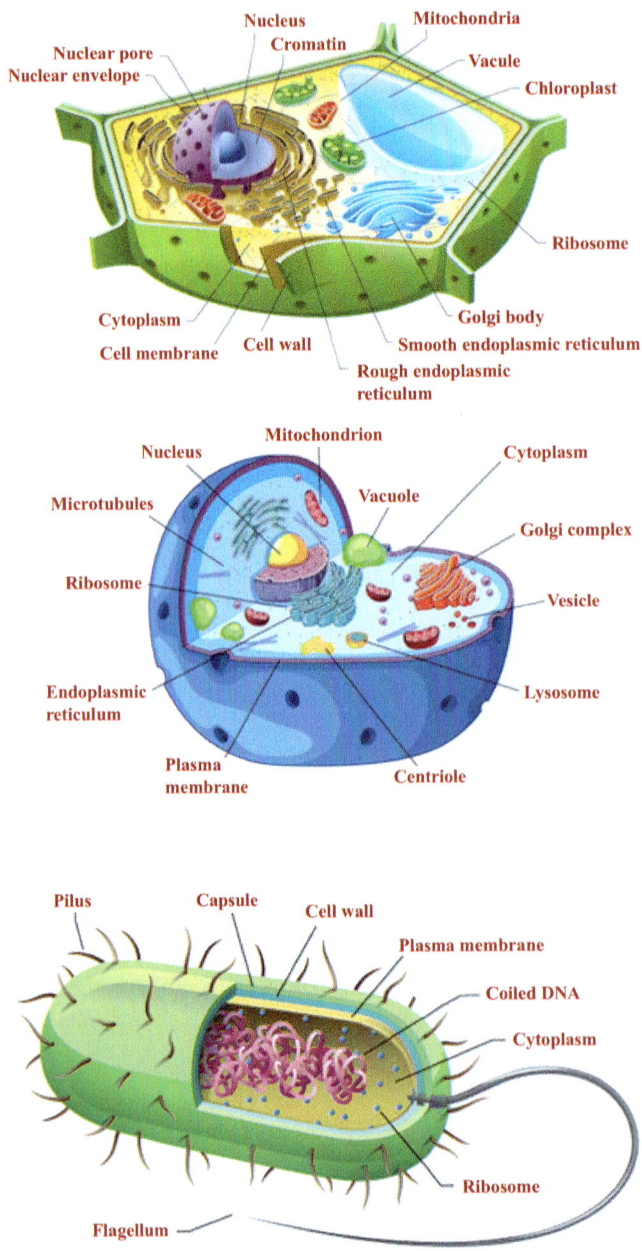

FIG. 4.1

Cross-section of plant cell; animal cell; and bacteria (from top to bottom) showing various components.

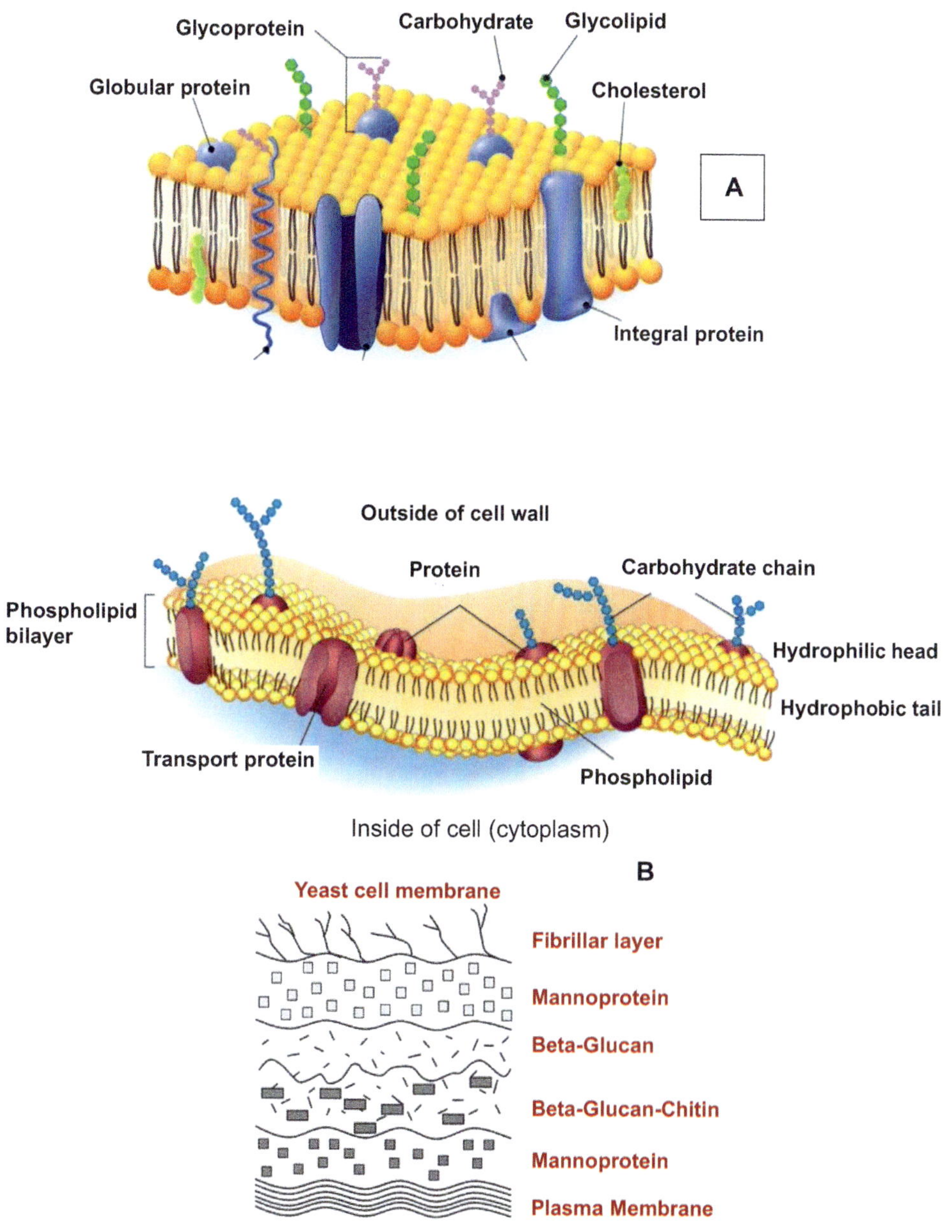

FIG. 4.2

(A) Membrane structure of animal cell and plant cell, from top to bottom. (B) Membrane structure of yeast and bacteria (top to bottom).

(Continued)

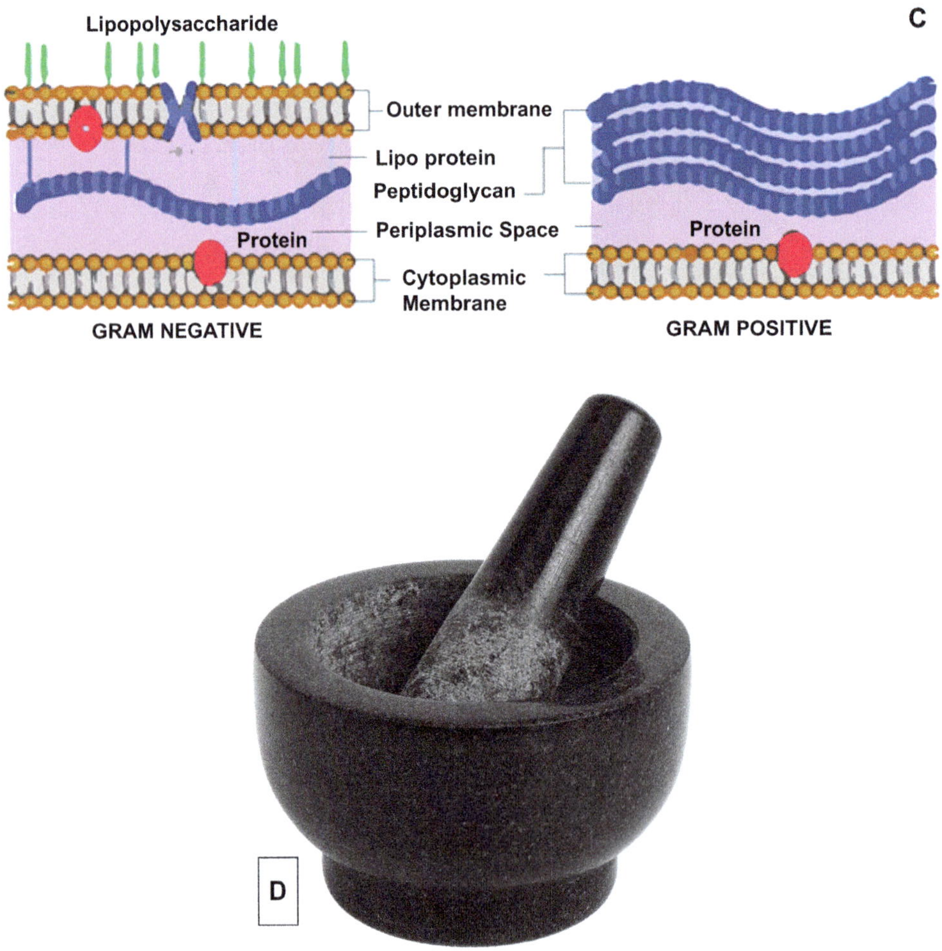

FIG. 4.2, Cont'd

(C) Membrane structural difference between Gram-positive bacteria and Gram-negative bacteria. (D) Cell lysis with motor-pestle after deep freezing in liquid nitrogen.

4.2.1 Bacterial cells

Bacterial cells may require different disruption methods depending on whether they are Gram-positive or Gram-negative, as the amount of peptidoglycan and presence of an envelope affects the overall process (Fig. 4.2B). Response of a bacterial system varies with different types of disintegration (digestive enzymes like lysozymes, osmotic shock (OS), chemical treatment). The peptidoglycan layer can withstand the osmotic pressure developed inside the cell due to the concentration difference of solutes across the membrane. For *Escherichia coli*, this pressure is estimated around 2 atm [1]. This layer is also responsible for maintaining the shape and rigidity of the cell. This layer consists of two sugar derivatives named N-acetylglucosamine and N-acetylmuramic acid as well as a small group of amino acids consisting of L-alanine, D-alanine, and D-glutamic acid.

In the Gram-positive bacteria, the cell boundary is mainly composed of a thick peptidoglycan layer. In addition, the bacterial cell membrane and plasma are made up of relatively fragile phospholipids and proteins. In certain cases, polysaccharide capsules may be present outside the cell wall. In Gram-positive bacteria, peptidoglycan layer makes up to 50%–80% of the envelope, and 10% of this layer is associated with teichoic acid, which provides a greater structural resistance to breakage [2]. In contrast, 10%–20% of the cell envelope of Gram-negative bacteria is composed of a 1.2- to 2.0-μm-thick peptidoglycan.

The cell wall of Gram-positive bacteria is particularly susceptible to lysis by the antibacterial enzymes and lysozyme. Unlike Gram-positive bacteria, Gram-negative bacteria do not have distinct cell walls but instead have multilayered envelopes. The peptidoglycan layer is significantly thinner than in Gram-positive bacteria. An external layer composed of lipopolysaccharides and proteins is usually present (Fig. 4.2C). Another difference with Gram-positive bacteria is the presence of the periplasm layers, which are two liquid filled gaps, one between the plasma membrane and the peptidoglycan layer and other between the peptidoglycan layer and the external lipopolysaccharides. Another periplasmic layer also exits in Gram-positive bacteria; these are important in bioprocessing since a large number of proteins, particularly recombinant proteins, are secreted into it.

4.2.2 Yeast/mold cells

Yeast cells are more rigid compared to that of bacterial cell membrane (Fig 4.2B). The yeasts are unicellular having 0.1–0.2 μm in thickness. These are mainly composed of polysaccharides such as glucans, mannans, and chitins. The plasma membrane in a yeast cell is composed of phospholipids and lipoproteins. Mold cells are largely similar to yeast cells in terms of cell wall and plasma membrane composition but are multicellular and filamentous.

Yeasts are more difficult to disrupt with enzymes and detergent than bacteria. Bead milling is a commonly acceptable method for rupturing yeast cells in isolating nucleic acids and some specific enzymes. A variety of equipment like Pulsing Vortex Mixer, HT Mini homogenizer, GenoGrinder, or HT Homogenizer with suitable size of bead can be used to break the outer boundary of yeast cells. The most effective bead size for *Saccharomyces* is 400-μm silica beads. However, smaller yeasts, such as *Pichia*, are best homogenized with 200-μm zirconium beads. A sample containing a very low number of cells should be disrupted using low binding beads so that analytes remain in the lysate instead of adsorbing into the beads.

4.2.3 Mammalians cells

Mammalian cells are easiest to disrupt as they lack a rigid cell wall (Fig. 4.2A). Commonly, a freeze-thaw method is used to disrupt bacterial and mammalian cells.

4.2.4 Plant cells

Plant cells are more difficult to disrupt than animal cells because of the very thick cellulosic walls mainly composed of cellulose and other polysaccharides (Fig. 4.2A). Cell walls, wherever present, are the main barriers that need to be disrupted to recover intracellular products.

A variety of options are available to disrupt the rigid plant cell wall. Commonly, manual grinding is the most common method used to disrupt plant cells. Tissue is frozen in liquid nitrogen and then crushed using mortar and pestle. Because of the tensile strength of the cellulose and other polysaccharides comprising the cell wall, this method is the fastest and most efficient way to access plant proteins and DNA (Fig. 4.2D)

4.3 Response of cell boundary to lysis

There are several varieties of mechanical or nonmechanical processes adapted to disrupt cells. Due to the variability in the cell boundary of plants, animals, fungi, and microorganisms (bacteria) (Fig. 4.2A–D), it is challenging to select a suitable cell disruption method for the release of cellular components without disturbing their structure and function.

Cell disruption is the initial step in the manufacturing of many biological products such as enzymes, nucleic acids, reporter assays, immunoassays, protein purification, antigens for vaccines, and viral vectors that deliver genes for next-generation vaccines or gene therapy. It is also the first step in conducting Western and Southern blotting to analyze specific proteins. Many mechanical and nonmechanical processes have been adapted to disrupt the cell (Fig. 4.3). The selection of method for lysis is based on its effect on safety of the structure and function of target product to be isolated [3]. Laboratory and industrial scale cell lysis methods have been developed and used for many years now. The global market for cell lysis is estimated at $2.35 billion in 2016 and is expected to reach $3.84 billion by 2021 [4].

Factors that influence the selection of disruption methods include:

1. Structural makeup of the cell boundary determines the suitability of a method to be applied for cell disruption. For example, yeast cells are particularly difficult to disrupt as the cell wall limits a solvent's access to the desired products.
2. The stability of the target product to be released from the cells depends on selection of a suitable method.

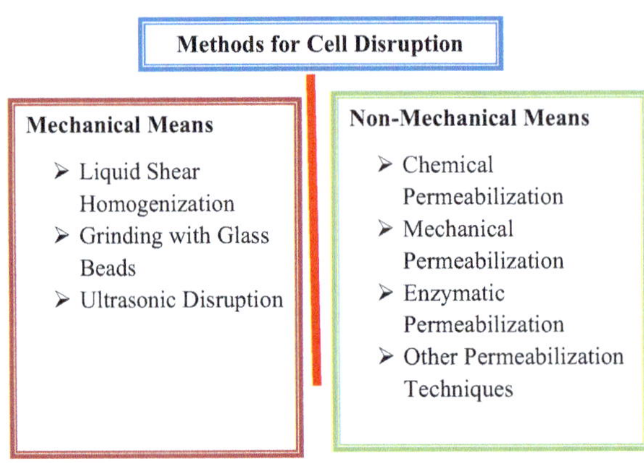

FIG. 4.3

Various methods used for cell disruption.

3. Mechanical methods produce heat during the process, so additional cooling systems are required when using mechanical cell disruption methods.
4. Extraction of cell debris is also one of the challenging issues while isolating a target product.
5. The economy of cell disruption mainly depends on the time needed for conducting the process and the quantity of energy input in the entire process.
6. In some cases, the drying of the cell mass enhances disruption methods and may help bring down the costs.

There are many methods of cellular disruption because there are many types of cells.

4.3.1 Cell lysis buffer

During cell lysis, the intracellular materials of the cell come out and mix with the surrounding medium of the harvested broth, causing a change in pH. The change in pH may bring instability to the target product. The primary purpose of lysis is isolating the product of interest and keeping it in a stable environment, and for such purposes, a suitable lysis is used. Most lysis buffers contain buffering salts like Tris-HCL and ionic salts like NaCl to monitor the pH and osmolarity of the lysate. The word *lysis* comes from the Greek word for "loosen." Cell lysis buffer can be used on both animal and plant tissue cells.

4.3.1.1 Selecting a buffer

Due to the differential nature of structure and function of therapeutic proteins, it is necessary to select the most suitable buffer based on the purpose and design of the project undertaken. The important factors to be considered are pH, ionic strength, usage of detergent, and protease inhibitor to prevent proteolytic processes. For example, Gram-negative bacteria need detergent for lysis, whereas Gram-positive bacteria can be disrupted without detergent. Commonly, a protease inhibitor is added to the lysis buffer, along with other enzyme inhibitors of choice, such as a phosphatase inhibitor when studying proteins with phosphorylation.

4.3.1.2 Composition of lysis buffer

A buffer is a chemical surrounding to protect a target protein. Each buffer has a specific pH range and should be compatible with the isolated protein. Commonly, HEPES-NaOH (7.2–8.2 pH), Tris-HCL (7.0–9.0 pH), and sodium dihydrogen phosphate or disodium hydrogen phosphate (5.8–8.0 pH) are used as lysis buffers.

4.3.1.3 Additive and detergent

Detergents are mostly used to maintain the ionic strength in the lysis buffer salts like NaCl, KCl, and $(NH_4)^2$ SO_4 and are used with a concentration range between 50 and 150 mM.

Detergents are organic amphiphilic surfactants having hydrophobic tail and a hydrophilic head, which are used to separate membrane proteins. The hydrophobic part of detergent can attach to a membrane protein and help in removing the membrane proteins. Detergents are, also, often categorized as nonionic, anionic, or zwitterionic based on their hydrophilic head. Nonionic detergents like Triton

X-100 and Zwitterionic detergents like CHAPS (3-[(3-cholarridoproyl)dimethylammonio]-1 propane sulfonate) are nondenaturing (will not disrupt protein function). Ionic detergents like sodium dodecyl sulphate (SDS) and cationic detergents like ethyl trimethyl ammonium bromide are denaturing (will disrupt protein function). The lysis strength is mainly dependent on detergents.

4.3.1.4 Commercial lysis buffer

There are several lysis buffers available commercially, mostly for laboratory use:

- NP-40 is a nonionic buffer widely used in the lysis of cells. This buffer can be a better substitute for replacement with other buffers but in different concentrations.
- RIPA (radioimmune precipitation assay) lysis buffer is commonly used as buffer for immunoprecipitation and general protein extraction from cell and tissues. It also disrupts most weak interaction between proteins.
- SDS lysis buffer is an ionic denaturing detergent. It is mostly used for solubilization of proteins at elevated temperature.
- ACK (ammonium-chloride-potassium) lysis buffer is commonly used for lysis of red blood cells in biological samples to obtain white blood cells.

4.3.2 Bacterial cells lysis

As stated earlier, bacterial cells may require different disruption methods, depending on whether they are Gram-positive or Gram-negative, as the amount of peptidoglycan and presence of an envelope affect the overall process. Response of a bacterial system varies with different types of disintegration (digestive enzymes like lysozymes, OS, chemical treatment). The peptidoglycan layer can withstand the osmotic pressure developed inside the cell due to the concentration difference of solutes across the membrane. For *E. coli*, this pressure is estimated around 2 atm [5]. This layer is also responsible for maintaining the shape and rigidity of the cell. This layer consists of two sugar derivatives named N-acetylglucosamine and N-acetylmuramic acid, as well as a small group of amino acids consisting of L-alanine, D-alanine, and D-glutamic acid.

In the Gram-positive bacteria, the cell boundary mainly composed of a thick peptidoglycan layer. In addition, the bacterial cell membrane and the plasma are made up of relatively fragile phospholipids and proteins. In certain cases, polysaccharide capsules may be present outside the cell wall. In Gram-positive bacteria, the peptidoglycan layer makes up 50%–80% of the envelope, and 10% of this layer is associated with teichoic acid, which provides a greater structural resistance to breakage [6]. In contrast, 10%–20% of the cell envelope of Gram-negative bacteria is composed of a 1.2- to 2.0-μm-thick peptidoglycan.

The cell wall of Gram-positive bacteria is particularly susceptible to lysis by the antibacterial enzyme and lysozyme. Unlike Gram-positive bacteria, Gram-negative bacteria do not have distinct cell walls but instead have multilayered envelopes. The peptidoglycan layer is significantly thinner than in Gram-positive bacteria. An external layer composed of lipopolysaccharides and proteins is usually present. Another difference with Gram-positive bacteria is the presence of the periplasm layers, which are two liquid-filled gaps, one between the plasma membrane and the peptidoglycan layer and other between the peptidoglycan layer and the external lipopolysaccharides. Another periplasmic layer also exits in Gram-positive bacteria, and these are important in bioprocessing since a large number of proteins, particularly recombinant proteins, are secreted into it.

4.3.3 Yeast/mold cell lysis

Yeast cells are more rigid compared to that of bacterial cell membrane (Fig. 4.2B). The yeasts are unicellular, having 0.1–0.2 μm in thickness. These are mainly composed of polysaccharides such as glucans, mannans, and chitins. The plasma membrane in a yeast cell is composed of phospholipids and lipoproteins. Mold cells are largely similar to yeast cells in terms of cell wall and plasma membrane composition but are multicellular and filamentous.

Yeasts are more difficult to lyse with enzymes and detergents than bacteria. Accordingly, bead beating is often used to crack open yeasts for both nucleic acid isolation and enzyme assays. Bead beating can be performed using a Pulsing Vortex Mixer, HT Mini homogenizer, GenoGrinder, or HT Homogenizer. Unfortunately, vortexing samples mixed with glass beads on basic laboratory vortexers are not that effective for disrupting cells.

The most effective bead size for *Saccharomyces* is 400-μm silica beads. However smaller yeasts, such as *Pichia*, are best homogenized with 200-μm zirconium beads. Samples containing very low numbers of cells should be disrupted using low binding beads so that analytes remain in the lysate instead of adsorbing to the beads.

4.3.4 Mammalian cells

Mammalian cells are the easiest to disrupt as they lack a cell wall. Mammalian cell boundary is known as the plasma membrane consisting of lipids (phospholipids and cholesterol), proteins, and carbohydrate groups that are attached to some of the lipids and proteins (Fig. 4.2A). Mammalian organisms are made of four major tissue types: connective, epithelial, nerve, and muscle. These can be further differentiated into specialized tissues, all of which have unique structures. So, suitability of a method for mammalian cell rupture may differ with tissue. The contents of mammalian cells are only separated from the extracellular environment by a thin membrane that forms a stable barrier between aqueous compartments. The plasma membrane is permeable to specific molecules. It allows nutrients and other essential elements to enter the cell and waste materials to leave the cell. Multiple cell rupture techniques exist, such as homogenization, mechanical shearing, freezing, and sonication. Commonly, primary and culture cells do not require physical disruption; however, detergent-based cell lysis is an easy alternative that can be combined with homogenization for optimal results. Detergents work to break apart the lipid bilayer, and homogenizers finalize particle breakage, so that all contents of interest can be accessed.

4.3.5 Plant cells lysis

Plant cells are more difficult to disrupt than animal cells because of the thick walls. The plant cell wall is mainly composed of cellulose and other polysaccharides (Fig. 4.2A). A variety of mechanical methods can be applicable to disrupt the cell wall. However, choosing a suitable chemical method is based on the specific product of interest.

A range of mechanical methods can be used to disrupt the cell wall. Chemical methods when used for cell disruption are based on specific targeting of key cell wall components. For example, lysozyme is used to disrupt the cell wall of Gram-positive bacteria since it degrades peptidoglycan, which is a key cell wall constituent. In Gram-negative bacteria, the peptidoglycan layer is less susceptible to lysis by lysozyme since it is shielded by a layer composed of lipopolysaccharides and proteins.

Irrespective of the nature of the cells belonging to mammal, yeast, bacteria, or virus, it is necessary to select the suitability of a method for the disruption of the cells of interest. In addition, it is important to consider the location of product of interest and stability in determining the best way to approach cell disruption. For example, certain proteins are located in specific organelles. Rupture methods that release the contents from every compartment make it incredibly difficult to achieve a high yield of a specific protein of interest. However, methods like high pressure homogenization will save precious time while still achieving a high yield of intracellular contents.

4.4 Methods for cell lysis (disruption)

A variety of mechanical and nonmechanical methods are used for cell disruption (lysis) (Fig. 4.3). The selection of a suitable method for cell disruption depends on the type of microorganisms or cells from other biological resources.

4.4.1 Mechanical methods

The two types of biomolecules of commercial value are available in an extracellular matrix or present as intracellular materials. The extracellular materials are produced within the cell but then excreted into the surrounding environment, whereas the intracellular materials remain inside the cell (Table 4.1)

Table 4.1 Some intercellular and extracellular enzymes of commercial importance.

Enzyme[a]	EC number[b]	Source	Intra (I)/ extracellular (E)[c]	Scale of production[d]	Use
Animal enzyme					
Catalyze	1.11.1.6	Liver	I	–	Food
Chymotrypsin	3.4.21.1	Pancreas	E	–	Leather
Lipase[e]	3.1.1.3	Pancreas	E	–	Food
Rennet[f]	3.4.23.4	Abomasum	E	+	Cheese
Trypsin	3.4.21.4	Pancreas	E	–	Leather
Plant enzyme					
Actinidin	3.4.22.14	Kiwi fruit	E	–	Food
α-Amylase	3.2.1.1	Malted barley	E	+++	Brewing
β-Amylase	3.2.1.2	Malted barley	E	+++	Brewing
Bromelin	3.4.22.4	Pineapple latex	E	–	Brewing
β-Glucanase[g]	3.2.1.6	Malted barley	E	++	Brewing
Ficin	3.4.22.3	Fix latex	E	–	Food
Lipoxygenase	1.13.11.12	Soybeans	I	–	Food
Papain	3.4.22.2	Pawpaw latex	E	++	Meat
Bacterial enzyme					
α-Amylase	3.2.1.1	*Bacillus*	E	+++	Starch
β-Amylase	3.2.1.2	*Bacillus*	E	+	Starch
Asparaginase	3.5.1.1	*Escherichia coli*	I	–	Health

Table 4.1 Some intercellular and extracellular enzymes of commercial importance—cont'd

Enzyme	EC number	Source	Intra (I)/ extracellular (E)	Scale of production	Use
Glucose isomerase[h]	5.3.1.5	*Bacillus*	I	++	Fructose syrup
Penicillin amidase	3.5.1.11	*Bacillus*	I	–	Pharma
Protease[i]	3.4.21.14	*Bacillus*	E	+++	Detergent
Pullulanase[j]		*Klebsiella*	E	–	Starch
Fungal enzymes					
α-Amylase	3.2.1.1	*Aspergillus*	E	++	Baking
Aminoacylase	3.5.1.14	*Aspergillus*	I	–	Pharma
Glucoamylase[k]	3.2.1.3	*Aspergillus*	E	+++	Starch
Catalase	1.11.1.6	*Aspergillus*	I	–	Food
Cellulose	3.2.1.4	*Trichoderma*	E	–	Waste
Dextranase	3.2.11	*Penicillium*	E	–	Food
Glucose oxidase	1.1.3.4	*Aspergillus*	I	–	Food
Lactase[l]	3.2.1.23	*Aspergillus*	E	–	Dairy
Lipase[e]	3.1.1.3	*Rhizopus*	E	–	Food
Rennet[m]	3.4.23.6	*Mucor miehei*	E	++	Drinks
Pectinase[n]	3.2.1.15	*Aspergillus*	E	–	Drinks
Pectin lyase	4.2.2.10	*Aspergillus*	E	+	Backing
Protease[m]	3.4.23.6	*Aspergillus*	E	–	Food
Raffinase[o]	3.2.1.22	*Mrtierella*	I	–	
Yeast enzymes					
Invertase[p]	3.2.1.26	*Saccharomyces*	I/E	–	Confectionary
Lactase[l]	3.2.1.23	*Kluyveromyces*	I/E	–	Dairy
Lipase[e]	3.1.1.3	*Candida*	E	–	Food
Raffinase[o]	3.2.1.22	*Saccharomyces*	I	–	Food

[a]*The name in common usage are given. As most industrial enzymes consist of mixture of enzymes, these names may vary from the recommended names of their principal component. Where appropriate, the recommended name of this principal component is given below.*
[b]*EC, number of the principal component.*
[c]*I, intracellular enzyme; E, extracellular enzyme.*
[d]*+++>100 ton/year; ++>10 ton/year; +1 ton/year; –< ton/year.*
[e]*Triacylglyceol lipase.*
[f]*Chymosin.*
[g]*Endo-1,3 (4)-β-glucanase.*
[h]*Xylose isomerise.*
[i]*Subtilisin.*
[j]*α-Dextrin endo-1.6-α-glucosidase.*
[k]*Glucan 1,4-α-glucosidase.*
[l]*β-Galactosidase.*
[m]*Microbial aspartic proteinase.*
[n]*Polygalacturonae.*
[o]*α-Galactosidase.*
[p]*β-Fructofuranosidase.*

[7–9]. The isolation of intracellular material requires that the cell either be genetically engineered, so what would normally be an intracellular product is extracted into the surrounding environment, or it must be disintegrated by physical, chemical, or enzymatic means to release contents into the surrounding medium. The genetic manipulation of microbial cells to make them leaky is limited in scope. Making the cell fully permeable to any significant fraction of intracellular products and enzymes would not only be difficult, but also implies discontinued existence of the cell. It is in this context that microbial cell disruption for intracellular product isolation becomes of increasing importance. However, few studies have been published on the subject, probably because of the high capital and operating costs of pilot plants for large-scale isolation of intracellular products and the requirement of sizable teams of scientists and technical staff to obtain meaningful biochemical engineering design data [9].

In the past, several intracellular enzymes have been produced industrially. Examples include glucose oxidase for food preservation, penicillin acylase for antibiotic conversion, and asparaginase for possible cancer therapy [10]. Other examples of intracellular microbial enzymes produced commercially are given in Table 4.1. The necessity of harvesting the producing cells to subsequently extract an internal constituent is a major economic disadvantage. Possible cost reductions, however, can be achieved by simultaneous isolation of a number of intracellular products following cell disruption [11–13]. This would be even more advantageous if extracellular product isolation preceded intracellular product isolation from the same fermentation batch.

As the biological materials are temperature- and other stress and strains-sensitive, it is essential to obtain intracellular contents (Fig. 4.1) from cell rupturing under controlled conditions. The mechanical operations with high-pressure homogenization (HOH) are most commonly used on the large scale due to their rapid handling of a variety of cell suspensions and ease of scalability. Problems like rate of fluid flow, variation in particle-particle interaction, and pressure drop can be overcome, while operating the process can be monitored by equipment modification. The energy dissipated as heat can cause structural damage of cellular components but can be managed with efficient cooling systems.

The effectiveness of a cell disruption method determines the accessibility of intracellular proteins for extraction and solubility. Different biological materials require different cell disruption strategies, which can be divided into two categories: gentle and harsher (Table 4.2).

(i) High-shear homogenization vs. high-pressure homogenization

High-shear mixers rely on rotor/stator shearing action, whereas high-pressure homogenizers utilize a high amount of energy to form pressure differentials to homogenize or reduce the size of ingredients. High-speed, high-shear rotor/stator mixers are able to close the gap between traditional rotor/stator mixers and high-pressure homogenizers into submicron size reduction.

(ii) High-shear homogenization

High-shear homogenizer (HSH) or mixers, also known as high-shear reactors (HSRs), rotor-stator mixers, and HSHs are used to create emulsions, suspensions, and granular product. It is commonly used in the food, pharmaceutical, and cosmetic industries to emulsify, homogenize, disperse, grind, and/or dissolve immiscible mixtures with components of the same or different phases. These machines have characteristics of high rotor tip speeds, high shear rates, localized energy dissipation rates, and higher power consumption than ordinary mixers. This equipment is available for laboratory purposes and industrial types (Fig. 4.4).

Table 4.2 Suitability of cell disruption methods for various ample types.

Technique	Description	Yeast	Plant cells	Mammalian cells	Bacteria
Gentle method					
Osmotic lysis	Suspension of cell in hypotonic solution; cell swell and burst Freezing in liquid nitrogen and subsequent thawing of cell	–	–		–
Freeze thaw process	Freezing in liquid nitrogen and subsequent thawing of cell	–	–		–
Detergent lysis	Suspension of cells in detergent solution to solubilize the cell membrane. This method is usually followed by another disruption method, such as sonication	–	–		–
Enzymatic lysis	Enzymatic digestion in iso-osmotic solution. Cellulose and pectinase is used for plant cells, and lyticase for yeast cells, and lysozyme for bacterial cells				
Harsher methods					
Sonication	Disruptor of cells is carried in cool water to avoid heating damage		–		
French press	Application of shear forces by forcing a cell suspension through a small orifice at high pressure		–		
Grinding	Breaking cells with a mortar and pestle			–	–
Mechanical homogenization	Mechanical homogenization with either a handheld device (e.g., Dounce and Potter-Elvehjem homogenizer), blender, or other motorized device; this method is best suited for soft, solid tissues	–		–	–
Glass-beads homogenization	Application of gentle abrasion by vortexing cells with glass beads		–		

Because high-shear mixing makes use of variable speed, the technique provides the ability to customize the amount of shear energy used for mixing applications. As a result, this technique can mix a substance, whether liquid or solid, into a material with which it typically would not easily mix. High-shear mixing often isn't as speedy as homogenizing, but this processes' higher flow can accommodate a wider range of viscosities, making it a great choice for applications including paint and ink mixing and emulsion for ceramics and adhesives. Also, high-shear mixing generates heat, which can negatively affect some temperature-sensitive mixtures and requires a cooling system for effective temperature control.

FIG. 4.4

Photo of High shear mixer (A) batch type for laboratory purpose. (B) Industrial type for large scale production. (C) Continuous type.

(A and B) With courtesy from: Tera Pak, High shear Mixer, Batch unit. (C) With courtesy from-Peerless.

(iii) High-pressure homogenizer

For two decades, wine-making industries have been using a variety of cell disruption techniques like ultrasound (US), pulsed electric fields (PEF), or high hydrostatic pressures (HHP) to increase the process efficiency in the extraction of grape juice [14–17]. However, in Europe, the use of these techniques are still not allowed at the winery scale [18], but some of them, e.g., US or PEF, are under active con sideration by the International Organization of Vine and Wine (OIV) [19].

In this method, cells in media are forced through an orifice valve using high pressure, causing the formation of very fine emulsion droplets. The origin of this shear is the sudden restriction of flow under high pressure through a restrictive valve [20]. Extreme shear and high-energy input have to be applied to reduce droplets from the microscale to the nanoscale range. The Laplace pressure, which causes the resistance to droplet deformation and breakup, needs to be overcome. This Laplace pressure increases when the droplet diameter decreases and is therefore larger for nanoemulsions than for ordinary emulsion [21]. In nanoemulsion, because of its high surface-to-volume ratio, there is a need for high amounts of surfactants (such as proteins) for their stabilization. The interaction of proteins with the newly formed droplet is improved under HPH. So, the basic target of homogenization is to create a reduction in particle size to a narrow size range [22] by applying physical force. Mostly, valve technology is being used to design a high-pressure homogenizer. In this system, a fluid is forced to pass through the homogenization valve, and the mechanical forces lead to cell disruption and release of product. This is due to an increase in surface area of disrupted cells and an increase in releasing of product in stable form. Mostly, homogenizers of 20–50 MPa is used in chemical, pharmaceutical, biotechnology, and food industries mainly to reduce the particle size of liquid dispersions, preventing phase separation during storage.

Currently, much higher pressure HPHs with a range of 150–400 MPa are available on the basis of this requirement [23]. Basically, during the HPH process, the fluid is forced to pass through a narrow gap in the homogenizer valve, where it is submitted to a rapid acceleration [24]. The overall structural design, including the valve of high-pressure homogenizer, even an ultrapressure homogenizer is made resist the intense mechanical stresses generated during the fluid passage. Presently, a variety of valve geometry made of different materials are commercially available [25].

HPH is a highly acceptable method of cell rupturing by industrialists. This is a purely mechanical process that involves flowing a fluidic product through a narrow gap (the homogenizing nozzle) at high pressure (150–200 MPa or 350–400 MPa) for ultrahigh pressure homogenization (Fig. 4.5) [3,26].

4.4.1.1 Functional components

In HPH, high-pressure pumps and disruption units are the main components responsible for cell disruption. In a pulsating flow pattern, the pumps produce positive displacement. In case of formation of any cavitation, provisions of dampers are attached to pipes. To reduce vibration noise, the dampers are attached to outer pipes. The high-pressure pump forces the cell suspension to pass through a very narrow but adjustable orifice to break the suspension into very fine emulsion droplets. The origin of this shear is the sudden restriction of flow under high pressure through a restrictive valve [27]. Extreme shear and high-energy input have to be applied to reduce droplets from the microscale to the nanoscale range. The Laplace pressure, which causes the resistance to droplet deformation and breakup, needs to be overcome. This Laplace pressure increases when the droplet diameter decreases and is therefore larger for nanoemulsions than for ordinary emulsion [28]. In nanoemulsion, because of their high

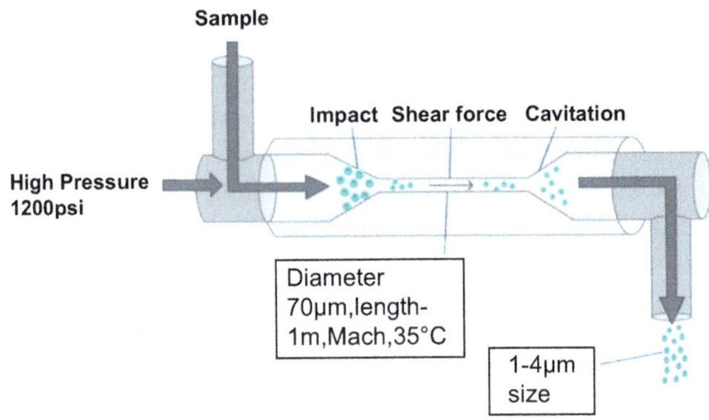

FIG. 4.5

Functional line diagram of liquid shear homogenizer.

surface-to-volume ratio, there is a need for high amounts of surfactants (such as proteins) for their stabilization. The interaction of proteins with the newly formed droplet is improved under HPH [29]. This is possible due to the impact ring at the end of narrow pathway, which provides efficient breakup of suspended particles. Passing through this narrow path due to high pressure, the particle tends to hit the impact ring of the homogenizer with high velocity, resulting in a pressure drop, turbulent eddies, and strong shearing forces produced by the high-pressure pump, causing the breaking of fat globules in milk (Fig. 4.6).

Basically, two theories (turbulent eddies and cavitation) mark this event. On the basis of turbulent eddies, high-pressure forces form small high-energy eddies in the stream that collides with the fat globules. This is mainly due to the turbulent flow of milk.

On the other hand, cavitation theory is based on generation of shockwaves in the stream of milk while it passes through the adjustable gap. This is mainly due to the pressure drop in the stream of milk. Vapor bubbles are formed due to high pressure during their turbulent flow, and when pressure is released, the bubbles collapse, which breaks apart the dispersed droplet.

The mechanism of droplet disruption depends on the operating parameters, including flow rate, pressure, number of cycles, temperature, and device parameters (e.g., nozzle geometry and impingement of design). The mechanism also depends on product parameters such as composition (fat, water, protein, etc.) and viscosity (although of less importance than in rotor-stators systems). HPH is a mature technology in milk processing. High-pressure homogenizers of piston-gap type can currently deliver pressures up to 150–200 MPa and even up to 300–400 MPa that collide with one another to disrupt particles. Oil emulsion can be performed by direct injection of the oily phase under pressure (up to 350 MPa) into a continuous aqueous phase at low pressure [30]. Microfluidization (up to 200 MPa) produces a tighter particle size distribution than that of traditional techniques. The scalability of the process depends on the formulation of the product. Multiple-pass HPH can pose a problem with regard to industrial implementation but can be overcome by applying higher pressure.

The overall design of HPH (Fig. 4.6) varies on the basis of high amount of solids, up to 50% of the feed. Heat generation is also high: 1.50 C/1000 psi [31]. Protein release is the first order with respect to the number of passes (N). The dependence of protein release on operating pressure (p 400–600 bar) can be expressed as a function of the pressure raised to an exponent [32].

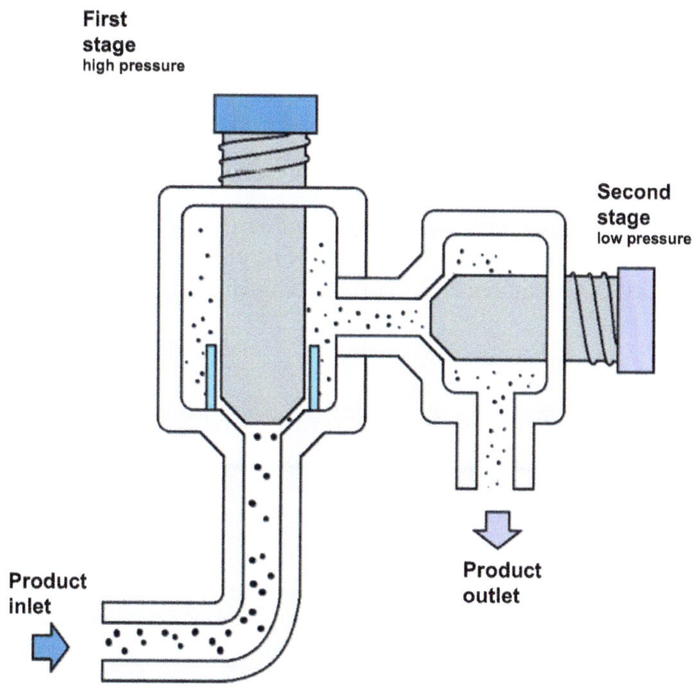

FIG. 4.6

Schematic diagram of a high pressure homogenizer.

In this connection, Sauer's model [33] for the amount of protein release is an ideal model (Eq. 4.1) to explain how the release of protein from *E. coli* is independent of biomass concentration, and higher concentration of cells can be disrupted at the same time.

But the main problem is to manage the generation of heat in this process to avoid degradation of some enzymes. This problem can be overcome with the provision of a cooling system [33] in this device. A combination of lysis methods, for example, chemical treatment with homogenization can also resolve the problem [34].

$$\text{In}\frac{Rm}{Rm-R}=kNP^{a} \tag{4.1}$$

The results can be combined in the correlation, where R is the protein released (kg protein/kg biomass), Rm is max protein available, k is rate constant and a function of temperature, and a is the pressure exponent.

4.4.1.2 Additional effects

The high pressures applied cause a temperature increase because of the heat of compression. This temperature increase is enhanced by shear effects and partial conversion of mechanical energy into heat. The result is a total temperature increase of 17–21°C per 100MPa. This needs to be controlled in the case of heat-sensitive food compounds. HPH also enhances product texture and mouth feels. Using nanoemulsion in food products can facilitate the use of less fat without compromising creaminess, thus offering the consumer a healthier option [27–29].

(iv) Grinding with glass beads

Basically, a glass bead mill or ball mill is a type of grinder used to grind or blend materials for use in mineral dressing processes, such as paints, pyrotechnics, and ceramics. For two decades [35], this method has been used for cell disruption, both in small scale and large scale level. This is one of the most acceptable methods both in laboratory purposes and large scale level in biopharmaceutical industries. The main principle requires a jacketed grinding chamber with a rotating shaft running in its center (Fig. 4.7).

4.4.1.3 How does it work?

Basically, the effective acceleration of a grind mill with beads is monitored by two methods:

(a) By shaking the entire container (i.e., cell disruption using roto-stator mill)

A rotor-stator ball mill (RSM) consists of a stationary block with a tapered cavity known as stator and a truncated cone-shaped rotating accessory device called the rotor (Fig. 4.8) having speeds in the 10,000–50,000 rpm range. A slurry containing cells is fed into the tiny gap between the rotating rotor and the fixed stator. The feed (slurry containing cells) is drawn in due to the rotation and expelled through the outlet due to centrifugal action. The high rate of shear generated in the space between the rotor and the stator, as well as the turbulence thus generated, are responsible for cell disruption. These mills are more commonly used for disruption of plant and animal tissue-based material and are operated in the multi-pass mode, i.e., the disrupted material is sent back into the device for more complete disruption. The cell disruption caused within the rotor-stator mill can be described using the equation discussed for a bead mill.

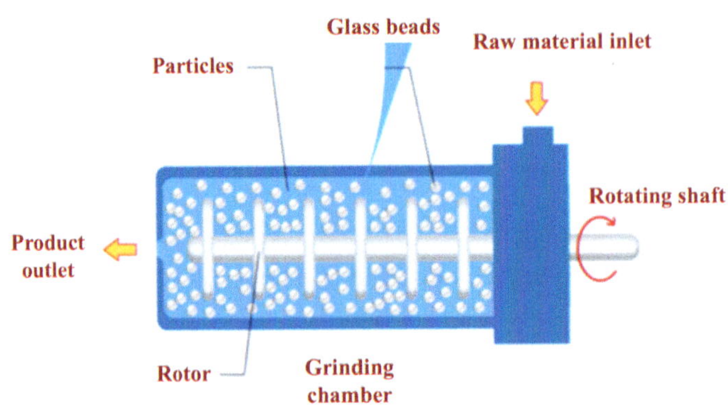

FIG. 4.7

Glass beads mill.

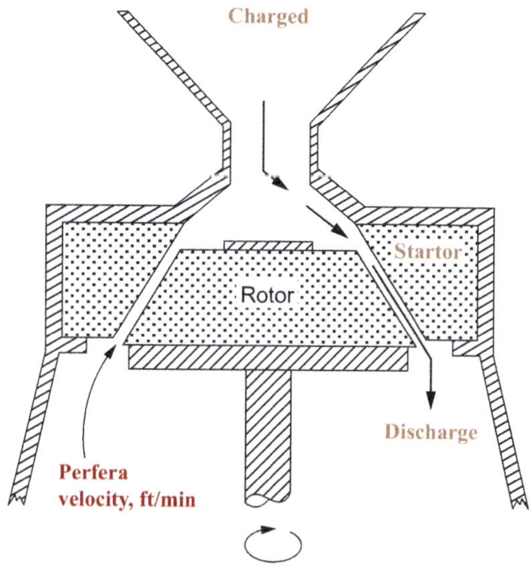

FIG. 4.8

Functional line diagram of a rotor-stator ball mill.

(b) By a spinning agitator within a container

The basic working principle is a shaft fitted with a series of rotating grinding discs and a charge of glass ballotine (Fig. 4.9). The mill-base is pumped through the grinding chamber and the disc rotations move the beads and mill-base mixture at high speed. The product is separated from the grinding beads at the outlet by a slotted screen. The optimum concentration of ballotine is, usually, maintained at about 70%–90% of the chamber geometry volume. The maximum bead size is usually about 0.2–0.5 mm (like, bacterial cell wall); however, in rigid cell boundary like yeast cells, it is 0.4–0.7 mm. The frequency and the strength of collisions during milling depend on the type of cells used [36]. Commonly, high-speed bead mills have been used to disrupt the cells of several yeasts, bacteria, and fungi [37,38]. Main issues related to bead mills are the high temperature rises with increase of bead volume, poor scale-up, and most importantly, there is a high chance of contamination [3].

In this process, a lot of heat is produced due to the friction of the glass beads and the inner wall of the cells' slurry container. To overcome the problem, it is advised to use liquid nitrogen along with the cell suspension. This process is referred to as cryogenic bead milling. An alternative approach is to use glycol-cooled equipment. Commonly, in the laboratory, the bead mill operates in small batch scale, and for industrial process, it is used at a turnkey level in continuous process. The bead mill is used for disrupting yeast cells and for grinding animal tissue. Using a small scale unit operation in a continuous mode, a few kilograms of yeast cells can be disrupted per hour. Large units can handle hundreds of kilograms of cells per hour.

FIG. 4.9

Schematic diagram of a grinding mill.

4.4.1.4 Disadvantages of mechanical disruption

Mechanical disruption methods suffer from several drawbacks. Because cells are broken completely, all intracellular materials are released. Therefore, the product of interest must be separated from a complex mixture of protein, nucleic acids, and cell fragments. Released nucleic acids can increase the viscosity of the solution and complicate subsequent processing steps or experimental procedures, especially chromatography. The cell debris produced by mechanical lysis often consists of small cell fragments, making the solution difficult to clarify. Complete product release often requires more than one pass through the disruption device, exacerbating the problem by further reducing the size of fragments. Small fragments are difficult to remove by continuous centrifugation because the throughput of the device is related to the square of the particle diameter. Filtration is complicated by the gelatinous nature of the homogenate and by its tendency to foul membranes; furthermore, mechanical methods expose the cells, and hence the extracted product, to very harsh conditions. Although it is now generally accepted that most proteins can tolerate the high pressure inside a homogenizer or ball mill, most proteins are denaturized by the heat generated unless the device is sufficiently cooled.

4.4.2 Nonmechanical means

Nonmechanical disruption, in contrast to mechanical, uses methods other than force to rupture a cell's wall and/or membrane. This can be categorized as physical, chemical, and biological, where each group is further classified based on the specific techniques and methods used for disruption. A detailed description of each type is presented as follows.

4.4.2.1 Physical disruption

Physical disruption is a noncontact method that uses external force to rupture the cell membrane. The different forces include heat, pressure, and sound energy, and on this basis, they can be grouped into as: thermal lysis, cavitation, and osmotic.

4.4.2.2 Thermal disruption

This technique involves freezing a cell suspension in a dry ice/ethanol bath or freezer and then thawing the material at room temperature or 37°C. This method of lysis causes cells to swell and ultimately break as ice crystals form during the freezing process and then contracting during thawing. This causes formation of ice on the cell membrane, which helps in breaking down the cell membrane. Multiple cycles are necessary for efficient lysis, and the process can be quite lengthy. However, freeze/thaw has been shown to effectively release recombinant protein located in the cytoplasm of bacteria and is recommended for the lysis of mammalian cells in some protocols. This method is time consuming and cannot be used for extracting cellular components sensitive to temperature. However, it has been also proven to isolate highly expressed recombinant protein from *E. coli* by such techniques [38]. Elevated temperature has also been shown to be capable of cell lysis [39,40]. However, heating for a long period may damage the DNA. But this method is highly expensive and not recommended for commercial use [41]. Raised temperature can also cause damage of target materials such as temperature labile proteins and enzymes [42]. Thermal lysis is an attractive method at the microscale used in many microfluidic devices.

4.4.2.3 Cavitation

In this technique, the local micro surrounding pressure is reduced by increasing the velocity, ultrasonic vibration, etc. The reduction in pressure results in collapsing the cavity or bubble. This pressure fluctuation is of the order of 1000 MPa [3]. During the collapse of a bubble, a large amount of mechanical energy is released in the form of a shockwave that propagates through the media. Since this shockwave has high energy, it has been used to disintegrate the cell membrane. Ultrasonic and hydrodynamic methods have been used for generating cavitation used to disrupt cells.

4.4.2.4 Ultrasonic disruption

Another widely applied method is cell lysis using high-frequency sound that is produced electronically and transported through a metallic tip to an appropriately concentrated cellular suspension. Because of the very high cost, it is only used at the laboratory scale for lysis of cells with less resistant cell walls, such as bacteria and fungi. Ultrasonic disruption of cells involves cavitations, microscopic bubbles, or cavities generated by pressure waves. It is performed by ultrasonic vibrators that produce high-frequency sound with a wave density of approximately 20 kHz [43]. A transducer converts the wave into mechanical oscillations via a titanium probe immersed in a concentrated cell suspension. However, this technique also generates heat, which can denature thermolabile proteins [44]. Rod-shaped bacteria are easier to break than cocci, and Gram-negative organisms are more easily disrupted than Gram-positive cells. Sonication is effective on a small scale but is not routinely used in large-scale operations because of problems with power transmission and heat dissipation. To overcome the problems associated with ultrasonic cavitation, such as high power requirement and high energy to dissipate heat, hydrodynamic cavitation has been used to disrupt the cell membrane [45]. Hydrodynamic cavitation is produced by pumping the cell suspension through a constricted channel that results in an increase in velocity [46]. The energy required for lipid extraction from cells using the hydrodynamic cavitation technique is about 3 MJ/kg, which is 10 times more efficient compared to sonication in terms of energy consumption [47].

4.4.2.5 Osmotic shock (OS)

This method is based on the sudden changing of the concentration of surrounding salt to create a difference in pressure between the inside and outside of the cell; the membrane becomes permeable to water due to osmosis. The entry of water inside the cell depends on the concentration of salt surrounding the cell. The concentration of surrounding salt is lower as compared to inside the cell. Water starts entering the cell, leading to swelling of the cell, and it subsequently bursts. This technique is suitable for mammalian cells due to the fragile structure of the membrane; however, periplasmic proteins may be released in the case of Gram-negative bacteria [48–50].

4.4.2.6 Chemical permeabilization (CP)

Many chemical methods have been employed to extract intracellular components from microorganisms by permeabilizing the outer wall barriers. Often used with plant cells (and sometimes in combination with shearing), organic solvents such as toluene, ether, benzene, butanol, chloroform, and methanol can be used to permeate cell walls by creating channels through the cell membrane. Chemical permeabilization can also be achieved using phenylethyl alcohol, DMSO, antibiotics, ionones, surfactants (triton, Brij, Duponal), chaotropic agents, and chelates. Ethylenediaminetetraacetic acid (EDTA) is widely used to disrupt Gram-negative microorganisms because it chelates cations, causing holes in the cell walls. Its effectiveness is a result of its ability to bond the divalent cations Ca^{2+} and Mg^{2+}, stabilizing the structure of outer membranes by bonding the lipopolysaccharides to each other. Once these cations are removed by EDTA, the lipopolysaccharides are released, resulting in areas of increased permeability in the outer walls.

Chaotropic agents such as urea and guanidine are capable of bringing some hydrophobic compounds into aqueous solution. They accomplish this by disrupting the structure of water (the hydrogen-bonding network), making it a less hydrophilic environment and weakening the hydrophobic interaction between solute molecules.

4.4.2.7 Alkaline lysis

SDS is an alkaline lysis buffer. This is used with sodium hydroxide. The OH^- ion reacts with the cell membrane and breaks the fatty acid-glycerol ester bonds and subsequently makes the cell membrane permeable, and SDS solubilizes the proteins and the membrane [36,37,51,52]. This is a very common method applied to all kind of cells, especially to isolate plasmid DNA from bacteria [53,54].

In alkaline lysis, OH^- ions are the main component used for lysing the cell membrane [55]. The lysis buffer consists of sodium hydroxide and SDS. At a pH range of 11.5–12.5, the OH^- ion reacts with the cell membrane, breaks the fatty acid-glycerol ester bonds, and subsequently makes the cell membrane permeable, and the SDS solubilizes the proteins and the membrane. The pH range of 11.5–12.5 is preferable for cell lysis [3,56]. Although this method is suitable for all kinds of cells, this process is very slow and takes about 6–12 h. This method is mostly used for isolating plasmid DNA from bacteria [57,58].

4.4.2.8 Enzymatic permeabilization

Enzymatic digestion is often used in the laboratory as a gentle method of cell rupture. One advantage of enzymatic lysis is its specificity. The cell walls of yeast and bacteria are digested with special enzymes, and then the unprotected cell can easily be ruptured by mild OS. Cell wall-degrading enzymes such as lysozyme, which degrades β-1,4 glycosidic linkages, is useful for Gram-positive organisms. Addition of EDTA to chelate metal ions improves the effectiveness of lysozyme. Yeast cell walls can be degraded by snail gut enzymes (containing a mixture of gluconates). Enzymatic rupture methods are, in general, too expensive for use in large-scale manufacturing. Most of the enzymes are available commercially and can be used for large scale lysis.

4.4.2.9 Other permeabilization techniques

Basic proteins such as protamine or the cationic polysaccharide chitosan can permeabilize yeast cells. Similarly, mammalian cells can be permeabilized by exposure to several natural substances such as streptolysin or even viruses. Electric discharges have also been used to permeabilize mammalian cells to study secretion by exocytosis.

4.4.2.10 Detergents

Detergents that are ionic in nature (e.g., cationic acetyl trimethyl ammonium bromide and anionic sodium lauryl sulphate) can denaturate membrane proteins and lyse the cells (Fig. 4.10). Nonionic detergents (although less reactive than ionic ones) are also used to some extent (e.g., TritonX-100 and Tween). Table 4.3 lists all the detergents according to their charge and properties [59].

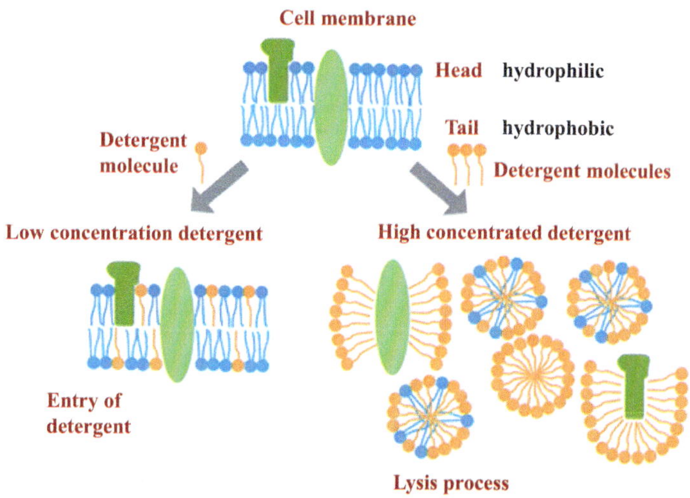

FIG. 4.10

Lysis of cell by detergent.

Table 4.3 List of some detergents and their properties [41].

Detergent	Charge	Properties
Sodium dodecyl sulphate (SDS)	Anionic	Strong lysis agent. Good for most cells. Not suitable for sensitive protein extraction
Triton X (100, 114)	Nonionic	Mild lysis agent. Good for protein analysis
NP-40	Nonionic	Mild lysis agent. Good for isolating cytoplasmic proteins but not nuclear proteins
Tween (20, 80)	Nonionic	Mild lysis agent. Good for cell lysis and protein isolation
Cetyltrimethylammonium bromide (CTAB)	Cationic	Generally used for isolating plant DNA
CHAPS, CHAPSO	Zwitterionic	Mild lysis agent. Good for protein isolation

The problem with the use of detergents is that they affect later purification steps, particularly salt precipitation. This limitation can be overcome by using ultrafiltration or ion-exchange chromatography for purification. Commonly, detergents are used for lysing mammalian cells. In the case of bacterial cells, prior to using detergent, the cell wall first has to be broken down by mechanical means to access the cell membrane.

References

[1] P. Grayson, A. Evilevitch, M.M. Inamdar, P.K. Purohit, W.M. Gelbart, C.M. Knobler, R. Phillips, The effect of genome length on ejection forces in bacteriophage lambda, Virology 348 (2006) 430–436, https://doi.org/10.1016/j.virol.2006.01.003.

[2] T.J. Silhavy, D. Kahne, S. Walker, The bacterial cell envelope, Cold Spring Harb. Perspect. Biol. 2 (2010), a000414, https://doi.org/10.1101/cshperspect.a000414.

[3] S.T.L. Harrison, Bacterial cell disruption: a key unit operation in the recovery of intracellular products, Biotechnol. Adv. 9 (1991) 217–240, https://doi.org/10.1016/0734-9750(91)90005-G.

[4] Markets and Markets Cell Lysis/Cell Fractionation Market-Global Forecasts to 2021, Available online: http://www.marketsandmarkets.com/Market-Reports/cell-lysis-market260138321.html.

[5] P. Grayson, A. Evilevitch, M.M. Inamdar, P.K. Purohit, W.M. Gelbart, C.M. Knobler, R. Phillips, The effect of genome length on ejection forces in bacteriophage lambda, Virology 348 (2006) 430–436, https://doi.org/10.1016/j.virol.2006.01.003.

[6] T.J. Silhavy, D. Kahne, S. Walker, The bacterial cell envelope, Cold Spring Harb. Perspect. Biol. 2 (2010), a000414, https://doi.org/10.1101/cshperspect.a000414.

[7] T.C. Ford, J.M. Graham, An Introduction to Centrifugation by, Bios Scientific Publishers, Oxford, 1991, p. 118.

[8] Baldwin & Murray Moo-Young, Disruption of a filamentous fungal organism (*N. sitophila*) using a bead mill of novel design. II. Increased recovery of cellulases, Biotechnol. Tech. 5 (1991) 337–342.

[9] A.R. Kleinig, A.P.J. Middelberg, On the mechanism of microbial cell disruption in high-pressure homogenisation, Chem. Eng. Sci. 53 (1998) 891–898.

[10] P. Dunnil, Biosafety in the large-scale isolation of intracellular microbial enzymes, Chem. Ind. 22 (1982) 270–y3.

[11] R. Cameron, P. Hambleton, J. Melling, Assessing microbiological safety of bioprocessing equipment, in: Proceedings 4th European Congress on Biotechnology, vol. 1, 1987, pp. 139–142.

[12] C.R. Engler, C.W. Robinson, Disruption of *Candida utilis* cells in high pressure flow devices, Biotechnol. Bioeng. 23 (1981) 765–780, https://doi.org/10.1002/bit.260230408.
[13] A.P.J. Middelberg, Process scale disruption of microorganisms, Biotechnol. Adv. 13 (1995) 491–551, https://doi.org/10.1016/0734-9750(95)02007-P.
[14] M. Corrales, S. Toepfl, P. Butz, D. Knorr, B. Tauscher, Extraction of anthocyanins from grape by-products assisted by ultrasonics, high hydrostatic pressure or pulsed electric fields: a comparison, Innov. Food Sci. Emerg. Technol. 9 (2008) 85–91.
[15] A. Morata, I. Loira, R. Vejarano, M.A. Bañuelos, P.D. Sanz, L. Otero, J.A. Suárez-Lepe, Grape processing by high hydrostatic pressure: effect on microbial populations, phenol extraction and wine quality, Food Bioprocess Technol. 8 (2015) 277–286.
[16] M.C. Santos, C. Nunes, J.A. Saraiva, M.A. Coimbra, Chemical and physical methodologies for the replacement/reduction of sulfur dioxide use during winemaking: review of their potentialities and limitations, Eur. Food Res. Technol. 234 (2012) 1–12.
[17] P. Ribéreau-Gayon, D. Dubourdieu, B. Donèche, A. Lonvaud, Handbook of Enology. Vol. 1. The Microbiology of Wine and Vinifications, second ed., John Wiley & Sons Ltd., West Sussex, UK, 2006.
[18] Commission Regulation (EC) No. 606, Laying down certain detailed rules for implementing Council Regulation (EC) No 479/2008 as regards the categories of grapevine products, oenological practices and the applicable restrictions, J. Eur. Commun. 193 (2009) 1–59.
[19] International Organization of Vine and Wine—OIV, International Code of Oenological Practices, Paris, France, OIV, 2018.
[20] P. Sanguansri, M.A. Augustin, Nanoscale materials development—a food industry perspective, Trends Food Sci. Technol. 17 (10) (2006) 547–556.
[21] M. Cortés-Muñoz, D. Chevalier-Lucia, E. Dumay, Characteristics of submicron emulsions prepared by ultra-high-pressure homogenisation. Effect of chilled or frozen storage, Food Hydrocoll. 23 (2009) 640–654.
[22] P. Halim, et al., Microalgal cell disruption for biofuel development, Appl. Energy 91 (2012) 116–121.
[23] P. Walstra, Principles of emulsion formation, Chem. Eng. Sci. 48 (1993) 333–349.
[24] J. Floury, J. Bellettre, J. Legrand, A. Desrumaux, Analysis of a new type of high pressure homogeniser. A study of the flow pattern, Chem. Eng. Sci. 59 (2004) 843–853.
[25] S.I. Martínez-Monteagudo, B. Yan, V.M. Balasubramaniam, Engineering process characterization of high-pressure homogenization—from laboratory to industrial scale, Food Eng. Rev. 9 (2017) 143–169.
[26] C.R. Engler, C.W. Robinson, Disruption of candida utilis cells in high pressure flow devices, Biotechnol. Bioeng. 23 (1981) 765–780, https://doi.org/10.1002/bit.260230408.
[27] P. Sanguansri, M.A. Augustin, Nanoscale materials development—a food industry perspective, Trends Food Sci. Technol. 17 (10) (2006) 547–556.
[28] M. Cortés-Muñoz, D. Chevalier-Lucia, E. Dumay, Characteristics of submicron emulsions prepared by ultra-high pressure homogenisation: effect of chilled or frozen storage, Food Hydrocoll. 23 (2009) 640–654.
[29] R. Halim, M.K. Danquah, P.A. Webley, Extraction of oil from microalgae for biodiesel production: a review, Biotechnol. Adv. 30 (3) (2012) 709–732, https://doi.org/10.1016/j.biotechadv.2012.01.001.
[30] S.P. Heffernan, A.L. Kelly, D.M. Mulvihill, U. Lambrich, H. Schuchmann, Efficiency of a range of homogenization as a process for emulsion formation, Chem. Eng. Technol. 27 (2011) 361–368.
[31] J. Geciova, D. Bury, P. Jelen, Methods for disruption of microbial cells for potential use in the dairy industry—a review, Int. Dairy J. 12 (2002) 541–553. S0958-6946(02)00038-9.
[32] A. Middelberg, Process-scale disruption of microorganisms, Biotechnol. Adv. 13 (3) (1995) 491–551. 0734-9750(95)02007-P.
[33] T. Sauer, C.W. Robinson, B.R. Glick, Disruption of native and recombinant *Escherichia coli* in a high-pressure homogenizer, Biotechnol. Bioeng. 33 (1989) 1330–1342, https://doi.org/10.1002/bit.260331016.
[34] D.C. Augenstein, K. Thrasher, A.J. Sinskey, D.I.C. Wang, Optimization in the recovery of a labile intracellular enzyme, Biotechnol. Bioeng. 16 (1974) 1433–1447, https://doi.org/10.1002/bit.260161102.

[35] Y. Chisti, M. Moo-Young, Disruption of microbial cells for intracellular products, Enzym. Microb. Technol. 8 (1986) 194–204, https://doi.org/10.1016/0141-0229(86)90087-6.

[36] R.M. Taskova, H. Zorn, U. Krings, H. Bouws, R.G. Berger, A comparison of cell wall disruption techniques for the isolation of intracellular metabolites from *Pleurotus* and *Lepista* sp, Z. Naturforsch. C J. Biosci. 61 (2006) 347–350, https://doi.org/10.1515/znc-2006-5-608.

[37] C.W. Ho, W.S. Tan, W.B. Yap, T.C. Ling, B.T. Tey, Comparative evaluation of different cell disruption methods for the release of recombinant hepatitis b core antigen from *Escherichia coli*, Biotechnol. Bioprocess Eng. 13 (2008) 577–583, https://doi.org/10.1007/s12257-008-0020-9.

[38] B.H. Johnson, M.H. Hecht, Cells by repeated cycles of freezing and thawing, Biotechnology 12 (1994) 1357, https://doi.org/10.1038/nbt1294-1357.

[39] M. Goodfellow, E. Stackebrandt, Nucleic Acid Techniques in Bacterial Systematics, vol. 5, John Wiley & Sons, Hoboken, NJ, USA, 1991.

[40] J.S. Watson, R.H. Cumming, G. Street, J.M. Tuffnell, Release of Intracellular Protein by Thermolysis, London, UK, Ellis Horwood, 1987, pp. 105–109.

[41] B. Wang, M. Merva, W.V. Williams, D.B. Weiner, Large-scale preparation of plasmid DNA by microwave lysis, Biotechniques 18 (1995) 554.

[42] K. Zhu, H. Jin, Z. He, Q. Zhu, B. Wang, A continuous method for the large-scale extraction of plasmid DNA by modified boiling lysis, Nat. Protoc. 1 (2007) 3088–3093, https://doi.org/10.1038/nprot.2006.452.

[43] D. Mark, S. Haeberle, G. Roth, F. von Stetten, R. Zengerle, Microfluidic lab-on-a-chip platforms: requirements, characteristics and applications, Chem. Soc. Rev. 39 (2010) 1153–1182, https://doi.org/10.1039/b820557b.

[44] M.D. Lilly, P. Dunnill, Fermentation Advances, Academic Press, London, UK, 1969, pp. 225–248. Isolation of intracellular enzymes from micro-organisms-the development of a continuous process.

[45] B. Balasundaram, S. Harrison, D.G. Bracewell, Advances in product release strategies and impact on bioprocess design, Trends Biotechnol. 27 (2009) 477–485, https://doi.org/10.1016/j.tibtech.2009.04.004.

[46] A.K. Lee, D.M. Lewis, P.J. Ashman, Microalgal cell disruption by hydrodynamic cavitation for the production of biofuels, J. Appl. Phycol. 27 (2015) 1881–1889, https://doi.org/10.1007/s10811-014-0483-3.

[47] M. Capocellia, M. Prisciandarob, A. Lanciac, D. Musmarraa, Comparison between hydrodynamic and acoustic cavitation in microbial cell disruption, Chem. Eng. (2014) 38, https://doi.org/10.3303/CET1438003.

[48] L.P. Fonseca, J. Cabral, Penicillin acylase release from *Escherichia coli* cells by mechanical cell disruption and permeabilization, J. Chem. Technol. Biotechnol. 77 (2002) 159–167, https://doi.org/10.1002/jctb.541.

[49] Y.-C. Chen, L.-A. Chen, S.-J. Chen, M.-C. Chang, T.-L. Chen, A modified osmotic shock for periplasmic release of a recombinant creatinase from *Escherichia coli*, Biochem. Eng. J. 19 (2004) 211–215, https://doi.org/10.1016/j.bej.2004.03.001.

[50] A.R. Byreddy, A. Gupta, C.J. Barrow, M. Puri, Comparison of cell disruption methods for improving lipid extraction from thraustochytrid strains, Mar. Drugs 13 (2015) 5111–5127, https://doi.org/10.3390/md13085111.

[51] H.C. Bimboim, J. Doly, A rapid alkaline extraction procedure for screening recombinant plasmid DNA, Nucleic Acids Res. 7 (1979) 1513–1523, https://doi.org/10.1093/nar/7.6.1513.

[52] P.F. Stanbury, A. Whitaker, Principles of Fermentation Technology, Pergamon Press, Oxford, UK, 1984.

[53] K. Tamura, T. Aotsuka, Rapid isolation method of animal mitochondrial DNA by the alkaline lysis procedure, Biochem. Genet. 26 (1988) 815–819, https://doi.org/10.1007/BF02395525.

[54] I. Feliciello, G. Chinali, A modified alkaline lysis method for the preparation of highly purified plasmid DNA from *Escherichia coli*, Anal. Biochem. 212 (1993) 394–401, https://doi.org/10.1006/abio.1993.1346.

[55] H.C. Bimboim, J. Doly, A rapid alkaline extraction procedure for screening recombinant plasmid DNA, Nucleic Acids Res. 7 (1979) 1513–1523, https://doi.org/10.1093/nar/7.6.1513.

[56] P.F. Stanbury, A. Whitaker, Principles of Fermentation Technology, Elsevier, Oxford, UK, 1984 (Google Scholar).

[57] K. Tamura, T. Aotsuka, Rapid isolation method of animal mitochondrial DNA by the alkaline lysis procedure, Biochem. Genet. 26 (1988) 815–819, https://doi.org/10.1007/BF02395525.
[58] I. Feliciello, G. Chinali, A modified alkaline lysis method for the preparation of highly purified plasmid DNA from *Escherichia coli*, Anal. Biochem. 212 (1993) 394–401, https://doi.org/10.1006/abio.1993.1346.
[59] PanReac AppliChem Detergents—More than Foam!. https://www.applichem.com/en/literature/brochures/brochures-biochemical-support/detergents/.

CHAPTER

Extraction of product of interest phase IV

5

Abbreviations

AC	affinity chromatography
AP	adsorption chromatography
ATPS	aqueous two-phase system
CMC	critical micelle concentration
GC	gas chromatography
GFC	gel filtration chromatography
GMP	good manufacture practice
GSC	gas-solid chromatography
IEC	ion-exchange chromatography
LC	liquid chromatography
LLC	liquid-liquid chromatography
LLE	liquid-liquid extraction
LSC	liquid-solid chromatography
MA	membrane absorbers
MF	membrane filtration
PC	packing the column
PC	partition chromatography
PEG	polyethylene glycol
SCF	supercritical fluid
TLC	thin-layer chromatography

5.1 Extraction

Extraction is physicochemical process to isolate biopharmaceuticals from a matrix (slurry). Common examples include liquid-liquid extraction (LLE) and solid-phase extraction. The distribution of a solute between two phases is an equilibrium condition described by partition theory. Several methods are being used in the extraction from plants, animals, or microorganisms such as maceration, infusion, decoction, percolation, digestion, Soxhlet extraction, superficial extraction, and ultrasound-assisted and microwave-assisted extraction. In the case of solvent extraction, the target compound transfers from one solvent to another owing to the difference in solubility or distribution coefficient between these two immiscible (or slightly soluble) solvents.

Techniques for Downstream process for Biologic Drugs and Vaccines. https://doi.org/10.1016/B978-0-443-19157-2.00001-7

FIG. 5.1

The process of liquid-liquid extraction.

5.2 Liquid-liquid extraction (LLE)

This process is referred to as the separation of a biological target product present in harvested broth through transferring the desired product from one liquid phase to another liquid phase, which can be further processed for concentration through a suitable evaporation process in the form of partial purification (Fig. 5.1).

LLE based on the principle that a solute or an analyte can be distributed in self in a certain ratio between two immiscible solvents, usually water (aqueous phase) and organic solvent (organic phase). The efficacy of extraction is dependent on the partition coefficient, i.e., the relative distribution of a substance between the two liquid phases. The components that need to be recovered are extracted from the feed stream with the help of an extractant (often called solvent). Both liquids have to be thoroughly mixed and subsequently separated from each other again (Fig. 5.2).

To achieve high purities and yields, it is necessary to operate with multiple stages and liquid-flowing countercurrent.

LLE is widely used in the production of organic compounds, the processing of perfumes, and the production of vegetable oils and biodiesel. It is among the most common initial separation techniques, though some difficulties result in extracting closely related functional groups. Various types of LLEs are used for target product recovery (Fig. 5.3).

Different types of well-established techniques are used for protein extraction from plants, animals, bacteria, etc. The water-soluble proteins can be separated by using a specific buffer. But some water-insoluble proteins are too rigid to be extracted by physical means, or they lose activity when extracted

FIG. 5.2

Liquid-liquid separation method for isolation of a target biological product.

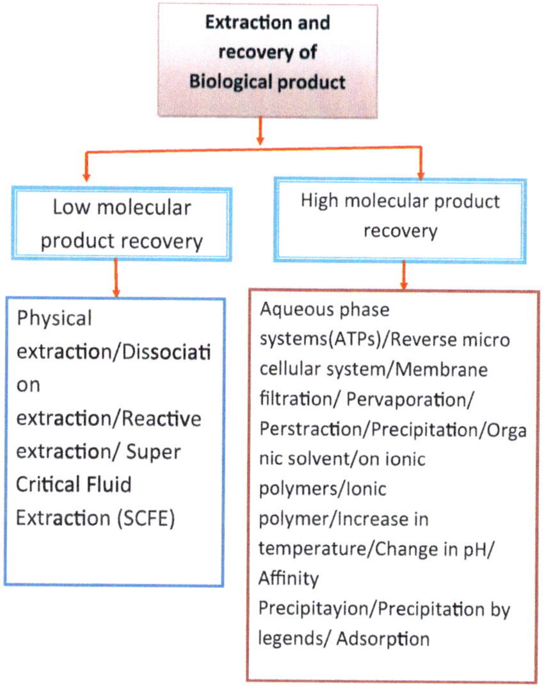

FIG. 5.3

Various extraction processes for biological product recovery.

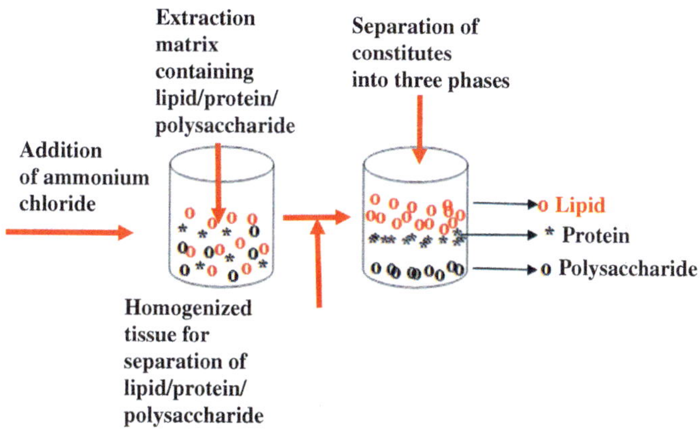

FIG. 5.4

Separation of protein from disrupted animal/plant/microbe matrix by organic solvent extraction.

by denaturants. The better alternative for the extraction of such proteins is using C4 and C5 solvent (butanol or pentanol). The addition of such an organic solvent should be followed by the addition of ammonium sulfate, resulting in three phases of suspended solution (Fig. 5.4).

One of the promising methods is to use organic solvents to extract proteins in their inactive state. C4 and C5 alcohols (butanol and pentanol) are mostly used for this purpose as they act as differentiating solvents. When ammonium sulfate is added to a slurry of disrupted cells of biological materials followed by butanol, the slurry separates into three phases. In upper organic phase and lower aqueous phase, depending on the concentration of ammonium sulfate, some proteins precipitate out and form an intermediate layer. Through this method, lipids, detergents, a few proteins, phenolic compounds, and some low molecular-weight contaminants can be separated from the target protein. This method is suitable to isolate proteins such as superoxide dismutase and catalase from human erythrocytes, leaving the major contaminant hemoglobin in the inorganic phase (t-butanol). Therapeutic proteins like interferon beta is extracted with 2-butanol.

5.2.1 Applications of liquid-liquid separation

In pharmaceutical industry, LLE methods are applied for the isolation and identification of proteins, antibiotics, amino acids, and enzymes. Additionally, professionals can use LLE ions to extract essential oils, which manufacturers can use in fragrances, flavoring, and pharmaceuticals.

By using organic solvents, the lipophilic compounds (low molecular weight products) can be conveniently extracted. However, it is quite difficult to extract hydrophilic compounds. The nonionizing compound becomes distributed between two liquid phases based on the physical properties.

An additional component has to be introduced as an extractant, which makes further separation steps necessary. Therefore, LLE is mostly used when the separation of components by distillation is either uneconomical or impossible. Some examples are as follows:

- Simultaneous separation of various components with widely differing boiling points.
- Separation of high-boiling products or pollutants that are present in only low concentrations.
- Separation of components with similar boiling points or components forming azeotropes.
- Separation of mixtures with thermally sensitive components.
- Selective separation of single components out of a multicomponent mixtures.
- Separation of components from electrolytic solutions.

5.3 Reverse micellar system

It is referred to as the smallest amphiphile aggregates, consisting of a fluid-like core formed by the nonpolar portion of the amphiphiles encased in a shell formed by the polar portion of the amphiphiles in contact with fluid water (Fig. 5.5).

Reverse micelles are thermodynamically stable, minute surfactants that hold water in their interior, surrounded by an organic phase. Only the protein of interest is entrapped in the micelle (Fig. 5.6A and B), whereas other impurities remain in the organic phase.

The reverse micellar system is an encouraging strategy for downstream processing. It is ideal for extracting biomolecules from diluted samples. Reverse micellar systems poses high sample loading capacity, are specific, and are easy to operate.

Ideally, micelles vary in size from 2 to 20 nm, depending on their number and composition. Besides the critical micelle concentration (CMC), there are other factors that influence the rate of micelle

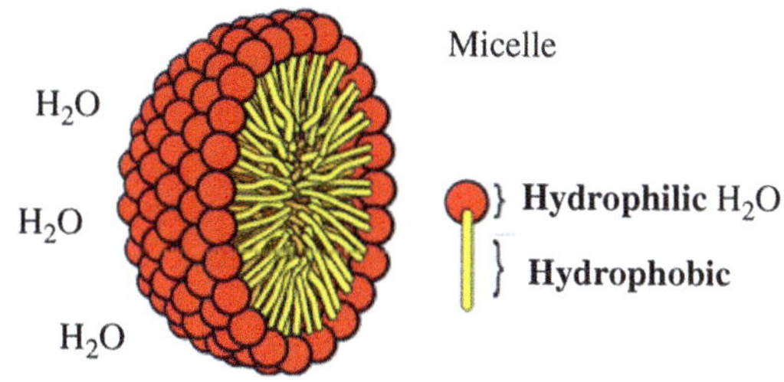

FIG. 5.5

Diagrammatic view of a micelle showing the hydrophilic head facing outside and hydrophobic tails concentrating toward inner core of the micelle.

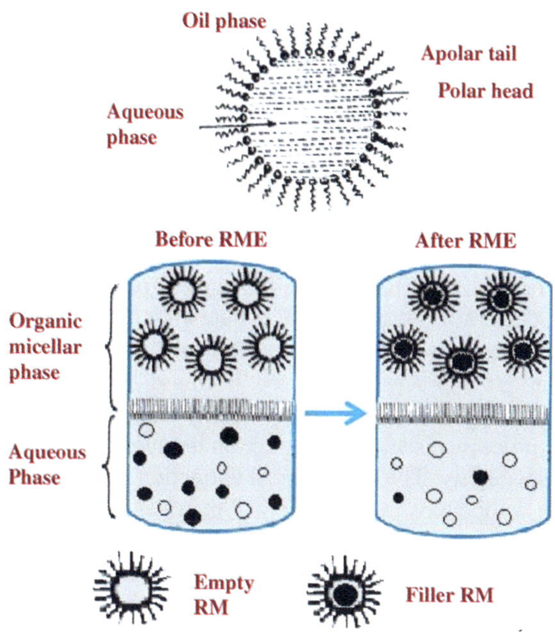

FIG. 5.6

Stages in the working of a reverse micellar system.

formation, such as temperature. Micelle formation can take place only above a particular temperature. The Krafft temperature is the minimum temperature at which the amphipathic molecules form micelles. Below this temperature, molecules remain in crystalline form.

In a polar medium like water, the number of counterions is always equal to their degree of binding to the micelle. As counterions bind to the polar head of amphipathic molecules, they decrease the repulsion and thus increase the stability of micelles. The CMC decreases with an increase in the number of counterions.

Detergents are capable of forming micelles that function as emulsifiers, dissolving substances that are usually insoluble in water. The hydrophobic tails of detergents interact with insoluble dirt and oil such that the dirt is entrapped in the hydrophobic center of these micelles. Washing clothes with water removes the dirt particles in micelles and cleans the surface.

A reverse micellar extraction cycle is basically composed of two steps: forward and back extraction. In the forward extraction process, biomolecules are transferred from the reverse micelles back to the aqueous stripping solution. The extraction can be performed batchwise on a bench scale, employing different procedures. Continuous extraction using reverse micellar systems is an efficient process with a reduced number of steps and is used in the purification of biomolecules.

5.4 Supercritical fluid extraction

This technique differs from the aforementioned procedures in that the materials used for extraction are supercritical fluids (SCFs). SCFs are intermediates between gases and liquids and exist as fluids above their critical temperature and pressure. Supercritical CO_2, with a low critical temperature and pressure, is commonly used in extractions. SCF extraction is relatively expensive and hence not widely used (SCF has been used for the extraction of caffeine from coffee beans and of pigments and flavor ingredients from biological material).

Replacing conventional organic solvents with SCFs in extraction procedures is a major advancement in today's pollution programs. SCFs in extraction allow separation and minimization of waste, as well as solvent recycling. Other advantages of SCF extraction include high efficiency, high extraction rates, and good selectivity. Processes such as supercritical CO_2 extraction [1–3] and microwave-assisted extraction [4,5] are green processes that have been used to obtain bioactive compounds from plants (phytochemicals).

The use of such green processes is attracting attention because of their sustainability and ability to increase target molecule specificity. The supercritical extraction procedure has been used extensively in the past to extract target compounds from a variety of matrices at laboratory and commercial scales. The most well-known examples of large-scale SCF extraction are the decaffeination of coffee beans [6,7] and the extraction of α-acids from hops to produce hop resins.

CO_2 is the main solvent used in this technique, especially when the target molecule is polar. It is safe, food grade, nonflammable, recyclable, and readily available at relatively low cost and high purity. In addition, CO_2 has a relative low critical temperature of 31.1°C and a low critical pressure of 7.4 MPa. At supercritical conditions, above its critical temperature and above its critical pressure (Fig. 5.7), CO_2 poses the properties of both a gas and a liquid.

Thus, it acquires very good mass-transfer properties because of the high, more gas-like, diffusion coefficients and low viscosity values. These properties provide effective penetration into solid matrices and help preserve the integrity and stability of more delicate compounds as a result of their relatively low surface tension. Solvents other than CO_2 are not necessarily required, and the expense of removing polluting organic solvents can be eliminated. Another advantage is its high selectivity and easy separation from the extracted compounds, without leaving toxic residues in the extracts.

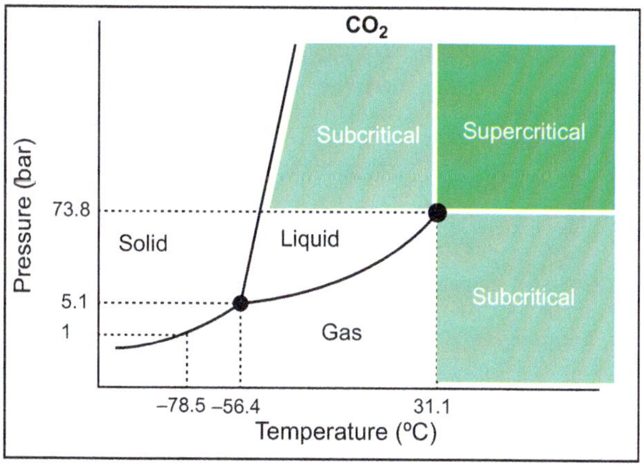

FIG. 5.7

Phase separation diagram for carbon dioxide.

5.5 Microwave-assisted extraction

Microwave-assisted extraction is also a novel green method for extracting soluble compounds into a fluid from a wide range of matrices. Microwaves are electromagnetic fields with a frequency of 0.3–300 GHz. They are made up of oscillating fields, electric and magnetic, that are perpendicular. These waves can penetrate the sample matrix and interact with polar molecules (such as water) to create heat. Consequently, microwaves can simultaneously heat the whole material to penetration depth. The extraction process involves the separation of solutes from active sites of the sample matrix under increased temperature and pressure, diffusion of solvent across the sample matrix, and release of solutes to solvent caused by expansion and breakdown of cell walls (Fig. 5.8).

The process thus partitions compounds of interest from the sample matrix into the solvent.

Furthermore, the migration of dissolved ions increases solvent penetration into the matrix and thus facilitates the release of phytochemicals. The effect of microwave energy is strongly dependent on the dielectric susceptibility of both the solvent and the solid plant matrix.

5.6 Membrane filtration (MF)

Membrane filtration (MF) has become a common separation technique in industrial biotechnology. It can be conveniently used for the separation of biomolecules and particles, and for the concentration of fluids. The MF technique basically involves the use of a semipermeable membrane that selectively retains particles or molecules that are bigger than the pore size, while smaller molecules pass through the membrane pores.

Membranes used in filtration are made up of polymeric materials such as polyethersulfone and polyvinyl difluorides. It is rather difficult to sterilize membrane filters. In recent years, microfilters and ultrafilters composed of ceramics and steel have become available and are more easily cleaned and sterilized. Following are a few commonly used MF techniques.

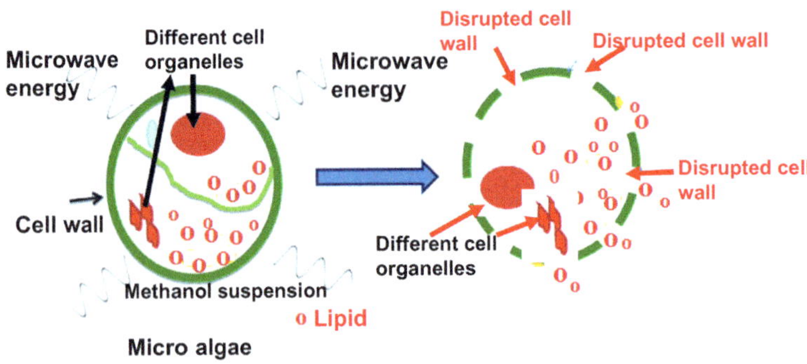

FIG. 5.8

Microwave-induced algal cell disruption process.

5.6.1 Membrane adsorbers (MA)

Membrane adsorbers are micro- or macroporous membranes with attached ion exchange groups and/or affinity ligands. Membrane adsorbers can bind to proteins and retain them. Methods and buffers known from conventional gel chromatography can be directly applied to separate target biologics.

5.6.1.1 Prevaporation

This is a technique in which volatile products can be separated by a process of permeation through a membrane, coupled with evaporation. Pervaporation is useful for the extraction, recovery, and concentration of volatile products. However, this procedure cannot be used for large-scale separation of volatile products because of its relatively high cost.

5.6.1.2 Perstraction

This is an advanced technique based on the principle of MF coupled with solvent extraction. Hydrophobic compounds can be recovered or concentrated by this method.

5.6.1.3 Precipitation

Precipitating is the most commonly used technique in industry for the concentration of macromolecules such as proteins and polysaccharides. Furthermore, precipitation can also be employed for the removal of certain unwanted by-products such as nucleic acids and pigments. Neutral salts, organic solvents, high-molecular-weight polymers (ionic or nonionic), and alternations in temperature and/pH are used in precipitation. In addition to these nonspecific protein precipitation reactions (i.e., the nature of the protein is unimportant), there are some protein-specific precipitation techniques such as affinity precipitation and ligand precipitation. Some details are as follows:

- Neutral salt: The most commonly used salt is ammonium sulfate because it is highly soluble, nontoxic to proteins, and cheap. Ammonium sulfate increases the hydrophobic interaction between protein molecules, resulting in their precipitation. The precipitation of proteins is dependent on several factors such as protein concentration, pH, and temperature.
- Organic solvents: Ethanol, acetone, and propanol are the commonly used organic solvents for protein precipitation. They reduce the dielectric constant of the medium and enhance the

electrostatic interactions between protein molecules that lead to precipitation. Because proteins are denaturated by organic solvents, the precipitation process has to be carried out below 0°C.

- Nonionic polymers: PEG is a high-molecular-weight nonionic polymer that can precipitate proteins. It reduces the quantity of water available for protein salvation and precipitates protein. PEG does not denaturate proteins and is nontoxic.
- Ionic polymers: Charged polymers such as polyacrylic acid and polyethyleneimine are used. They form complexes with oppositely charged protein molecules, resulting in charge neutralization and precipitation.
- Increase in temperature: Heat-sensitive proteins can be precipitated by increasing the temperature.
- Change in PH: Alterations in pH can also lead to protein precipitation.
- Affinity precipitation: Affinity interaction (e.g., between antigen and antibody) can be exploited for the precipitation of proteins.
- Precipitation by ligands: Ligands with specific binding sites for proteins have been successfully used for selective precipitation.

5.6.1.4 Adsorption

The biological products of fermentation can be concentrated by using solid adsorbent particles. In the early days, activated charcoal was used as the adsorbent material. More recently, cellulose-based adsorbents have been employed for protein concentration. The matrices used for the concentration of low-molecular-weight compounds (vitamins, antibiotics, and peptides) are generally based on polystyrene, methacrylate, and acrylate. The process of adsorption can be carried out by making a bed of adsorbent columns and passing the culture through them. The desired product, held by the adsorbent, can be eluted.

5.7 Chromatography

Chromatography is referred to as a technique for the separation of different components of specific properties from a mixture of fluid solvent in the gaseous or liquid phase, known as the mobile phase (Table 5.1).

5.7.1 Functioning of adsorption chromatography

In any chromatographic technique, the stationary phase is usually a solid, thick liquid, or bonded coating that stays fixed in one place.

A sample to be separated, when placed on the stationary phase, gradually moves along in the same direction as the mobile phase. If a sample compound (or analyte) has no interaction with the stationary phase, it runs right through and comes out of the system (as elute) at the same rate as the mobile phase (Figs. 5.9 and 5.10). On the other hand, if an analyte has no interaction with the mobile phase, it sticks directly to the stationary phase and never elutes [8].

In a well-designed chromatography process, precautions are taken so that the stationary and mobile phases both have at least some interaction with the analytes. Any individual sample molecule interacts first with one phase and then the other, back and forth repeatedly, but the fraction of each analyte in

Table 5.1 Chromatography separation of biomolecules on the basis of specific properties.

Property	Technique[a]
Biorecognition (legend specificity)	Affinity chromatography
Charge	Ion-exchange chromatography
Size	Gel filtration or size-exclusion chromatography
Hydrophobicity	Hydrophobic interaction chromatography/Reverse phase chromatography

[a]*Expanded bed adsorption is a technique used for large-scale purification. Proteins can be purified from crude samples without the need for separate clarification, concentration, and initial purification to remove particulate matter.*

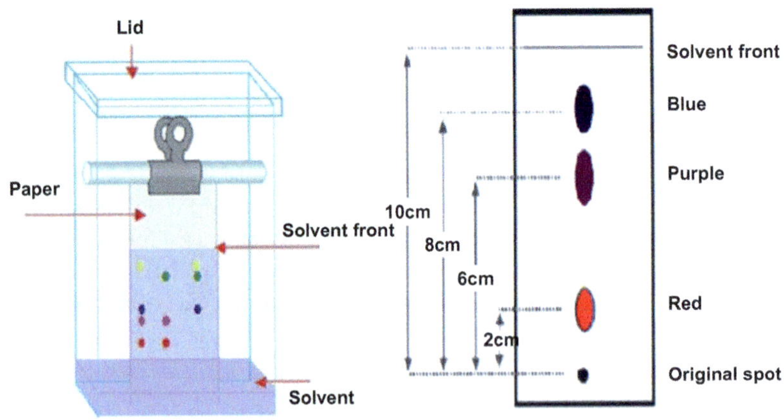

FIG. 5.9

Demonstrative diagram of paper chromatography showing the movement of target molecules along with the mobile phase.

each phase remains constant. This distribution ratio between the selected phases must differ for each analyte in order for them to be separated.

5.7.2 Types of chromatography

Various chromatography methods have been developed to separate different types of biomolecules present in a mixture. Some of them include column chromatography, thin-layer chromatography (TLC), paper chromatography, gas chromatography, ion-exchange chromatography (IEC), gel permeation chromatography, high-pressure liquid chromatography, and affinity chromatography (AC).

Mainly the chromatography can be categorized into two types:

(i) Adsorption chromatography (AC)
(ii) Partition chromatography (PC)

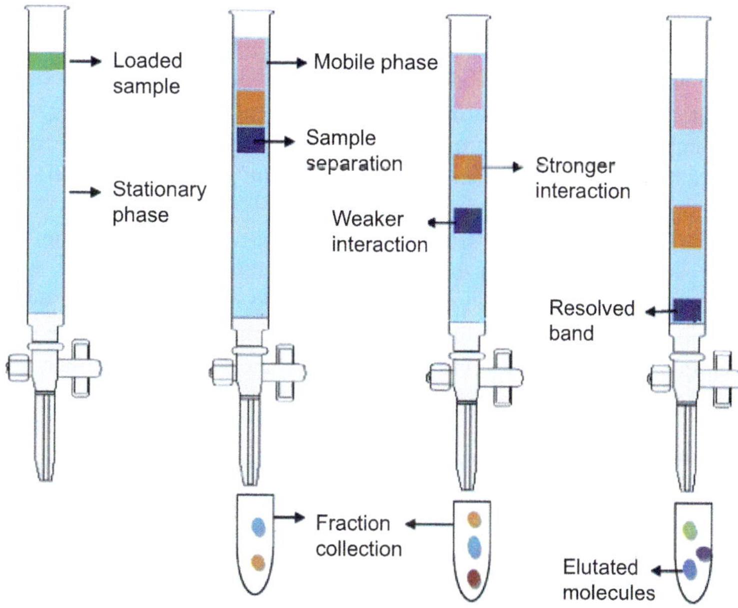

FIG. 5.10

Column packed with silica slurry and sample to be eluted. The colored bands travel down the columns the compound is eluted.

5.7.2.1 Adsorption chromatography

AC is the oldest type of chromatography technique. It makes use of a mobile phase, which is either in liquid or gaseous form. The mobile phase is adsorbed onto the surface of the stationary phase.

Components of AC

Before going through the functional aspects of AC, it is necessary to understand its different components.

Stationary phase: Adsorbent is the stationary phase in adsorption chromatography. The forces involved help to remove solutes from the adsorbent, so that they can move with the mobile phase.

Mobile phase: Either a liquid or a gas is used as a mobile phase in adsorption chromatography. Forces involved to remove solute from the adsorbent so that they can move with the mobile phase. When a liquid is used as a mobile phase, it is called liquid-solid chromatography (LSC). When a gas is used as a mobile phase, it is called gas-solid chromatography (GSC).

The following apparatuses are used in adsorption chromatography:

Chromatography jar: The glass jar has a lid. It helps to maintain a proper environment during separation.

Thin-layer chromatograph plate: Borosilicate glass plate with size 20×20 cm, 20×10 cm, 20×5 cm.

Capillary tube: Sample mixture is applied to TLC with the help of this tube.

Mobile phase: Mobile phases generally consist of a water/organic solvent mixture, an aqueous buffer, or a mixture of organic solvents with or without modifiers. In isocratic mode, the composition of the mobile phase remains unchanged throughout the analytical run, whereas in gradient elution mode the mobile phase composition changes through the programming of the pump. There are many factors related to the selection of the mobile phase, such as the solubility and ionization of the analyte, eluent, and whether to use normal or reverse phase mode.

Stationary phase: The stationary phase (adsorbents) is the solid media or medium through which the unknown sample or solution to be analyzed is passed. As the term implies, it is stationary and substances in the unknown sample that have affinity for it (or to reagents in it) do not travel along the column, but stay with the stationary medium, thereby allowing separation of the substances present in the unknown sample or solution. Alumina (Al_2O_3) and silica (siO_2xH_2O) are the two most common stationary phases in column chromatography and TLC.

Packing the column (PC)

There are several methods used in packing a column. The two main types of slurry and dry packing methods are being used. The slurry method is often used for macroscale separations. The solid stationary phase is combined with a small amount of nonpolar solvent in a beaker. They are thoroughly mixed until a consistent paste is formed that is still capable of flowing. This homogeneous mixture is poured into the column as carefully as possible using a spatula to scrape out the solid. The slurry method normally gives the best column packing, but is the more difficult technique to master (Fig. 5.10).

Adsorption chromatography can be grouped into the following categories:

(a) Thin-layer chromatography: In this technique, the mobile phase travels over an adsorbent. The adsorbent is a thin layer that is supported by a solid surface for the separation of components. The separation takes place through differential migrations, which occurs when the solvent moves along the powder spread on the glass plates.

(b) Paper chromatography: The technique that uses paper sheets or strips as the adsorbent, the stationary phase through which a solution is made to pass, is called paper chromatography. The solid surface of the paper is the stationary phase, and the liquid phase is the mobile phase.

The major difference between TLC and paper chromatography is the stationary phase used in paper chromatography is made of cellulose filter paper containing water in its pore, whereas in TLC the stationary phase used is a glass plate coated with silica gel (Fig. 5.11).

(c) Column chromatography: It is the technique in which the solutes of a solution are allowed to travel down a column where the individual components are absorbed by the stationary phase (Fig. 5.12).

Based on the affinity toward the adsorbent, the components take position on the column. The most strongly adsorbed component is seen at the top of the column.

(d) Gas-solid chromatography: The principle of separation in GSC is adsorption (Fig. 5.13). It is used for solutes that have less solubility in the stationary phase.

This type of chromatography technique has a very limited number of stationary phases available, and therefore GSC is not commonly used.

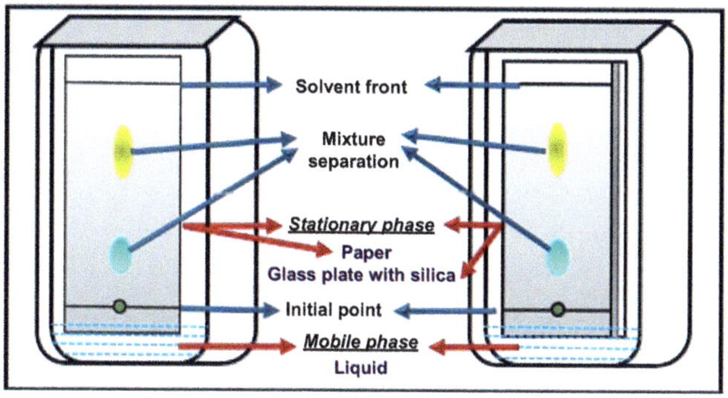

FIG. 5.11

Difference between thin layer chromatography and paper chromatography.

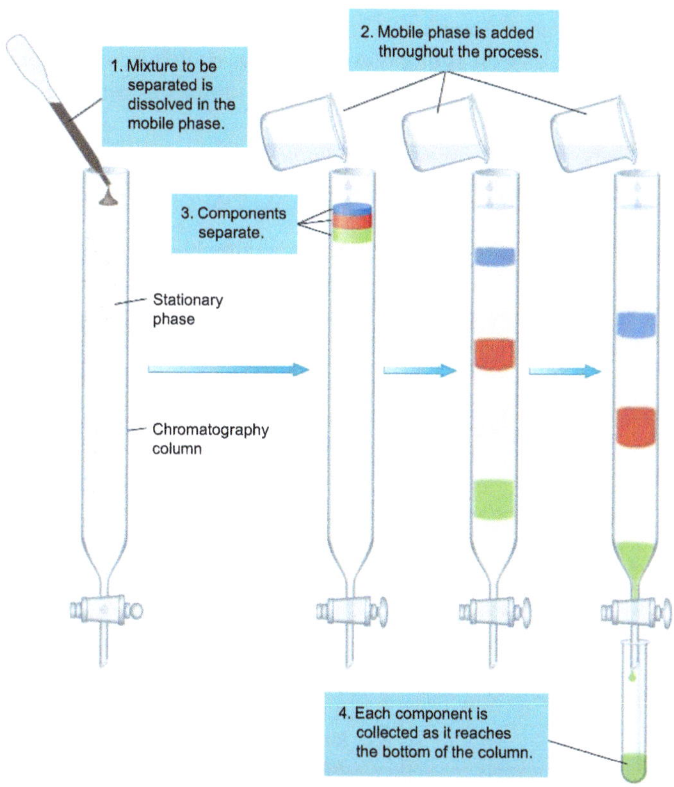

FIG. 5.12

Functioning of a column chromatography.

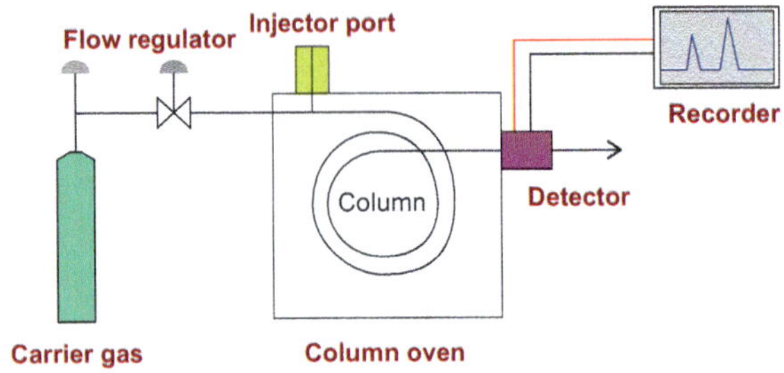

FIG. 5.13

Functional line diagram of gas-solid chromatography.

Adsorption chromatography is used for the separation of amino acids, isolation of antibiotics, separation of carbohydrates, separation of fat and fatty acids, and isolation of peptides and proteins

5.7.2.2 Partition chromatography (PC)

In partition or LLC, a wide range of compounds such as water-soluble and oil soluble compounds and ionic and nonionic compounds, as well as biopolymers such as nucleic acids and proteins, can be easily separated [9]. The basic of LLC is the distribution of sample molecules between two immiscible liquid phases, a stationary phase and a mobile phase (Fig. 5.14).

In conventional LLC, the stationary phase is mechanically held to a support by adsorption. In liquid chromatography (LC), the most widely used supports are those with chemically bonded organic stationary phases.

In 1940, Martin and Synge developed the LLC chromatographic separation process in which target molecule(s) can be separated by using two liquid phases. The target molecule(s) are located between the two liquid phases. Similar to the separator funnel liquid-liquid separation dynamic [10].

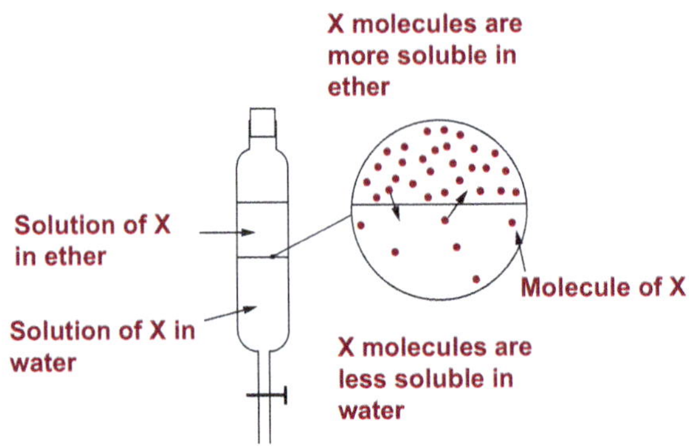

FIG. 5.14

Application of partition chromatography for the separation of "X" molecule.

5.7.3 Gas chromatography (GC)

The credit for the invention of gas chromatography goes to Anthony T. James and Archer J.P. in 1952 [11,12]. Gas chromatography is based on the basic principle of partition chromatography rather than adsorption chromatography.

Like all chromatographic techniques, a mobile and stationary phase are required for gas chromatography. The mobile phase (carrier gas) is an inert gas (i.e., helium, argon, or nitrogen). The stationary phase consists of a packed column in which the packing or solid support acts as a stationary phase or is coated with a liquid stationary phase (high boiling polymer). Most analytical GCs use a capillary column in which the stationary phase directly coats the walls of the small-diameter tube (i.e., 0.25-μm film in a 0.32-mm tube).

The separation of compounds is based on the different strengths of interaction of the compounds with the stationary phase ("like-dissolve-like" rule): The stronger the interaction, the longer the compound interacts with the stationary phase, and the more time it takes to migrate through the column (longer retention time). In Fig. 5.15, heptane interacts more strongly with the stationary phase and therefore lags behind pentane in its movement through the column. As a result, pentane has a much shorter retention time than heptane.

5.7.3.1 Practical procedure

The sample solution injected into the instrument enters a gas stream, which transports the sample into a separation tube known as the "column." The various components are separated inside the column. The detector measures the quantity of the components that exit the column. To measure a sample with an unknown concentration, a standard sample of known concentration is injected into the instrument. The standard sample of known concentration is injected into the instrument. The standard sample peak

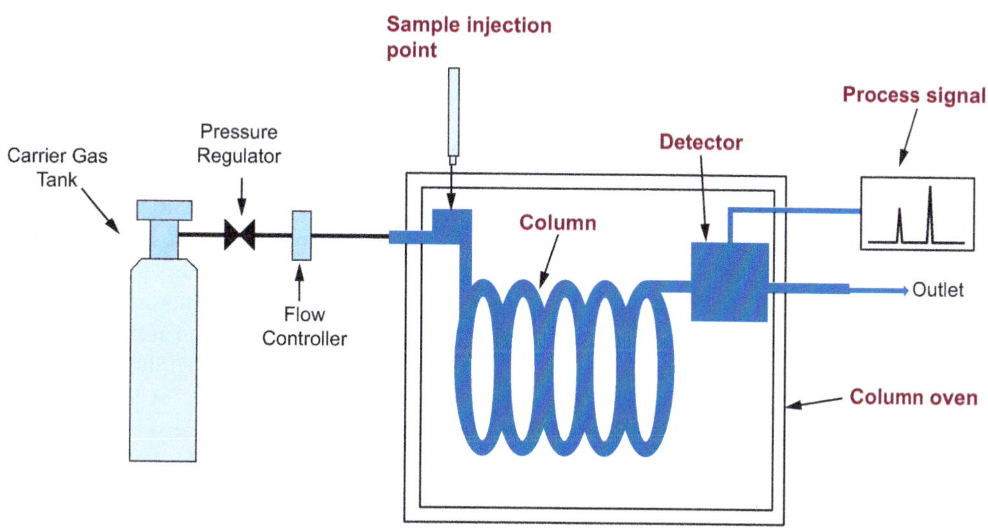

FIG. 5.15

Line diagram of gas chromatography showing functional aspects.

retention time, appearance time, and area are compared with those of the test sample to calculate the concentration. Six factors that influence the separation of components are the following:

(i) Amount of material inject

Ideally, the peaks in a chromatogram display a symmetric shape (Gaussian curve). If too much sample is injected, the peaks show significant tailing, which causes poorer separation. Most detectors are relatively sensitive and do not need a lot of material to produce a detectable signal. Strictly speaking, under standard conditions, only 1%–2% of the compound injected into the injection passes through the column, because most GC instruments are operated in split-mode to prevent overloading of the column detector. The splitless mode is only used if the sample has an extremely low analyte concentration.

(ii) Stationary compound selection

The polarity of the solute is crucial for the choice of a stationary compound, which in an optimal case has similar polarity to the solute. Common stationary phases in open tubular columns are cyanopropylphenyl dimethyl polysiloxane, carbowax polyethylene glycol, biscyanopropyl polysiloxane, and diphenyl dimethyl polysiloxane. For packed columns, more options are available.

(iii) Vapor pressure

The boiling point of a compound is often related to its polarity. The lower the boiling point, the higher the vapor pressure of the compound and the shorter the retention time because the compound spends more time in the gas phase. That is one of the main reasons why low boiling solvents (i.e., diethyl ether and dichloromethane) are used as solvents to dissolve the sample. The temperature of the column does not have to be above the boiling point because every compound has a nonzero vapor pressure at any given temperature, even solids. That is the reason why we can smell compounds such as camphor (0.065 mmHg/25°C). However, their vapor pressures are low compared with those of liquids (e.g., water, 24 mmHg/25°C; ethyl acetate, 95 mmHg/25°C; and diethyl ether, 520 mmHg/25°C).

(iv) Column temperature

The column(s) in a GC are contained in an oven, the temperature of which is precisely controlled electrically. The rate at which a sample passes through the column is directly proportional to the temperature of the column. The higher the column, the faster the sample moves through the column. However, the faster a sample moves through the column, the less it interacts with the stationary phase and the less the analytes are separated. In general, the column temperature is selected as a compromise between the length of the analysis and the level of separation. A method that holds the column at the same temperature for the entire analysis is called "isothermal." However, if the column temperature increases (the temperature "ramp"), the final temperature is called the "temperature program." A temperature program allows analytes that elute early in the analysis to separate adequately, while shortening the time it takes for late-eluting analytes to pass through the column.

(v) Carrier gas flow rate

A high flow rate reduces retention times, but results in poor separation because components have very little time to interact with the stationary phase and are just being pushed through the column.

(vi) Column length

Longer columns provide more time during which partitioning and, thus, separation can occur. This is especially needed for the separation of compounds with similar functional groups and weights.

5.7.4 Gel filtration chromatography (GFC)

Gel filtration chromatography (GFC) refers to the molecular size and the hydrodynamic volume of the components. The molecules are separated by the differential exclusion or inclusion of solutes as they pass through the stationary phase containing heterosporous cross-linking polymeric gels or beads. This technique has also been known as gel-exclusion, size-exclusion, and molecular sieve. In this technique, the separation of molecules is based on the size, shape, and molecular weight of the sample rather than on chemical interaction with the gel beads. Gel filtration resin beads contain pores of a defined size range. The sponge-like gel beads with pores serve as molecular sieves for the separation of smaller and bigger molecules. Large proteins, which cannot enter the pores, pass around the outside of the beads. Similar molecules enter the gel beads through their pores and become trapped (Fig. 5.16).

Gel filtration is mainly carried out in columns. The column is filled with beads of gel, which contain pores. The gel is made up of dextrans (Sephadex), agarose, or polyacrylamide. The sample is dissolved

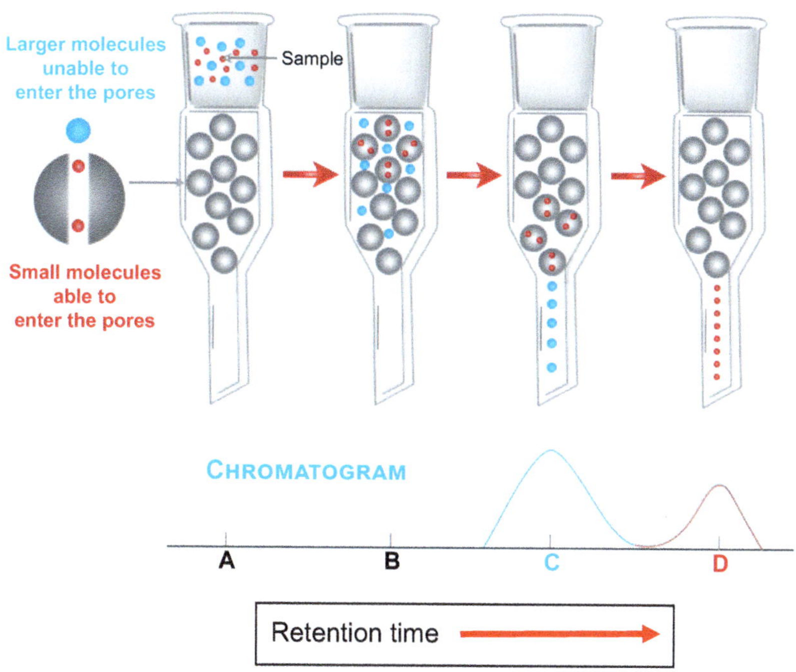

FIG. 5.16

A mixture of different sizes is injected onto the column. The larger sizes cannot enter as many pores as the smaller sizes and thus elute first. This technique is known as size-exclusion chromatography (SEC).

in buffer and allowed to flow through the column. Larger molecules cannot enter the pores of the beads, so they are excluded and pass down the column rapidly. Small molecules can enter into the pores of the beads, so their speed is restarted and they move down slowly.

5.7.4.1 Types of gels

The separating gels used for gel chromatography can be divided into swelling and nonswelling materials. Additional cross-linking of linear polymers or weak cross-linking during polymerization produces gels that can take up great amounts of solvent and swell appreciably. Dextran gels belong to this type and have proved very valuable for measuring and separating proteins and other water-soluble polymers. Sephadex is the most popular gel for the separation of proteins and most biomolecules. When *Leuconostoc mesenteroides* indulges in sucrose fermentation, large polymers of glucose are the result. These polymers are known as dextrans and are used to prepare Sephadex, which is a cross-linked polymer. Sephadex gels are insoluble in water and stable in bases, weak acids, and mild reducing and oxidizing agents. They cannot be used to separate biopolymers larger than 300,000 Da. Some identified gel serial numbers are G-25, G-50, G-75, G-100, and G-200.

Very rigid polymer gels of large pore sizes are also available, such as those prepared by copolymerization of styrene and divinylbenzene in the presence of different diluents (e.g., in a mixture of toluene and n-dodecane). The pore size of this gel depends on the composition of the reaction mixture, but all gels are rigid and swell only very little. For completely nonaqueous separations, a gel that swells in an organic solvent is required. Styragel provides this option. It is a rigid, cross-linked polystyrene gel. The gel structure is unaffected by temperatures as high as 150°C. The gel can be used with solvents such as tetrahydrofuran, cresol, dimethyl sulfoxide, chloroform, and carbon tetrachloride. Similar gels with large pores and greater rigidity are also prepared from vinyl acetate and divinyl adipate in suitable diluents.

Silica gels are principally applicable to gel chromatography. Recently it became possible to produce this material as beads with well-defined diameters and pore sizes. It is commercially available as Spherosile or Porasilc and has already been used extensively in gel chromatography. A glass powder of uniform pore size is also available. These fine glass spheres are manufactured from borosilicate glass to contain large number of pores with a very narrow size distribution. The glass spheres have a molecular exclusion limit ranging from 3000 to 9×10^6 Da. Manufacture involves the use of sintered borosilicate glass, which is thermally treated to produce heterogeneous regions. The glass is crushed, and the silica-poor phase is leached out to leave the desired pore system. The powder is well suited as a packing for polymers. Polyacrylamide is also a very popular medium and is produced by polymerizing acrylamide into bead form. Polyacrylamide gel can be used to separate molecules of up to 300,000 Da. This gel is insoluble in water, but common organic solvents in the pH range of 2–11 can be used. Some common gels are Bio-Gel P-100, Bio-Gel P-200 and Bio-Gel P-300.

5.7.4.2 Application of gel filtration chromatography

- Separation and purification of molecules, viruses, nucleic acids, hormones, enzymes, proteins, and antibodies.
- Separation of vitamins, steroids, neuropeptides, and drugs can be achieved very quickly.
- Determination of molecular weight of molecules.
- Study of protein receptor binding.
- Separation of 4S tRNA.
- Separation of DNA (from bacteria, usually Gram-positive) from the invariable contaminants, the teichoic acids.

5.7.5 Ion-exchange chromatography (IEC)

IEC refers to a process for the separation of charged molecules (ion and polar molecules) based on their affinity to the ion exchanger (Fig. 5.17).

An ion is a charged atom or molecule. It is charged because the number of electrons does not equal the number of protons in the atom or molecule. An atom can acquire a positive or negative charge depending on whether the number of electrons in an atom is greater or less than the number of protons in the atom. It can be applicable to any kind of charged molecules, including large proteins, small nucleotides, and amino acids [13].

Almost all proteins contain charged amino acids. Nucleic acids are highly negatively charged compared with proteins. But most lipids are nonpolar (having no charged areas) or only slightly polar, with a very few charged areas. Water mixes with hydrophilic (water-loving) compounds by sticking to their charged groups. Since lipids lack charged groups, the water molecules have nothing to stick to and don't mix with them.

Ion chromatography can be applied to any kind of charged molecules including large proteins, small nucleotides, and amino acids. However, ion chromatography must be done in conditions that are one unit away from the isoelectric of a protein.

IEC can be broadly classified as anion exchange and cation exchange. In cation-exchange chromatography, the molecule of interest is positively charged. In this type of chromatography, the stationary phase is negatively charged, and positively charged molecules are loaded to be attracted to it. In anion-exchange chromatography, the stationary phase is positively charged, and negative charged molecules are loaded to be attracted to it (Fig. 5.17) [14].

It is often used in protein purification, water analysis [15,16], and quality control. In IEC, water-soluble and charged molecules such as proteins, amino acids, and peptides bind to moieties that are

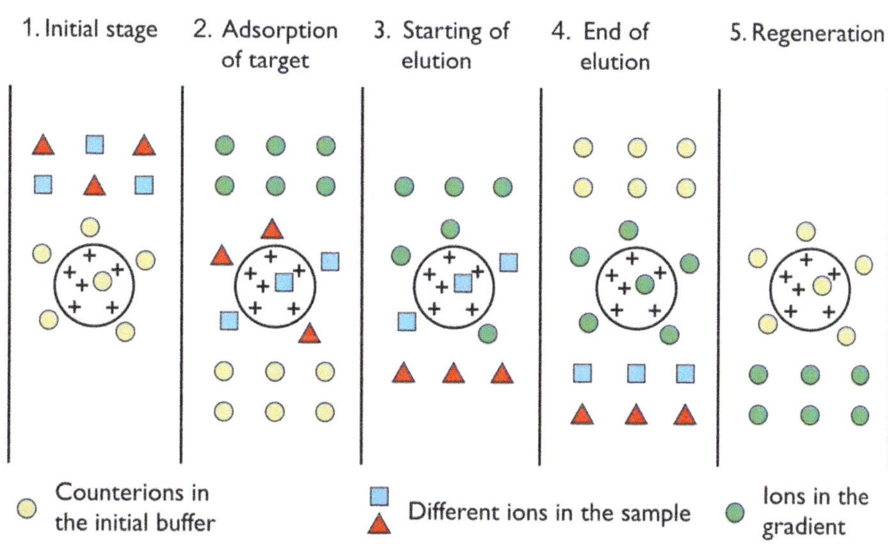

FIG. 5.17

Line diagram of ion-exchange chromatography showing functional aspects.

oppositely charged by forming ionic bonds to the insoluble stationary phase [17]. A cationic stationary phase is used to separate anions, and an anion stationary phase is used to separate cations [18].

5.7.5.1 Working condition

As a rule, the pH of the mobile phase buffer must be between the pl (isoelectric point) or pK_a (acid dissociation constant) of the charged molecule and the pK_a of the charged group on the solid support. For example, in cation-exchange chromatography, using a functional group of the solid with a pK_a of 1.2, a sample molecule with a pl of 8.2 can be run in a mobile phase buffer of pH 6.0. In anion-exchange chromatography, a molecule with a pl of 6.8 can be run in a mobile phase buffer at pH 8.0 when the pK_a of the solid support is 10.3. As in most other modes of chromatography, a protein sample is injected onto the column under conditions where it is strongly retained. A gradient of linearly increasing salt concentration is then applied to elute the sample components from the column. An alternative to using a linear gradient is to use a step gradient. This requires less complicated equipment and can be very effective for eluting different fractions if the appropriate concentration of salt is known, usually from linear gradient experiments (Fig. 5.18).

Many chromatographers also use changes in pH to affect a separation. In cation-exchange chromatography, raising the pH of the mobile phase buffer causes the protein molecule to become less protonated and, hence, less positively charged. The result is that the protein can be no longer for ionic interaction with the negatively charged solid support, which ultimately results in the molecule eluting from ionic interactions with the negatively charged solid support, which ultimately results in the molecule eluting from the column. In anion-exchange chromatography, lowering the pH of the mobile phase buffer causes the molecule to become more protonated and, hence, more positively (and less negatively) charged. The result, as earlier, is that the protein can no longer form an ionic interaction with positively charged solid support, which causes the molecules to elute from the column.

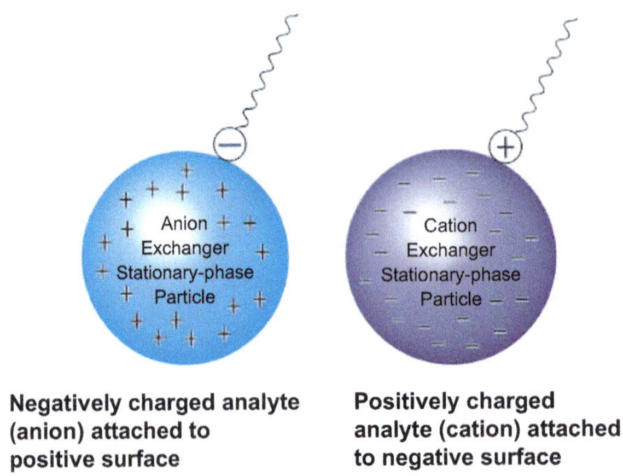

FIG. 5.18

Left: In cation exchange chromatography, positively charged molecules are attracted to a negatively charged slid support. Right: In anion exchange chromatography, negatively charged molecule are attracted to a positively charged solid support.

5.7.6 Affinity chromatography (AC)

AC refers to a method of isolating a product of interest based on a specific binding interaction between an immobilized ligand and its binding partner. Examples include antibody/antigen, enzyme/substrate, and enzyme/inhibitor interactions. The degree of purification can be quite high depending on the specificity of the interaction, and, consequently, it is generally the first, if not the only, step in a purification strategy.

AC was originally developed for the purification of enzymes but was later extended to nucleotides, nucleic acids, and immunoglobulin. For the purification of enzymes, the ligand attached to the adsorbent is usually a powerful inhibitor that binds to only one enzyme in a complex mixture. Affinity purification and many assay techniques depend on the reversible binding interactions of target biomolecules in liquid samples with specific chemical groups or macromolecules that are immobilized on a solid material. The latter class of molecules are called affinity ligands, and they can be attached (immobilized) to the solid material (stationary phase) in a variety of ways. This process involves three main steps: loading, separation, and elution (Fig. 5.19).

Most affinity purification strategies, especially those involving antibodies and other proteins, depend on covalent chemical conjugation of the ligand to the solid support matrix (Fig. 5.20) (Table 5.2).

5.7.6.1 Basic principle of functioning of affinity chromatography

The process of AC is mainly based on interactions between the analyte of interest (normally dissolved in the mobile phase) and a binding partner or ligand (immobilized on the stationary phase).

In a typical AC experiment, the ligand is attached to a solid, insoluble matrix (usually a chemically modified polymer such as agarose or polyacrylamide). The stationary phase is first filled into a column into which the mobile phase is poured slowly. The legend is blinded with a target molecule associated with the stationary phase. The nontarget biomolecules loosely linked with stationary phase are washed with a washing buffer and removed from the column. An elution buffer is used to remove the biomolecule of interest.

Affinity ligands that have broad applicability are commercially available in a variety of ready-to-use, preimmobilized forms. Examples include protein A agarose resin for general antibody purification involving biotinylated molecules. This method finds its widest application in the field of protein purification using streptavidin magnetic beads for, involving biotinylated molecules. This method finds its widest application in the field of protein purification and has been useful in studying quantitative aspects of protein-ligand interaction.

5.7.6.2 Components of affinity chromatography

(i) Supporting matrix

The most commonly used supporting matrices are polyacrylamide and agarose due to their unique properties like the nonabsorption property of molecules in substantial quantity, coupled with ligands without changing the binding properties and stability in a variety of experimental conditions. A variety of materials are used in the matrix (Table 5.3).

(ii) Space arm

To avoid steric repulsion between the ligands or activated groups of the matrix and the macromolecules of the matrix backbone, a space is introduced between activated groups of the support and the ligand.

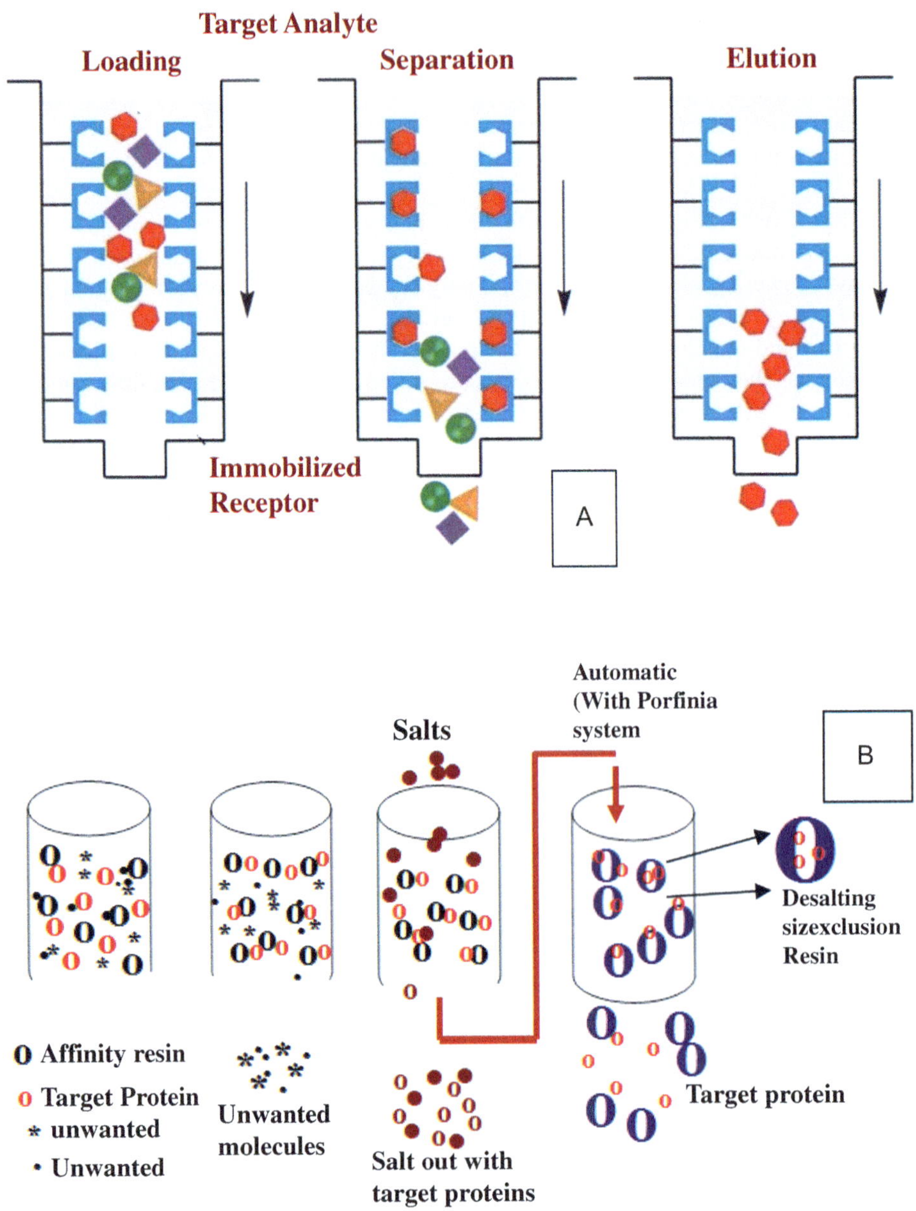

FIG. 5.19

(A) Affinity chromatography involves three major steps: loading, separation, and elution for the separation of target product with salt. (B) Desalting by size exclusion resin.

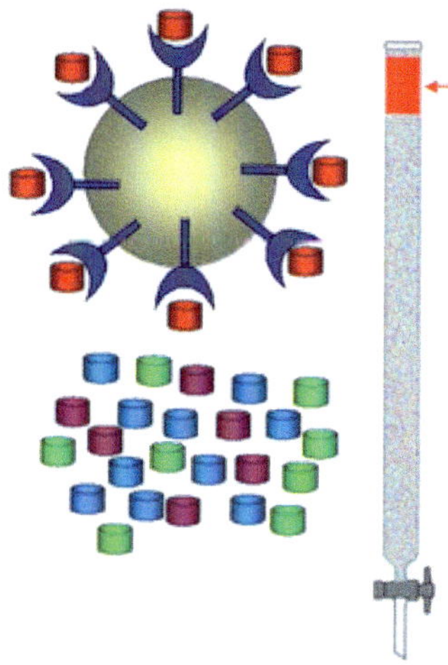

FIG. 5.20

Binding of ligand and macromolecular complex (matrix).

Table 5.2 Most commonly binding ligands and their respective target.

Ligand	Target
Antibody	Antigen
Iron and aluminum ions	Phosphoproteins
Avidin	Biotin
Glutathione	GST
Chelator + Ni and Co ions	His-tagged proteins

This spacer is known as "arm" made of a variety of chemicals (Table 5.4). The ligand then projects out from the macromolecule to prevent repulsion and allows unimpeded access to the target protein. Examples of compounds used as space arms include.

(iii) Activation of the matrix functional group

The agarose gel used as a matrix is available in different activated forms:

(a) CNBr-agarose (it is easy to use for the attachment of amines, but does not have a long space between the gel beads and the bound ligand).

(b) Aminohexanoic acid-agarose (CH-agarose, for reaction with amines) and diaminohexane-agarose (AH-agarose, for reaction with carboxylic acids) have relatively long methylene chains that keep the ligand a significant distance from the gel beads.

Table 5.3 List of variety of material used in the matrix.

	Features
Agarose	Hydrophilic, almost no unspecific bonds, and the gold standard for protein purification
Silica gel	Nanoporous (leads to unspecific bonds), functionalized via silane, and silanes are washed away by alkaline buffer → reduced stability; applications: bound nucleic acids chaotropic
Aluminum oxide	Acidic surface, binds amines irreversibly, and used to reduce the amount of specific substances
Acrylate	Partially hydrophobic (unspecific bounds possible), monodisperse particles, and used for cell separation
Organic polymers	Partially hydrophobic (unspecific bounds possible) and monodisperse particles can be used for ligand coupling, not recommended for protein purification because of unspecific bounds

Table 5.4 Commonly used coupling compounds used as space arms in affinity chromatography.

Chemical	Chain length
Cyanogen bromide	C1
Epoxide	C3
Epoxide with C6 acid	C10
Diamine	C10

5.7.6.3 Application of affinity chromatography

- Affinity chromatography has been used to purify a wide range of enzymes and proteins, including immunoglobulin and DNA-related enzymes. Heparin-agarose is used to purify the heparin binding domain of platelet thrombospondin.
- Messenger RNAs are routinely isolated using poly (U) Sepharose 4B.
- Clinical applications include immunoassay, therapeutic drug monitoring, and allergy testing.
- It has also been applied for the concentration of dilute proteins and for the storage of unstable protein in an immobilized form.

5.8 Formulation and polishing

Various formulations have been developed for biologic drugs, either in solid form or in liquid form. This is to save the biologics from damage due to ultraviolet radiation, rain, pH, and high temperatures. Conventional formulations have given way to advanced versions such as microencapsulation and microgranules. The formulation of low molecular weight products (solvents and organic acids) can be achieved by concentrating them through the removal of most of the water. For certain small molecules (antibiotics and citric acid), formulation involves crystallization by adding salts. Proteins are highly susceptible to loss of biological activity; hence, their formulation requires special care. Certain stabilizing additives can be added to protein formulations to prolong their shelf life. Stabilization includes sugars (sucrose and lactose), salt (sodium chloride and ammonium sulfate), polymers (polyethylene glycol), and polyhydric alcohols (glycerol). Protein can be formulated in the form of solutions, suspensions, or dry powders.

The improvements in techniques for biological control and the increasingly widespread use of metabolites produced by microorganisms to obtain products that can be applied in various fields are pushing scientists to provide and monitor economically viable supply chains for products of microbial origin. Generally, economically viable supply chains for products that can be incorporated into the supply chain must be free of moisture to prevent deterioration and maintain stability. There are several processes that can be used for drying, but spray drying and freeze drying are the most commonly used for biopharmaceuticals.

5.8.1 Spray drying

Spray drying is a physical process that involves the automization of a solution, slurry, or emulsion containing one or more components of the desired product into droplets by spraying followed by the rapid evaporation of the spray droplets into solids powered by hot air at a certain temperature and pressure (Fig. 5.21).

Spray drying can be used in a wide range of applications where the production of a free-flowing powder is required.

Almost all other methods of drying, including the use of ovens, freeze dryers, or rotary evaporators, produce a mass of material that requires further processing (e.g., grinding and filtering) and, therefore, produces particles of irregular size and shape. Spray drying, on the other hand, offers very flexible control over the powered particle properties such as density, size, flow characteristics, and moisture content.

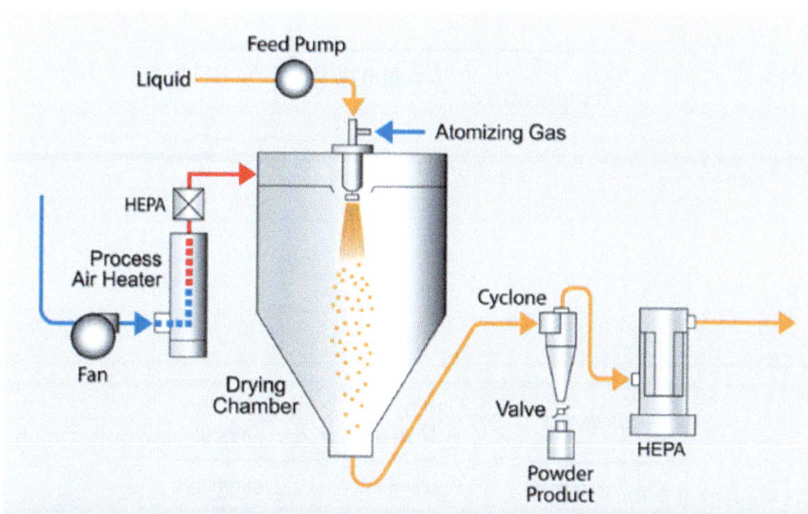

FIG. 5.21

Line diagram of spay drying technique and the functional aspects.

5.8.2 Freeze drying or lyophilization

Freeze drying is a process in which a completely frozen sample is placed under a vacuum in order to remove water or other solvents from a wide range of products, including pharmaceuticals, foodstuffs, diagnostics, bacteria, and viruses. This is mainly because freeze drying does not usually cause the loss of biological activity in the desired product (Fig. 5.22).

In this technique, the ice-frozen sample is allowed to change directly from solid to vapor without passing through liquid phase. There are three stages in the process: freezing, sublimation drying, and desorption drying (Fig. 5.23). The freezing phase is the most critical in the entire freeze-drying process. It is important to freeze biological samples rapidly in order to avoid the formation of large ice crystals, which deteriorate the final product's quality. Examples of freeze-dried products include antibiotics, bacteria, sera, vaccines, diagnostic medications, protein-containing products, biologic drugs, cells, tissues, and other biopharmaceutical products.

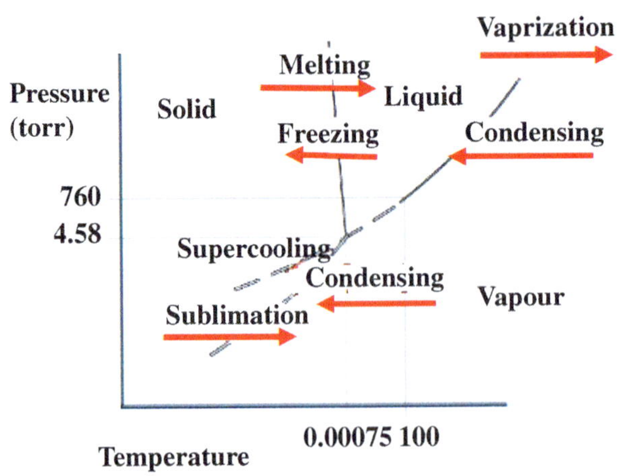

FIG. 5.22

Phase diagram of water.

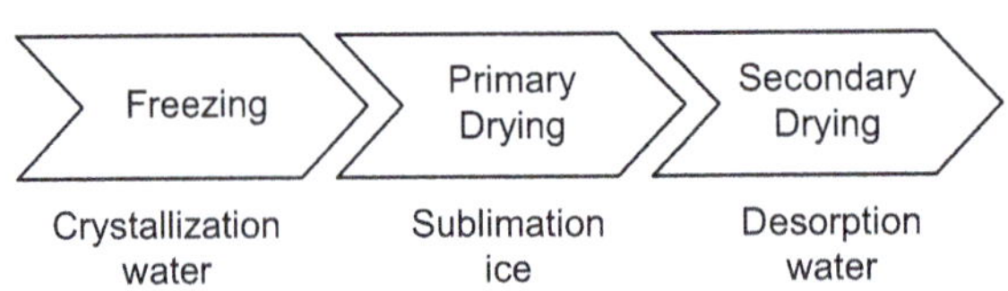

FIG. 5.23

Three processes of freezing.

5.9 Integration of different processes for biologic production

It is essential to develop strategies to integrate fermentation (upstream process) and downstream processes to finally obtain the product of interest of the biologic molecule. However, this has not been practicable for various reasons. Integration of certain stages in upgrading downstream processing for purification of the product has met with some success. For instance, protein concentration by extraction into two-phase systems combined with clarification and purification can be done together. Microbial process development and integration of unit operations during biopharmaceutical development for a typical microbial or mammalian cell culture process are important. Most biopharmaceutical companies have their own in-house process development programs. In addition, several CMOs are involved in process development and the production of material for preclinical, phases I–III, or even GMP manufacturing.

Despite all this, there is often a lack of vision regarding process development, especially from small research companies, resulting in unscalable and highly expensive processes. Developing a scalable, robust, and regulatory-compliant process requires time, sophisticated equipment, skilled scientists and engineers, and, above all, financial resources. Process development can only be justified if the developed process is scalable, cost-effective, and validated. Integration and selection of unit operations from the research phase is of prime importance, and making the right decision early on can definitely save time and money in the later stage of product development.

References

[1] J. Ver, M.R. Kula, Extraction of protein from biological raw material using aqueous(polyethylene) glycol-citrate, Biotechnol. Appl. Biochem. 12 (1990) 397–404.

[2] F. Adam, et al., "Solvent free" ultrasound-assisted extraction of lipids from fresh microalgal cells: a green, clean and scalable process, Bioresour. Technol. 114 (2012) 457–465.

[3] J. Azmir, et al., Techniques for extraction of bioactive compounds from plant material: a review, J. Food Sci. Emerg. Technol. 17 (2013) 106–113.

[4] Y. Chavan, R.S. Singhal, Ultrasound-assisted extraction (UAE) of bioactives from arecanut (*Areca catechu* L.) and optimization study using response methodology, Innov. Food Sci. Emerg. Technol. 17 (2013) 106–113.

[5] M. Herrero, et al., Compressed fluid for the extraction of bioactive compounds, Trends Anal. Chem. 43 (2013) 67–83.

[6] D. Hidalgo, et al., Integrated and sustainable system fir multi-waste valorization, Environ. Eng. Manag. J. 13 (2014) 2467–2475.

[7] L.B. Roseio, et al., Supercritical, ultrasound and conventional extracts from carob (*Ceratonis siliqua* L.) biomass effect on the phenolic profile and antiproliferative activity, Indus Crop Prod. 47 (2013) 132–138.

[8] J. McMurry, Organic Chemistry: With Biological Applications, second ed., Brooks/Cole, Belmont, CA, 2011, p. 395, ISBN: 9780495391470.

[9] H. Schindler, Notes on the history of the separatory funnel, J. Chem. Educ. 34 (11) (November 1957) 528. Bibcode:1957JChEd..34..528S https://doi.org/10.1021/ed034p528.

[10] C.W. Gehrke, R.L. Wixom, E. Bayer (Eds.), Chromatography-A Century of Discovery 1900-2000, The Bridge to The Sciences/Technology. Elsevier, 2001, ISBN: 978-0-08-047650-6.

[11] K.D. Bartle, P. Myers, History of gas chromatography, TrAC Trends Anal. Chem. 21 (9) (September 2002) 547–557, https://doi.org/10.1016/S0165-9936(02)00806-3.

[12] A.T. James, A.J.P. Martin, Gas-liquid partition chromatography: the separation and micro-estimation of volatile fatty acids from formic acid to dodecanoic acid, Biochem. J. 50 (5) (March 1952) 679–690, https://doi.org/10.1042/bj0500679.
[13] A.J. Ninfa, D.P. Ballou, M. Benore, Fundamental Laboratory Approaches for Biochemistry and Biotechnology, John Wiley, Hoboken, NJ, 2010.
[14] Principles of Ion Exchange Chromatography. separations.us.tosohbioscience.com. (Retrieved 1 May 2018).
[15] Handbook for Monitoring Industrial Wastewater, Environmental Protection Agency (USA), August 1973 (Retrieved 30 July 2016).
[16] A. Da¸browski, Z. Hubicki, P. Podkościelny, E. Robens, Selective removal of the heavy metal ions from waters and industrial wastewaters by ion-exchange method, Chemosphere 56 (2) (2004) 91–106.
[17] M. Luqman, Inamuddin, Ion Exchange Technology II, Springer Netherlands, 2012, p. 1, ISBN: 978-94-007-4026-6.
[18] J.S. Fritz, Ion chromatography, Anal. Chem. 59 (4) (1987) 33185A–33344A, https://doi.org/10.1021/ac00131a002.

CHAPTER

Viral vectors for vaccine production

6

Abbreviations

AAV	adeno-associated viruses
AVs	adenoviruses
cGMP	current good manufacturing practice
DNA	deoxyribonucleic acid
FDA	Food Development Authority
HIV	immunodeficiency virus
LV	lentiviral vector
mAbs	monoclonal antibodies
MLV	murine leukemia virus
PV	plant virus
RCPs	replication-competent particles
RNA	ribonucleic acid
RVs	retroviruses
VSV	vesicular stomatitis virus
WHO	World Health Organization

6.1 Viral vectors for vaccine production

A viral vector is a carrier of fragments of nucleic acid coding for antigen for another infectious agent [1]. Viral vector vaccines are harmless due to their lack of genetic material causing harmful effects [2], and the genetic material they discharge remain free in the host cells without integrating with a person's genome [3].

Currently, six viral vectors are used as biological tools to deliver the genetic material into cells [4]. These include retroviruses (RVs), adenoviruses (AVs), lentiviral vector (LV), adeno-associated viruses (AAV), and plant virus (PV) and hybrid. Viruses are intracellular parasites; they rely on their host cells for energy, macromolecular synthesis machinery, and work benches for genome replication and particle assembly. So, viral propagation is dependent on the transcription of viral messenger RNA (mRNA) and, in turn, translation of viral proteins as well as genome replication [1]. The specific receptors and/or other components on the host cell membrane recognize the virus particles and result in penetration and virion internalization. Virion uncoating releases the viral genome and thereafter

Techniques for Downstream process for Biologic Drugs and Vaccines. https://doi.org/10.1016/B978-0-443-19157-2.00006-6

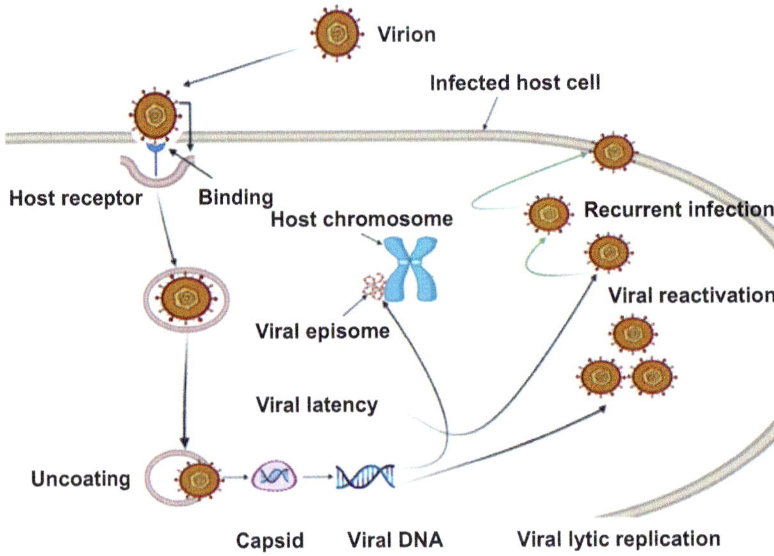

FIG. 6.1

General model of eukaryotic viral replication.

becomes ready for transcription and translation. Poxviruses and RNA viruses (with the exception of RVs) replicate in the cytoplasm. Transcription of all other DNA viruses takes place in the nucleus. Transcription and genome replication are followed by the cytoplasmic stages of mRNA translation and virion assembly. Release of mature virions occurs as a result of membrane lysis and the death of the host cell (Fig. 6.1) [5].

Highly informative reviews are available on virus genome transcription and replication [6–12]. Although viral vector vaccines lack the potential for antigen expression within cells, they induce a robust cytotoxic T-cell response, as noticed in subunit vaccines that are only restricted to humoral immunity [13,14]. Humoral immunity is referred to as the immunity that is mediated by macromolecules, including secreted antibodies, complement proteins, and certain antimicrobial peptides located in extracellular fluids.

Basically, viral vectors are designed to lack the potential for replication due to the removal of the necessary genes involved [13]. This is made to maintain high safety level in viral vector. So, it is preferable to select none or low-pathogenic viruses for developing viral vectors that are replication defective [13].

Replication-deficient viral vectors need higher dose injections and booster doses to maintain continuity in the immune system [15]. Viral vectors are, relatively, more stable as compared to other viral vectors and also require less sophisticated storage and handling systems. An example is Ad26.COV2·S, an adenoviral-based vector vaccine [16].

6.2 Some important viral vectors

In 1972, the first viral vector created from the SV40 virus was genetically engineered for human use [15,17]. During the later phase, other viruses like AVs, poxviruses, herpes viruses, vesicular stomatitis virus (VSV), cytomegalovirus, and lentiviruses have been introduced on the market for human use.

6.2.1 Adenovirus

Commonly, AVs are responsible for respiratory illnesses. The illnesses can range from the common cold to pneumonia, croup, and bronchitis. The advantages of AVs are their high transduction efficiency, transgenic expression, broad viral tropism, and that they can infect both dividing and nondividing cells. The only disadvantage with this is that many people have preexisting immunity to AVs due to previous exposure [18–20].

Most AV vectors are replication-defective because of the deletion of the E1A and E1B viral gene regions. Following are the four AV vector vaccines for COVID-19 that have been authorized in April 2021.

- The Oxford-AstraZeneca vaccine uses the modified chimpanzee AV ChAdOx1 [21–23].
- Sputnik V uses human AV serotype 26 for the first shot and serotype 5 for the second [24,25].
- The Janssen vaccine uses serotype [26–29].
- Convidecia uses serotype [30,31].

6.2.2 Lentiviruses

Lentiviruses are a subclass of RVs. These viruses can be integrated into the genomes of nondividing cells, whereas other RVs can infect only dividing cells. Lentiviruses are primarily a research tool used as viral vectors to transfer a specific genetic sequence into target cells. Efforts are going on to use lentiviruses to block the expression of a specific gene using RNA interference technology in high-throughput formats.

6.2.3 Vesicular stomatitis virus

Vesicular stomatitis is a disease of cattle, pigs, and horses in the Americas and is important because of its significance in the differential diagnosis of foot. In the early 1990s, it was introduced as a vaccine vector [32]. In 2019, FDA approved rVSV-ZEBOV vaccine (Ebola vaccine), known as Ervebo, based on a VSV. It is a recombinant, replication-competent vaccine consisting of VSV genetically engineered so that the gene for the natural VSV envelope glycoprotein is replaced with that from the Kiwit 1995 Zaire strain Ebola virus [33–35]. In most VSV vaccine vectors, attenuation provides safety against its virulence.

6.2.4 Poxvirus

Poxviruses are brick or oval shaped with large double-stranded DNA genomes. Poxviruses exist throughout the world and cause disease in humans and many types of animals. Due to its large size, the viral genome's capacity for foreign gene insertion is high. Poxvirus vectors are one of the most commonly used viral vectors for the development of preventive vaccines against infectious diseases and therapeutic cancer vaccines. These vectors are typically attenuated or engineered to safely deliver genes encoding antigens of different pathogens. Modified vaccinia strain Ankara (MVA) is a highly attenuated strain derived from the vaccinia strain Ankara. Mvabea, the second dose of the Zabdeno/Mvabea Ebola vaccine, is a modified vaccine vector, a type of poxvirus. Both doses are nonreplicating vectors and carry the genetic code of several Ebola virus proteins.

6.3 Types of nucleic acids associated with viral vectors

To understand more detail on viral vector vaccine production, it is necessary to know about the types of nucleic acids associated with the virus vectors.

The Baltimore system is based on how viruses use that machinery. Using this classification system, mRNA is at the center, and different pathways to mRNA from DNA or RNA genomes denote the different classes. The Baltimore Classification System initially included six classes of viruses. However, a seventh class was added to accommodate the gapped DNA genome of *Hepadnaviridae* (hepatitis B virus) (Fig. 6.2).

There are two ways that can be chosen to perform this technique: in vivo or in vitro condition. The former process is performed inside a living organism, and the later process is applicable in cell culture (Fig. 6.3).

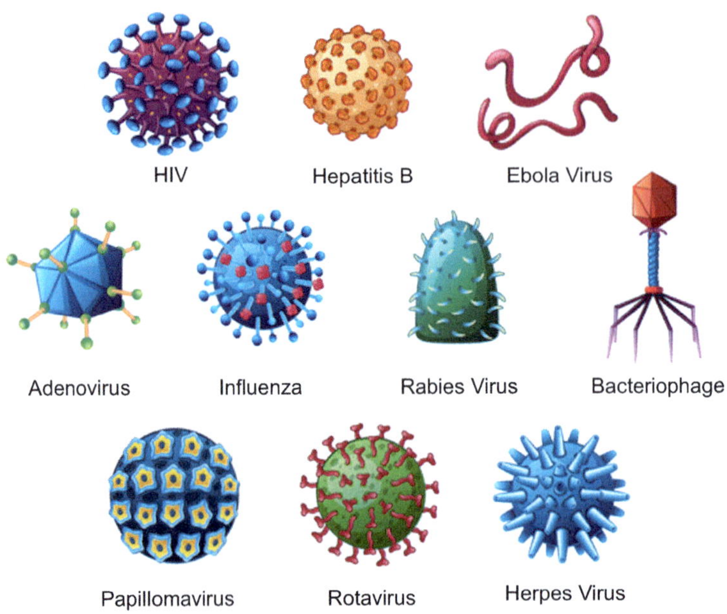

FIG. 6.2

Types of viral vectors.

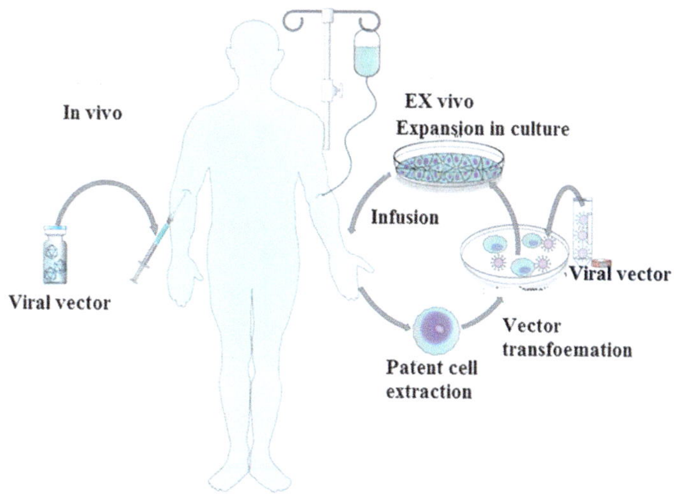

FIG. 6.3

Viral gene therapy modalities

Viruses have a special biological potential to deliver genes or other genetic material into a cell, and this process is known as transduction; the infected cells are described as transduced. In 1970, Paul Berg used a modified SV40 virus containing DNA from the bacteriophage λ to infect monkey kidney cells maintained in culture [36].

6.4 Key properties of viral vectors

From a safety point of view, virus vectors are tailored accordingly for human use.

Safety: Irrespective of the origin of the viral vector, pathogenic or nonpathogenic, their structure and function are amended in such a way as to cause the minimum risk for the body.
Low toxicity: The viral vector should have a minimal effect on the physiology of the cell it infects.
Stability: Some viruses are genetically unstable and can rapidly rearrange their genomes. So, it is essential for predictability and reproducibility of the work conducted using a viral vector and is avoided in their design.
Cell type specificity: Viral vectors are employed mainly as a gene delivery system. A great variety of viral expression systems have been developed and assessed for their ability to transfer genes into somatic cells. In particular, retroviral- and adenoviral-mediated gene transfer have been extensively studied and improved. The viral receptor can be modified to target the virus to a specific kind of cell. Viruses modified in this manner are said to be pseudotyped.
Identification: Viral vectors are often given certain genes that help identify which cells took up the viral genes. These genes are called markers. A common marker is resistance to a certain antibiotic. The cells can then be isolated easily, as those that have not taken up the viral vector genes do not have antibiotic resistance and so cannot grow in a culture with the relevant antibiotic present.

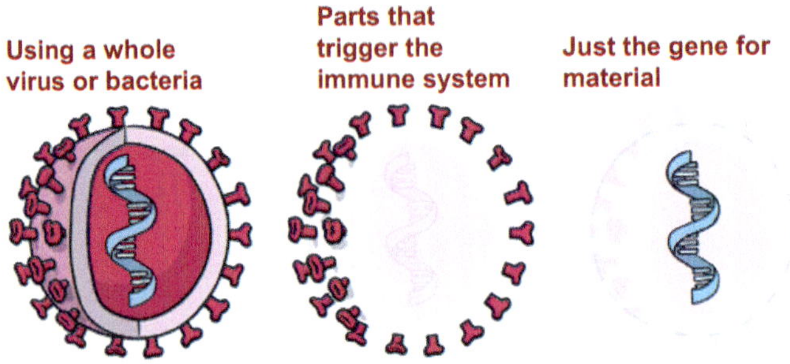

FIG. 6.4

Three main approaches for vaccine production for human use.

6.5 Development of viral vectors

A suitable vaccine for human use is screened from a variety of viruses. Roughly 7 out of every 100 will be chosen for clinical trials, after the necessary structural amendment of a pathogenic virus. Out of the vaccines that do make it to clinical trials, just one in five is successful. Mainly, there are three ways to develop vaccines for human use (Fig. 6.4). Their differences lie in whether they use a whole virus or bacterium, just the part of the germ that triggers the immune system, or just the genetic material that provides the instruction for making specific proteins and not the whole virus.

6.6 Types of vaccines for human use

Viral vector-based vaccines use a harmless virus to smuggle the instructions for making antigens from the disease-causing virus into cells, triggering protective immunity against it. Vaccines currently being used for humans are, mainly, divided into two groups: (i) virus-based (Fig. 6.5) and (ii) protein-based (Fig. 6.7) vaccines.

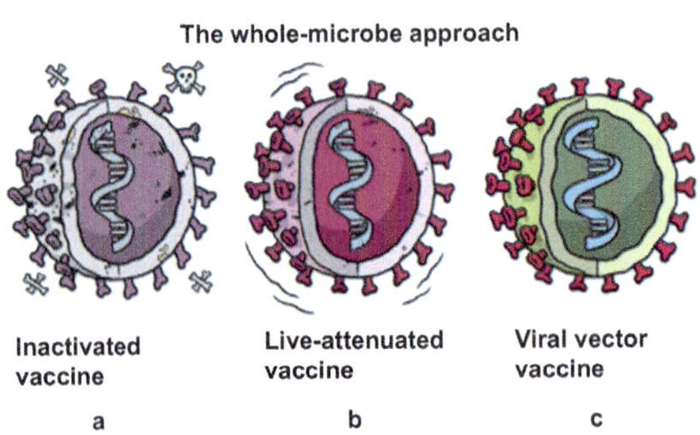

FIG. 6.5

Viral-based vectors for human use.

6.6.1 Virus-based vaccine (the whole-microbe approach)

A pathogenic virus or bacterium or one very similar to it is inactivated or killed. The inactivated step usually involves heat, radiation, or chemicals to destroy the pathogen's genetic materials, which stop it from replicating. Activated vaccines can trigger a strong immune reaction, but it is usually not as strong as the reaction that live attenuated vaccines can produce (Fig. 6.5). There are many existing vaccines, including those for yellow fever and measles (live attenuated vaccine), or seasonal influenza and hepatitis A (inactivated vaccines) belonging to this group. Bacterial-attenuated vaccines also exist, such as the BCG vaccine for tuberculosis.

Virus removal is separate and distinct from viral inactivation as it is a complementary process that further enhances viral safety by eliminating or separating the virus from the protein(s) or product of interest.

On the basis of size, type, and liability for therapeutics, multiple methods of viral inactivation are possible. The most common are as follows:

- Low pH-based methods, especially for biotherapeutic products like monoclonal antibodies (mAbs).
- Solvent- or surfactant-based methods usually for polysaccharides.

Less frequently hydroxide has been shown to be effective at activating lipid-envelope viruses such as human immunodeficiency virus and pseudorabies virus and against nonenveloped viruses such as porcine circovirus type 2 and minute virus in mice. Pasteurization, dry heat, and vapor heat methods are also less frequently used.

6.6.1.1 Inactivated virus vaccine

A virulent virus is exposed to chemical or physical agents for inactivation. The most commonly used inactivating agent, formalin, is known to induce irreversible changes in many viral antigens. The use of β-propiolactone in the manufacture of some human rabies vaccines has advantages in that those proteins are not damaged and the inactivating agent is completely hydrolyzed within hours to nontoxic products. Nonionic detergents such as polyoxylene ether are used in the case of enveloped viruses to solubilize virions and release glycoprotein peplomers and other envelope proteins. Differential centrifugation or ultrafiltration is used to semipurify the solubilized glycoproteins prior to formulation for use as so-called "split" vaccines. Some commonly used influenza vaccines are produced by this procedure.

The sources of virulent viruses are from infected animals like mouse brain, but infected cell cultures provide cleaner starting material. Commonly, inactivated virus vaccines are not elicit sufficient antibodies in the body's cells. With such vaccines, the primary vaccination course comprises two or three injections; further "booster" may be required at intervals to maintain protective immunity. Very critical dosages of chemical or physical treatment are necessary, or else there would be chances of damaging enough to modify immunogenicity, which would result in an immune response shorter in duration, narrower in spectrum for viral antigens, weaker cell-mediated and mucosal immune responses, and possibly less effective in preventing viral entry.

Advantages and disadvantages

(a) The standard protocol is well established and practiced at industrial level.
(b) Even, people having compromised immune system can avail safely.
(c) There is no risk of the vaccine triggering diseases due to its lack of any live components.
(d) Comparatively stable.
(e) There may be need of booster doses.

6.6.1.2 Live attenuated vaccine

Live vaccines are derived from "wild" viruses or bacteria. These wild viruses or bacteria are attenuated (weakened) in a laboratory, usually by repeated culturing. For example, the measles virus used as a vaccine today was isolated from a child with measles disease in 1945.

Advantages and disadvantages

(i) The standard protocol is well established and has been used in large-scale production.
(ii) Strong immune response.
(iii) Both the B cells and T cells are being stimulated for immune response.
(iv) Manufacturing process is simple.
(v) People having compromised immune system may not respond.
(vi) It may cause triggering for any hidden diseases in the body.

6.6.1.3 Viral vector vaccine

Since long, we have been producing vaccines on an empirical basis, relying mostly on the attenuation or inactivation of pathogens. But increased knowledge in recombinant DNA technology, immunology, functional genomics, and proteomics has added a new dimension to the vaccinology field. The use of recombinant proteins allows the targeting of immune responses focused against few protective antigens. A variety of expression systems can be availed for the production of large quantities of proteins depending on the required characteristics. Live recombinant bacteria or viral vectors equally stimulate the immune system as in the case of a natural infection. Currently, nucleic acid vaccines, subunit vaccines, and recombinant vaccines are also developed, based on DNA, mRNA, or proteins.

In the case of a viral vector vaccine, a safe virus is used to deliver specific subpart(s) in the form of proteins of the germ of interest so that it can trigger an immune response without causing disease (Fig. 6.6).

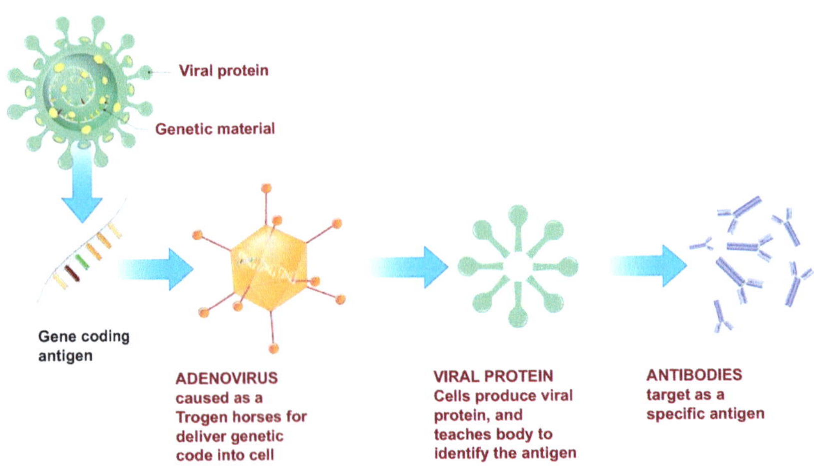

FIG. 6.6

Mechanism of action of viral vector vaccine.

To do this, the instructions for making particular parts of the pathogen of interest are inserted into a safe virus. The safe virus then serves as a platform or vector to deliver the protein into the body. The protein triggers the immune response. The Ebola vaccine is a viral vector vaccine, and this type can be developed rapidly.

6.6.2 Protein subunit-based vector

A subunit vaccine is one that only uses the very specific parts (the subunits) of a virus or bacterium that the immune system needs to recognize (Fig. 6.7).

The subunit may be proteins or sugars. It doesn't contain the whole microbe or use a safe virus as a vector (Fig. 6.8).

Because these vaccines contain only the essential antigens and not all the other molecules that make up the germ, side effects are less common. Most of the vaccines on the childhood schedule are subunit

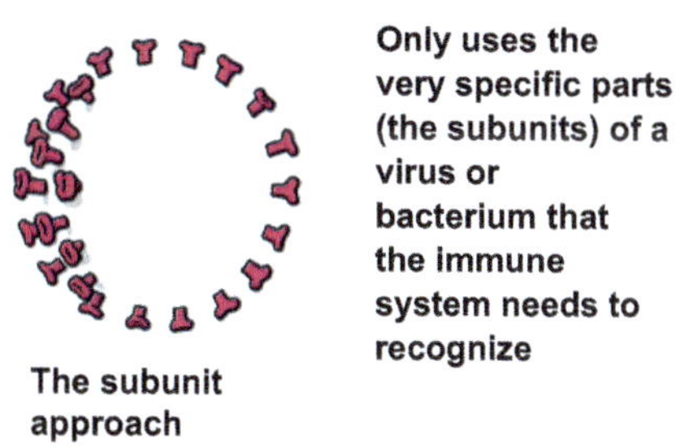

FIG. 6.7

Protein subunit-based vector.

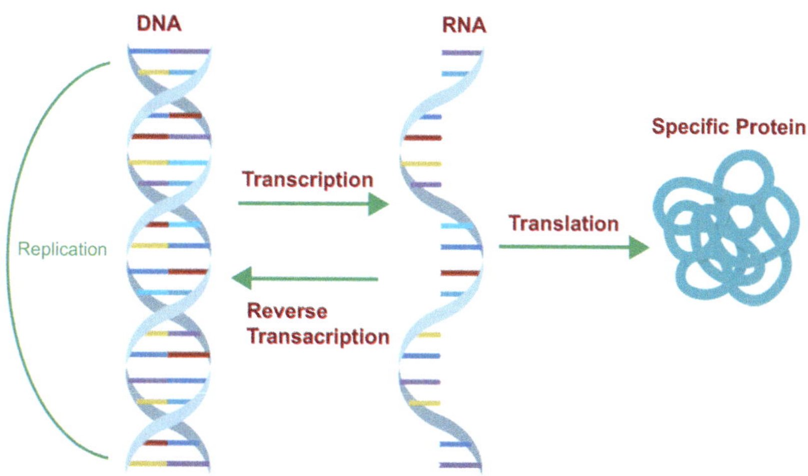

FIG. 6.8

Use of nucleic acid for specific proteins—the DNA or RNA.

vaccines, protecting people from diseases such as whooping cough, tetanus, diphtheria, and meningococcal meningitis. Unlike vaccine approaches that use either a weakened or dead whole microbe or parts of one, a nucleic acid vaccine just uses a section of genetic material that provides the instructions for specific proteins, not the whole microbe. DNA and RNA are the instructions our cells use to make proteins. In our cells, DNA is first turned into mRNA, which is then used as the blueprint to make specific proteins.

Instead of injecting a whole pathogen to trigger an immune response, a selective subunit of a pathogen, known as subunit vaccines (sometimes called acellular vaccines) containing purified pieces of it, can be used for stimulating immune cells. These protein subunits lack the potential to cause disease. So, subunit vaccines are considered very safe. Like protein subunit vaccines, polysaccharide vaccines contain chains of sugar molecules on the surface of the cell walls of some bacteria. The conjugate subunit vaccines bind a polysaccharide chain to a carrier protein to try and boost the immune response. Only protein subunit vaccines are being developed against the virus that causes COVID-19.

Presently, varieties of subunit vaccines are on the market for human use. For instance: hepatitis B and acellular pertussis vaccines (protein subunit), the pneumococcal polysaccharide vaccine, and the MenACWY vaccine from bacteria (causing meningococcal disease) containing surface polysaccharides joined to diphtheria or tetanus toxoid (conjugate subunit). The advantages and disadvantages of subunit vaccines are as follows:

- The standard protocol of large-scale production is well established.
- People having compromised immune systems can safely use.
- Due to the lack of living components, there will be very less chance of vaccine triggering disease.
- Relatively stable.
- Manufacturing system is complex.
- For improving entry into cells, adjuvants are required.
- Booster dose is needed to keep continuity of immune system.
- Determining the best antigen combination takes time.

6.6.2.1 Mood of action

As stated earlier, subunit vaccines contain fragments of protein and/or polysaccharide from the pathogen, which, in suitable combinations, give biological stimulation for an effective immune response. By restricting the immune system's access to the pathogen in this way, the risk of side effects can be reduced. Comparatively with other vaccines, the subunit vaccines are more stable and relatively cheap and easy to produce. In subunit vaccines, the antigen does not infect cells, but mainly triggers antibody-mediated immune responses. In order to accelerate the effective functional aspects, subunit vaccines are sometimes delivered alongside adjuvants and booster doses may be required.

6.7 Next-generation vaccines

With the modernization of tools and techniques in genetic engineering and the conceptual development of knowledge on genomics and proteomics, lot of efforts have been going on to develop next-generation vaccines with high efficacy, durability, or protection, reducing the number of doses needed. Hence, WHO is keen to encourage product characteristics for next-generation vaccines.

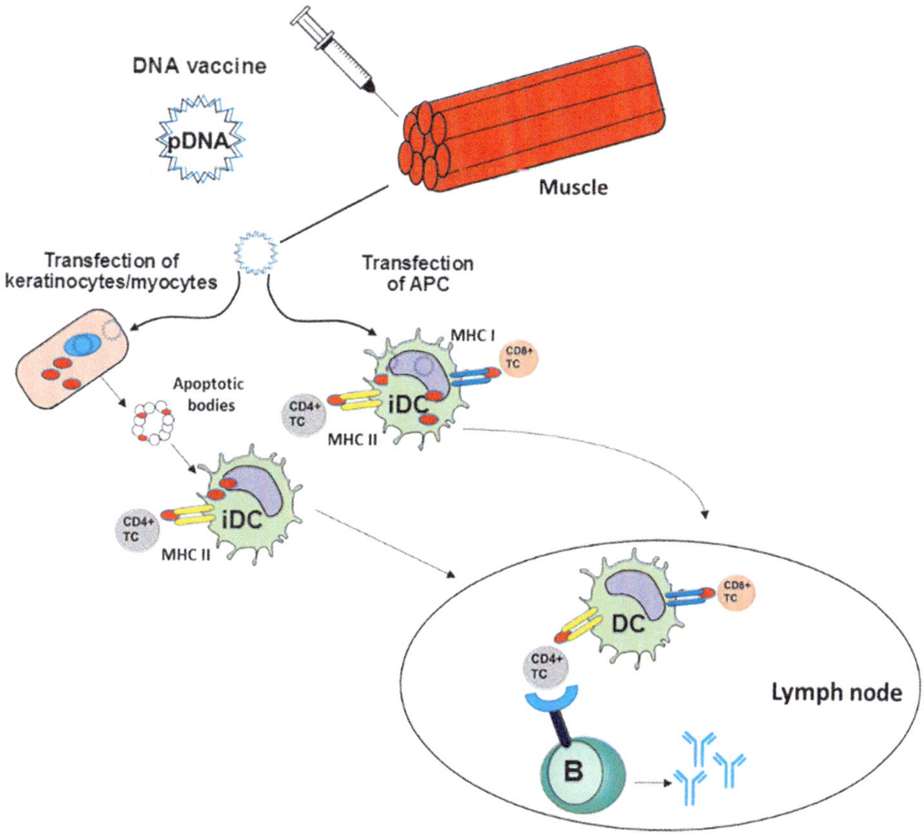

FIG. 6.9

Functional aspects of DNA-based vaccine.

6.7.1 DNA-based vaccine

DNA vaccines are nucleic acid vaccines that are genetically engineered into a nonreplicated circular plasmid. The DNA sequence codes for the antigen against which immunity is sought. In this case, the plasmid is transfected into the host cell and incorporated into the host genome (Fig. 6.9).

Since the protein is a bacterial or viral sequence, it is recognized as a foreign body by the host's machinery, and as a result, immune response is activated. The main advantages of DNA vaccines are their easier development, cost-effectiveness, and longer immune response to the host cells. This type of vaccine, also, stimulates both B- and T-Cell responses and improves vaccine stability. Currently, DNA vaccination is an emerging field. Use of adjuncts has catalyzed the functional aspects of DNA vaccines to assist in their entry into specific cells and stimulate or direct the immune response. Still, many aspects of the immune response generated by DNA vaccines are not clear.

In 2021, Zydus Cadila has received the Emergency Use Authorization (EUA) from the Drug Controller General of India for ZyCoV-D, the world's first circular plasmid DNA vaccine for COVID-19. Many types of DNA vaccines are under process for human use (Table 6.1).

Table 6.1 Selected DNA-based COVID-19 vaccines in development process.

Developer	Vaccine	Description	Status
Zydus Cadila; Department of Biotechnology, Government of India	ZyCoV-D	Three-dose DNA plasmid vaccine encoding SARS-CoV-2S protein and IgE signal peptide delivered intradermally by the needle-free PharmaJet Tropis device	Received Emergency Use Authorization in India, August 20, 2021
AnGes, Osaka University, Japan Agency for Medical Research and Development	AG0302-COVID19	Two-dose DNA plasmid vaccine encoding SARS-CoV-2S protein delivered by intramuscular injection	Phase 2/3
Inovio Pharmaceuticals, Advaccine Biopharmaceuticals (Suzhou, China)	INO-4800	Two-dose sequence-optimized DNA plasmid vaccine encoding a SARS-CoV-2S protein, delivered by intradermal injection followed by electroporation at the injection site with the Cellectra 2000 device	Phase 2/3
Genexine	GX-19N	Two-dose DNA plasmid vaccine encoding SARS-CoV-2S and N proteins	Phase 2/3
Entos Pharmaceuticals	Covigenix VAX-001	Two-dose proteolipid-vehicle-formulated DNA vaccine encoding the SARS-CoV-2S protein and two genetic adjuvants, delivered by intramuscular injection	Phase 1/2
Rottapharm Biotech (Monza, Italy), Takis (Rome)	COVID-eVax	Two-dose DNA plasmid vaccine encoding the receptor-binding domain of the SARS-CoV-2S protein, delivered by intramuscular injection followed by electroporation	Phase 1/2
Scancell (Nottingham, United Kingdom)	Covidity	Four-dose DNA vaccine regimen, delivered by intradermal and intramuscular injection with a PharmaJet needle-free device, comprising two doses each of SCOV1 and SCOV2 plasmid DNA, which encode the receptor-binding domain of the SARS-CoV-2S and N proteins expressed by the original pandemic strain and by a variant strain, respectively	Phase 1
Symvivo	bacTRL-Spike	Single oral dose of *Bifidobacterium longum* bacteria engineered to deliver plasmid DNA encoding SARS-CoV-2S protein	Phase 1

Sources: Company websites; ClinicalTrials.gov; PubMed.

Table 6.2 DNA vaccine vs recombinant vaccine.

DNA vaccine	Recombinant vaccine
Description	
DNA vaccines are prepared by inserting a DNA sequence (that codes for the antigen) into a plasmid and transfected to the host cell	Recombinant vaccines are protein or DNA recombinants that are inserted into the host cell to trigger immune response
DNA/protein	
DNA based	Both DNA and protein based
Incorporation	
The DNA sequence gets incorporated into the host genome	A gene of interest is incorporated into a vector which in turn is included in the host genome
Approval	
ZyCoV-D is a DNA vaccine approved for humans	Hepatitis B and HPV vaccines have been approved and are readily used

6.7.2 Recombinant vaccine

Yeast or bacterial cells are used for the production of recombinant vaccines. These protein or DNA recombinants are introduced into the host cells where they are identified as a foreign material, and as a result, an immune response is triggered. The advantage of recombinant vaccines is that they can be used for people with weakened immune systems as well. However, one shortcoming is that a booster dose is required to maintain the immune effect. Table 6.2 shows the differences between DNA vaccines and Recombinant vaccines.

6.7.3 Toxoid vaccine

It is used to prevent toxins (poisons) from producing bacteria. In toxoid vaccine, the toxin-producing site of bacterial vectors is weakened so they cannot cause illness. Weakened toxins are called toxoids. When the immune system receives a vaccine containing a toxoid, it learns how to fight off the natural toxin. The DTaP vaccine contains diphtheria and tetanus toxoids. For an instance, tetanus vaccines are based on inactivated tetanus toxin.

Toxigenic strains of *C. tetani* are grown in liquid media; the toxin is purified and then inactivated by treatment with formaldehyde to produce the toxoid antigen. After purification and sterilization, tetanus toxoid is formulated with aluminum or calcium salts and administered by intramuscular injection.

6.7.4 Conjugate vaccines

Conjugate vaccines fight a type of bacteria that has antigens. These bacteria have antigens with an outer coating of sugar-like substances called polysaccharides. This type of coating disguises the antigen, making it hard for a young child's immature immune system to recognize it and respond to it. Conjugate vaccines are effective for these types of bacteria because they connect (or conjugate) the polysaccharides to antigens that the immune system responds to very well. This linkage helps the immature system react to the coating and develop an immune response. An example of this type of vaccine would be the *Haemophilus influenzae* type B (Hib) vaccine and the mRNA COVID-19 vaccine.

6.8 Immune system triggering by whole virus

Both the live attenuated and inactivated vaccines contain the whole or part of the disease-causing pathogen and trigger immune system, but without causing illness. Inactivated vaccines contain viruses whose genetic material has been destroyed by heat, chemicals, or radiation so they cannot infect cells and replicate, but can still trigger an immune response.

When a live attenuated vaccine is injected, it will infect cells and replicate, but cause no or only very mild diseases. This short of vaccine may not be suitable for people with compromised immune systems (e.g., those with HIV) and pregnant women. In very rare cases, live attenuated vaccines can revert to a more pathogenic form, triggering disease in vaccinated individuals or their contacts. Evidence in such cases has been witnessed, for example, vaccine-derived poliovirus associated with the oral polio vaccine. In order to understand how vaccines work, it is necessary to know how the body fights illness. When germs, such as bacteria or viruses, invade the body, they attack and multiply. This invasion, called an infection, is what causes disease. The immune system uses white blood cells to fight infections. White blood cells are part of the body's immune system. They help the body fight infections and other diseases. There are 5 types of white blood cells:

- Neutrophils are the first responder of immune cells.
- Basophils release histamine to mount a nonspecific immune response.
- Eosinophils fight bacteria and parasites but also provoke allergy symptoms.
- Lymphocytes are B and T cells that defend against specific invaders.
- Monocytes clean up dead cells.

The immune system response to these vaccines is like that of the infection of natural pathogens; they induce stimulation of a range of white blood cells, including killer T cells, helper T cells, and antibody-producing B cells. This immune response continues until the virus is cleared from the body. In order to fastening the biological system for immune development, adjuvants and booster doses are given.

6.9 Upstream process for viral vector production

Growing therapeutic viral vector production is the most complex and challenging. Without the benefit of established process templates, manufacturers must work quickly within compressed timeframes and make efficient upstream decisions concerning raw materials, cell culture media preparation, and more. Decisions without proper risk mitigation may prove to have serious and irreversible consequences.

Critical areas typically include the following:

- Cell line and cell culture—Phase I
- Viral vector development—Phase II
- Turnkey scale therapeutic viral vector production—Phase III
- Therapeutic viral vector purification/polishing/formulation—Phase IV
- Virus vector packing for marketing—Phase V

Table 6.3 List of variety of suspension cell lines.

Cell line	Meaning	Organism	Origin tissue	Morphology	Links
CHO	Chinese Hamster Ovary	Hamster	Ovary	Epithelial	ECACC Cellosaurus
HeLa	"Henrietta Lacks"	Human	Cervix epithelium	Cervical carcinoma	ECACC Cellosaurus
H-9		Human	Embryonic stem cells	Lymphoblast	Cellosaurus
Jurkat		Human	White blood cells	Lymphoblast	ECACC Cellosaurus
C6/36		Insect—Asian tiger mosquito	Larval tissue		ECACC Cellosaurus
High Five		Insect (moth)—*Trichoplusia ni*	Ovary		Cellosaurus
S2	Schneider 2	Insect—*Drosophila melanogaster*	Late stage (20–24h old) embryos		ATCC Cellosaurus
Sf21	*Spodoptera frugiperda* 21	Insect (moth)—*S. frugiperda*	Ovary		ECACC Cellosaurus
Sf9	*S. frugiperda* 9	Insect (moth)—*S. frugiperda*	Ovary	Epithelial	ECACC Cellosaurus
SH-SY5Y		Human	Bone marrow	Epithelial	Cellosaurus
PC-3		Human	Prostate	Epithelial	Cellosaurus

6.9.1 Cell line and cell culture—Phase-I

In 1885, Wilhelm Roux initiated the groundwork for tissue culture techniques by developing a saline buffer system to culture chicken embryo cells [37]. Subsequently Ross Granville Harrison in 1907 developed the in vitro cell culture technique, by modifying the hanging drop technique for nerve cell culture [38]. Afterward, in 1910, Montrose Thomas Burrows adapted Harrison's technique in collaboration with Alexis Carrel and developed multiple tissue cultures by using fresh plasma in combination with saline solution [39]. Subsequently, Carrel developed a cell line culture derived from chicken embryo heart as an immortal cell line [40]. In 1952, George Otto Gay in association with his assistant Mary Kubicek cultured the first human-derived immortalized cell line-HeLa in a suspension cell culture system [41].

A variety of equipment such as magnetic stir plate, orbital shakers, incubators, and flasks (e.g., culture flasks, spinner flasks, and shaker flasks) with proper agitation systems are used for suspension cell culture. Currently, varieties of viral vectors are used as expression systems for delivering genes of interest into host cells (Table 6.3).

6.9.2 Therapeutic viral vector development—Phase II

Basically, the conceptual development of virus or viral vector production through fermentation technology was initiated for vaccine production, but subsequently the biopharmaceutical industries have taken this opportunity in other areas such as protein expression, gene therapy, treatment of cancer, and other biologics development [42–44]. The in vivo method of virus production is limited to a certain extent. Commonly, the treatments through virus particles involve doses of 100–1000 mL with a virus density of at least 106–107 infective units per mL. So, it is important to design highly efficient upstream production followed by an effective downstream process to manufacture sufficient quantities to meet the demand of the market. The overall production of a virus or virus vector depends on the virus/host cell interactions and the kinetics of virus particle synthesis and virus release.

Nonreplicating viral vectors are typically expressed in packing cells containing only the target of interest structural virus genes [45]. In packing cell lines, there is a continuous release of RNA-containing vector particles into the medium. Well-established packing cell lines can produce retroviral particles over a long period depending on their growth characteristics and are suitable for pilot-scale production in a suitable bioreactor.

The production of a virus or virus vector is mainly dependent on the expression systems that are, still, in constant evolution. Commonly, there are three expression systems: (i) Hek 293 cells (immortalized human embryonic kidney cells), (ii) an insect cell-based system based on infection by baculovirus of the Sf9 cell line, and (iii) packing/producer cells. The gene of interest from these expression systems is stably integrated into a host cell genome with a helper virus to trigger virus production. Currently, the majority of virus production requires transient transfection.

(i) HEK293 are immortalized human embryonic kidney cells. Originally isolated in the 1970s by Alex Van der Eb, a Dutch biologist, it was postdoc Frank Graham who transformed the cell line with sheared adenovirus 5 (Ad5). The "293" refers to the fact that it was Graham's 293rd experiment.

On December 21, 2020, the Roman Catholic Congregation for Doctrine of the Faith, with the approval of the Pope, declared the use of HEk293 as base material for therapeutic viral vectors for COVID-19 vaccines that have used cell lines from aborted fetuses in research and further processes, but only where other alternatives are nonexistent or there is the risk of a worse danger. During the COVID-19 pandemic, antivaccination activists noted that HEK 293 cells are used in the manufacturing of the Oxford-AstraZeneca COVID-19 vaccine (AKAAZD1222).

(ii) Insect cell line-based viral vector

Most commonly, in the baculovirus expression system, insect cells are used as hosts for the baculovirus and as core cells for the expression of exogenous proteins. The virus's amplification mainly depends on the type of insect cells used. So, the selection of host insect cells is very important for the production and quality of target proteins. The commonly used insect cell lines include *Trichoplusia ni* and Sf9 and Sf21 cell lines separated from *Spodoptera frugiperda*. Since the advent of the baculovirus expression system, the development of these three cell lines has been relatively mature and can be used for stable protein expression.

(iii) Package/producer cells for viral vector production

Package/producer cells refer to the efficacy of the transfer of genes and component materials from RV or any similar viruses into a variety of cell types from many animal species. It is a challenging task to develop RV packaging cells that allow the production of retroviral vectors in the absence of replication-competent viruses. So, retroviral vectors are the primary candidates for gene transfer into human somatic cells. These vectors have the potential to challenge efficiently for the treatment of disease by the transfer of functional genes, or gene therapy. A critical element in the production of the components to carry out retroviral-mediated gene transfer is the cell that generates the retroviral particles carrying the gene to be transferred. The cells are called "packaging cells" because they "package" the retroviral vector (which carries the gene of interest) into a delivery vehicle (the retroviral particles).

Production of viral vectors is the most delicate process involved in the requirement of the most sophisticated laboratory/manufacturing units under the control of regulatory norms of internationally approved agencies. This is because of to meet the stringent standards of purity, identity, potency scalable and commercially viable process to manufacture sufficient quantities of viral vector. Microchip-based sophisticated bioreactor is needed for this purpose.

A cell suspension or suspension culture is referred to a type of cell culture in which single cells or aggregates of cells are allowed to function and multiply in a bioreactor having suitable broth.

6.9.3 Cell isolation and culture process initiation—Phase III

Most commonly, primary cells are isolated from tissues by an enzymatic treatment process and suspended in a suitable medium for further culture. But some of the mammalian cells need some material to adhering to a surface. In the case of white blood cells, they can be cultured as a suspension, since they naturally exist in suspension in blood. So, the primary cell culture systems can be divided into two groups: (i) suspension cell culture and (ii) adherent cell culture.

(i) Suspension cell culture

Suspension cell culture (Fig. 6.10) is referred to as a type of cell culture in which single cells or small aggregates of cells are allowed to function and multiply in an agitated growth medium, thus forming suspension. For this purpose, the cells are either derived from homogenized tissues or from heterogeneous cell suspensions.

Commercially, a variety of immortalized mammalian cell lines, plant cells, and insect cells can be obtained cryopreserved and used to start a suspension culture.

FIG. 6.10

Suspension CHO recombinant cells.

There are many similarities and dissimilarities between adherent and suspension cell culture systems (Table 6.4). For laboratory purposes, spinner flasks with magnetic stirrers are used for small-scale cultures. Shaker flaks are also used for suspension cultures and appear similar to typical Erlenmeyer flasks but have a semipermeable lid to allow gas exchange. Commercially, varieties of shaker racks with the provision of holding different sizes of flasks are available for suspension culture (Fig. 6.11).

During suspension cell culturing, shaker flasks are loaded with cells and the appropriate media before they are placed on an orbital shaker. To optimize cell culture proliferation, the revolutions per minute of the orbital shaker must be adjusted within an acceptable range depending on the cells

Table 6.4 Similarity and dissimilarity between suspension and adherent cell culture system.

Suspension cell culture	Adherent cell culture
Required specialized nutrient-containing media	As in suspension cell culture
Needs containers for gas transfer	As in suspension cell culture
Needs complete aseptic condition to avoid cross-contamination	Same as suspension cell culture
Suspension culture needs to be agitated to avoid settling to the bottom of the flask	Adherent cell culture needs flat surface to promote lot of cells adherent
Suspension culture may contain a surfactant to protect cells from shear forces in addition to the amino acids, vitamins, and salt solution	Surfactant not necessary

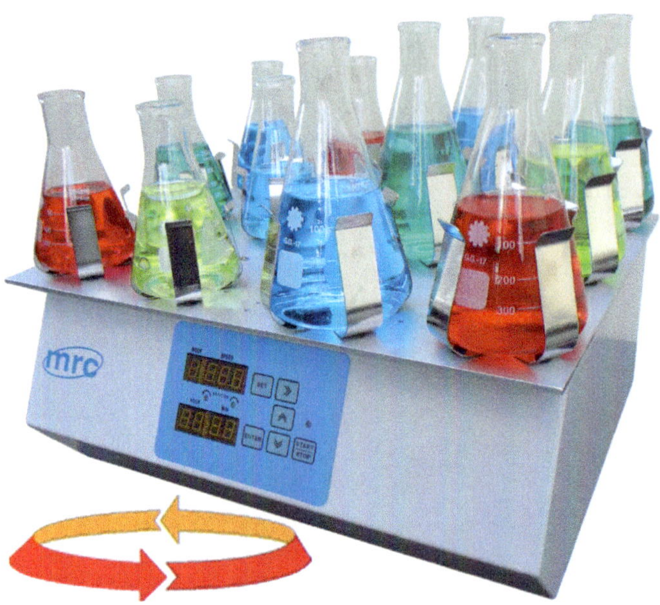

FIG. 6.11

Digital shaker flask equipment for suspension cell culture.

and media used. The media must be allowed to stir, but should not disturb the cells too much, causing them excessive stress. Shaker flasks are often used for fermentation cultures with microorganisms such as yeast and bacteria.

Subculture of suspension cells is a direct process, but in the case of adherent cells, which are first removed from the flask surface by enzyme treatment and processed further. The sample obtained is analyzed to determine the ratio of living to dead cells and the total concentration of cells in a flask. On the basis of the results obtained, a portion of the suspension culture will be transferred to an anther containing fresh medium. Media changes and subculturing are important to maintain cell lines, since cells will consume nutrients in the media to expand. Cells will keep on growing till a suspended medium is exhausted with nutrients, and attain an extreme pH.

After the first subculture, the primary cells start to become a cell line or subclone. Cell lines derived from primary cultures have a limited lifespan as a consequence of being outside of their niche. In case of continuous cell culture, cells are immortalized and can be grown for many passages with no significant loss of viability.

(ii) Adherent cell culture

Adherent cells are referred to as cells that grow while adhering to the culture vessel. Based on cell morphology, adherent cells can be divided into two main groups:

- Fibroblast-like: elongated shape and usually migrate on the dish
- Epithelial-like: polygonal shape, stationary, and grow in patches

Most vertebrate-derived cells (with the exception of hematopoietic cells) can be cultured and require a 2-dimensional monolayer to facilitate cell adhesion and spreading. Cell samples can be taken from tissue explants or cell suspension cultures. Adherent cell cultures with an adequate nutrient-containing growth medium will continue to grow until they cover an available surface area. Unlike suspension cultures, adherent cell cultures require regular passaging performed using mechanical or enzymatic dissociation.

In the beginning, viral vaccine production, cell therapies, and gene therapies used to be dependent heavily on adherent cell culture. But in later stages, producing high enough quantity of adherent cells to meet the capacity requirement has been a challenge. To grow adherent cell lines, manufacturers were for a long time relying on roller bottles or cell trays. These processes are sufficient for Phase I/II trials, which are typically only a few patients. Scale-up to Phase III/commercial scale requires large-scale production, which cannot be managed by adherent cell culture methods. Growing cells by adherent cell culture methods under current good manufacturing practice (cGMP) is a tougher problem. So, currently, well-designed large-scale bioreactors have been used to meet the challenges of growing suspension cell lines and to produce mAbs and other biologics. A variety of single-use bioreactors that have been specially designed to produce high quantities of cGMP-grade virus for viral vaccines are available on the market (Fig. 6.12). This unique technology provides the cells with a three-dimensional controlled growth environment with low shear stress, resulting in excellent cell growth and high virus productivity. It has provision for 500 m^2 of growth surface area, equivalent to 790 10-layer multitray stacks, in a small footprint. It also possess similar automation and control options compared to other bioreactors with an integrated pH, dissolve oxygen (DO), and optional biomass sensors. This technology has been gaining popularity within the biopharmaceutical industry for vaccine and gene therapy manufacturing purposes.

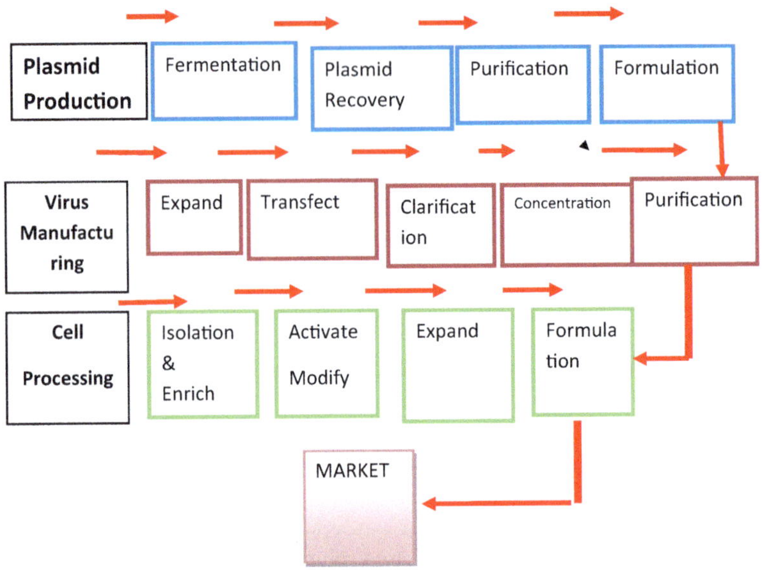

FIG. 6.12

Example of 3 manufacturing platforms for the generation of modified cells for ex vivo gene therapy via viral vectors produced by transfection.

In brief, prior to manufacturing viral vectors, several materials are needed for vaccine production for human use. These materials include plasmids encoding helper virus functions and the therapeutic gene, cell lines used to manufacture the vector, and others. In some cases, helper transducing viruses may be substituted for plasmids, which also need to be generated via cells. There are established packaging cell lines for some vector systems, such as gamma-retrovirus, while others are in development. One advantage to the use of stable lines for packaging is that they can reduce or eliminate transfection and/or transduction steps, simplify the production process, and lower the costs.

The overall target in viral vector design is to efficiently pack the therapeutic gene or nucleotides into infectious viral particles and avoid the generation of wild-type particles or empty particles. So, for safety purposes and to avoid toxicity, many vectors are manufactured by transient co-transfection of multiple plasmids. Commonly, AAV vector components and therapeutic genes are transfected as separate expression cassettes from different plasmids. Expression cassettes are expressed within the cell, resulting in viral proteins and a genomic ssDNA containing the expression cassette for the therapeutic gene(s). In the AAV system, viral particles containing the transgene assemble in the cytoplasm. Particles are released to the media via cellular lysis prior to further purification and characterization. Although retroviral and adenoviral vectors have been used in vaccine production and gene therapy, vector derived from AAV and LVs have gained more acceptance in vaccine production due to their nonpathogenic nature and well-tolerated habits. In addition, AAV does not integrate with the cell genome, as in case of lentivirus. But, LVs are now the most commonly used.

Table 6.5 Overview of common viruses used for generating gene therapy viral vectors.

Parameters	Retrovirus	Lentivirus	AAV	Adenovirus
Coat	Enveloped	Enveloped	Nonenveloped	Nonenveloped
Packaging capacity (Kb)	8	8	−4.5	7.5
Tropism/ infection	Dividing cells	Broad	Broad excluding hematopoietic stem cell	Broad
Inflammatory potential	Reduced	Reduced	Reduced	High
Host genome interaction	Integrating	Integrating	Integrating/nonintegrating	Nonintegrating
Transgene	Long lasting	Long lasting	Potentially long lasting	Transient or long lasting depending on immunogenicity

Table 6.5 summarizes some physical properties of the various viruses for their suitability in gene delivery system.

6.9.4 Trunkey scale therapeutic viral vector production—Phase-IV

The manufacturing process of viral vectors undergoes several complex and sophisticated phases. The initial stage of manufacturing processes begins with the selection of a suitable virus, which may be pathogenic, to convert it into a therapeutic viral vector by deleting the necessary gene coding responsible for the virulent characters. The viral vector carries the therapeutic gene into target cells, which can be modified either in vivo or ex vivo. In the ex vivo application, cells are modified in culture, which also allows for cell expansion and analytical characterization prior to reinfusion of the treated cells. These cells are examined pretreatment and posttreatment for viability, density, expression level, etc. The ex vivo process can sometimes result in more efficient transduction. Currently, a variety of viral vectors are in use on the basis of tissue tropism, desire for integrating or nonreplicating, and safety. In order to economize the process, high-efficiency transduction and robust levels of transgene expression are necessary.

The most important concern is designing virulent, free viral vectors from wild types. In addition, sufficient care is to be taken so that there would be no chances of reversible in forming wild type. In therapeutic viral vectors, the expressible units are typically separate plasmids introduced to producer cells either by transfection or through "helper" transduction viruses (Fig. 6.12). In most of the therapeutic viral vectors, the functional wild characters are disabled.

Commonly, two methods are used for viral vector production:

(i) Transient production system: It involves either (a) transfection of one or several plasmids encoding the helper virus functions (Fig. 6.13) or (b) viruses to provide helper function.

(ii) Stable producer cell line production system: It is the second major mode of vector production. Stable producer lines do not require transfection or transduction to enable vector packaging. Currently, the majority of vectors are generated by transient production systems due to convenience, lack of a stable producer cell line, or other reasons.

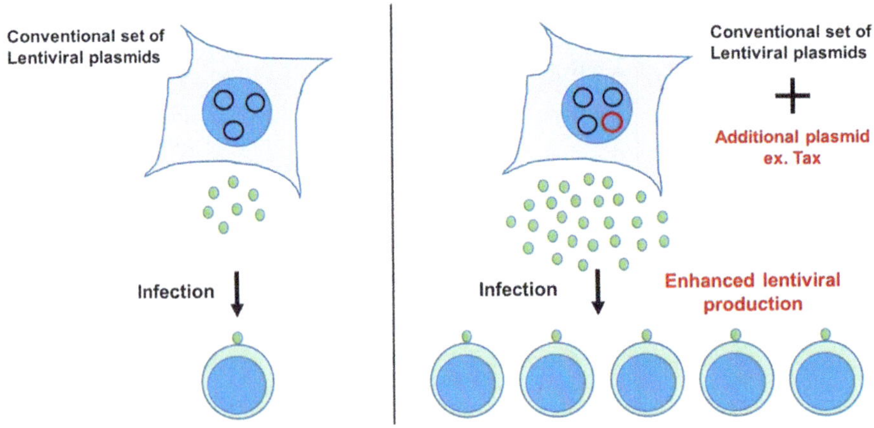

FIG. 6.13

Showing how virus production boosted in cells to generate more vectors for gene transfer.

6.9.5 Viral vector purification and formulation—Phase-V

6.9.5.1 Purification

Most stringent rules have been prescribed by international regulatory authorities such as Food and Drug Administration (FDA) and European Agency for the Evaluation of Medical Products (EMEA) and made mandatory to follow well-approved protocols for safe and secure therapeutic virus production by quality production. Like influenza viral vaccines, manufacturers begin to switch from egg cultivation to mammalian cell culture systems. Also within the purification procedure, a clear trend from classical purification methods like sucrose gradient centrifugation toward more sophisticated techniques like tangential flow filtration and liquid chromatography can be observed.

As the viral vectors are most delicate in structural stability and function, it is necessary to have the best formulation of the viral vector under suitable temperature conditions. So, it is necessary to have an intimate understanding of the physical, chemical, and biological characteristics of viruses that can be exploited in these processes. Structurally, virus particles have more dimensions of proteins used in biopharmaceutical applications with respect to weight and size. So, it is not simply a matter of "plugging" viruses into existing protein purification schemes and expecting quality output.

Commonly, downstream operations are greatly influenced by the presence of impurities and contaminants in the virus-containing lysates funneled into them, and because they account for as much as 70% of production costs, it is important that all phases of process development be designed in tandem. Virus particles are comparatively larger in size than the proteins, peptides, glycoproteins, sugars, and nucleic acids that are mostly present in the lysates harvested from virus-infected cells. While proteins typically range from 0.005 to 0.08×10^6 Da, viruses generally exceed 5×10^6 Da and can have a size of up to 1000 nm.

The process of virus purification depends on pI, surface hydrophobicity, the presence of an envelope, hydrodynamic diameter, and the liability of virus particles.

A variety of methods can be applicable for viral vector purification, which include density gradient ultracentrifugation, ultrafiltration, precipitation, two-phase extraction systems, and size-exclusive chromatography. Most commonly, high-speed preparative centrifugation (ultracentrifugation) using cesium chloride or sucrose gradient has been in application for viral vector purification. On the basis of suitability, iodixanol gradients have also been used for AV purification. Sometimes, virus particles may lose potential due to infectivity during the extended period of time required for fractionation of the gradients. Some viruses can be aggregated by precipitation with chemicals such as polyethylene glycol, ammonium sulfate, or calcium sulfate, which can be easily separated by low-speed centrifugation. But the problem with such a process is coprecipitation along with viral particles, and as a result, there is a chance of losing potential due to changes in osmotic pressure. To overcome such a problem, it is advisable to avail an aqueous two-phase separation system with PEG, dextran, or polyvinyl alcohol for viral vector purification. But the high cost of recycling the organic solutions has restricted the technique to be acceptable, in general. Currently, membrane-based tangential flow filtration has been gaining acceptance for the purification of viruses for vaccine production. The hollow-fiber crossflow membrane system is having good quality for the safe handling of virus particles and shear-sensitive proteins, so the structural entity of a virus vector (like RV) intact. Although membrane filtration is widely acceptable, due to its fouling at high pressure, loss of viral infectivity is possible due to shear forces that may act on the viral envelope.

Size exclusion chromatography (SEC) is also a suitable alternate method for virus vector purification. For example, RVs, which are typically in the order of 100 nm in diameter, have been separated from smaller biomolecules using crosslinked agarose SEC media.

A variety of chromatography-based virus purification methods are in practice. Adsorption of virus particles to solid phases, in fact, provides a convenient and practical choice for fractionating and recovering viruses from cell and media contaminants. Chromatographies also have certain advantages in large-scale virus purification: maintenance of high flow rates; biological activity of labile viruses is often preserved; scale-up is relatively easy; large volumes of cell lysates can be processed; and the overall cost of operation is relatively low and is easily monitored.

Affinity chromatography is an alternate method for the separation of two molecules based on specific and reversible binding properties. In immunoaffinity chromatography, binding occurs between an antigen and a matrix-bound antibody that will recognize a structural surface component on the virus. This procedure can render a high viral yield in a single purification step, thus simplifying the following downstream process. However, the high costs associated with antibody purification and immobilization may be prohibitive for industrial-scale use. But several viral vectors like adeno-AAV can be purified up to 99% through affinity chromatography with a lower number of transfection-deficient viral vectors.

For better results, the chromatography purification process is associated with some other methods, on the basis of a size-based separation procedure. For example, by a sequence of membrane ultrafiltration, SEC in group separation mode, followed by anion exchange chromatography in a negative mode, human influenza virus could be purified to a level of purity that is compatible with traditional preparations based on density-gradient centrifugation.

6.9.5.2 Formulation of vaccine

Formulation, final filtration, and filling are the last steps in viral vector production. For viral vector stability, formulation design developed by optimizing solution conditions and types of high-quality buffers, salts, and stabilizers, low in bioburdern and high-quality excipients, can significantly increase viral vector stability and self-life. For this purpose, high-quality buffers, salts, and stabilizers, low in bioburdern and endotoxins, are specially developed for high-risk applications and can ensure successful downstream processing and formulation of viral vectors. The active ingredients of vaccines help in stimulating immune response in the host cells of the body against harmful viruses, bacteria, and other pathogens. In order to bring structure and function stability and accelerate the response, all excipients to be incorporated in vaccine formulation are subjected to rigorous assessment to ensure the substances are safe in the quantities used. The main ingredient in most vaccines is water for injection. The other components used in vaccine formulation are used in very small quantities. Following are the different types of ingredients and the roles they play.

6.9.5.3 Active ingredients

Antigens

Antigens are the core that make vaccines work in triggering immune response. In the case of vaccines, the antigen can be the whole-inactivated or weakened virus or bacterium or any part (proteins/nucleic acids) of them.

Adjuvants

Vaccine adjuvants boost the immune system's response to the antigen. An adjuvant is an ingredient used in some vaccines, which helps create a stronger immune response in people receiving the vaccine. In other words, adjuvants help vaccines work better by keeping the antigen at the injection site for longer or by stimulating nearby immune cells. For instance, many vaccines contain aluminum salts, which slow down the release of the antigens from the vaccine once it is injected, strengthening and lengthening the immune response. They also help prevent proteins in the vaccine from sticking to container walls during storage. The amount of aluminum present in vaccines is small and well below the maximum levels considered safe for humans. Aluminum salts are commonly added to foods and other medicines.

Preservatives

The necessary preservatives added in the vaccines are limited to a single dose only. This is to avoid contamination with other pathogens. Commonly, single-dose vial does not require preservatives, but in the case of multidose vials a specific preservative is used. They play a critical role in global vaccination efforts, particularly in routine immunization programs and vaccination campaigns in low- and middle-income countries, where cold storage space for the vaccines is often limited. This is because multidose vials occupy less space and often need to be administered to large numbers of people in a short space of time. Examples of preservatives include 2-phenoxyethanol and thiomersal. The latter, which is also called thimerosal, has in the past raised public concern because it contains a mercury-derived compound called ethylmercury. However, unlike methylmercury, which is extremely toxic and can accumulate in the body, ethylmercury is quickly flushed out of the body, and extensive research has shown it to be safe to use in vaccines. Despite the overwhelming scientific evidence in its favor, some governments have removed it from vaccines to reassure the public.

Emulsifiers and stabilizers

Emulsifiers like polysorbate 80 help in keeping the vaccine formulation ingredients in suspension state. Some emulsifiers can also act as adjuvants by helping to boost the immune response to the vaccine. Stabilizers, such as sorbitol, which is found naturally in the body, as well as fruits and barriers, are used to protect the active ingredients in vaccine components from sticking to the walls of their containers. Other examples include gelatin, sugars, such as lactose or sucrose, amino acids, such as glycine, or proteins, such as recombinant human albumin, which is manufactured in yeast.

Residuals

In vaccine manufacturing process, residual trace amounts of materials used are left and later removed. These include cell culture media, such as egg white, inactivating ingredients used to kill viruses or make toxins safe, such as formaldehyde, or antibiotics used to prevent contamination.

6.10 Types of bioreactors used in virus/virus vector culture

Varieties of bioreactors are available for laboratory-scale and turnkey-level production of viruses or viral vectors. The selection of a bioreactor depends on the characteristics of the host cell. Commonly, animal/mammalian cells grow either in suspension or as adherent cells, and several bioreactors have been available at the commercial level for this purpose. On the basis of operational protocols, the bioreactors can be divided into three categories: (i) static systems, (ii) semidynamic systems, and (iii) dynamic systems (Fig. 6.14). Static cultivation systems are commonly used for laboratory purposes. This system is mainly used for screening purposes rather than production. Semidynamic systems retain the cells while supplying them with fresh medium. This system has the provision of power input, supply of nutrients, and agitation controlling device. The most frequently used and well-characterized bioreactors are dynamic cultivation systems. These systems are well equipped with accessories such as stirring system, pH, oxygen bubbling, etc. Adherent cell lines can be shielded from shear forces by using porous carriers, and macroporous carriers are advantageous despite mass transfer limitations. For virus culture, the selection of a bioreactor depends on the nature of the virus and host cell, their growth characteristics, and virus infection kinetics.

6.11 Case studies

Following are a few important and selective examples of virus/viral vector production through bioreactors.

6.11.1 Retroviral/LV production

Manufacturing large-scale, higher-titer, clinical-grade retroviral vectors under cGMP is a prerequisite for the implementation of Phase I/II clinical trials using cell engineering methods. In 1990, the first viral vector employed in gene therapy was introduced as gamma-retroviral vectors. But, at a later phase, interest was developed to use LVs, derived from complex RVs such as the human immunodeficiency virus (HIV) [46,47]. Retroviral and LVs gained more acceptance as gene transfer tools due to their high transducing efficiency in vivo and in vitro and the ability to permanently modify the genetic content of the target cell, sustaining a long-term expression of the delivered gene [48,49]. During the 2010s, retroviral and LVs represented 23% of all the vector types and 33% of the viral vectors used in gene therapy clinical trials. About a decade ago, retroviral vectors were the blockbuster vectors for the

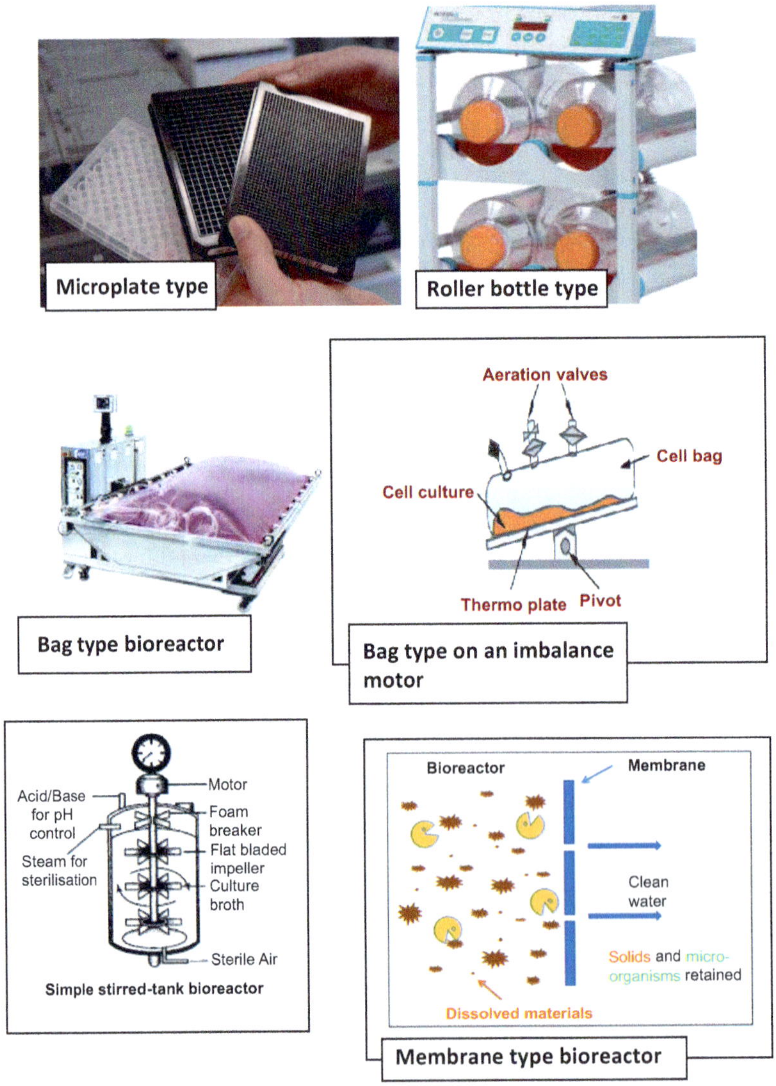

FIG. 6.14

Different types of bioreactors for virus/viral vector culture.

treatment of monogenic and infectious diseases and gene-marking clinical trials. The establishment of retroviral and lentiviral producer cells, named packing cell limes, has been based on the physical separation of the viral genome into different transcriptional units to minimize the risk of generating replication-competent particles (RCPs).

Currently, stirred tank bioreactors are widely used for suspension culture of RVs and lentiviruses. As compared to static/adherent conditions, the most important advantage of a stirred tank bioreactor is its scalability for higher productivity since suspension culture in the stirred system presents increased ratios of cell numbers per volume of culture medium. In addition, it is convenient to provide provision for a well-agitated system for homogeneous cell suspension and maintaining other physical necessities such as pH, oxygen, carbon dioxide, and temperature.

High-titer retroviral vector production by suspension systems was based on a T-lymphoblastoid cell line using a third-generation packing construct. This was based on murine leukemia virus (MLV)-derived retroviral vectors pseudotyped with an amphotropic envelope: CEM FLY cells [1,50]. After CHEMFLYA, many other viral vectors were produced with high-titer value with or without serum-containing media [5,51]. This was followed by RV and LV manufacturing through suspension and under serum-free conditions [6–8,52–54]. For clinical trial purposes, in the culture of retroviral and LVs, adherent static systems (T-flasks and roller bottles) are more acceptable than suspension system.

The stability of biological vectors during mass scale culture mainly depends on physicochemical parameters such as pH, temperature, osmolarity, and O_2 and CO_2 concentration. The optimal cell culture parameters have shown to be producer cell line and viral vector dependent.

The supplementation of mammalian cell culture media with animal sera has been common practice in biomedical and biotechnological research, since it provides critical nutrients and factors that support cell growth and proliferation. But due to its ill-defined composition and high batch-to-batch variability of serum, it is not commonly recommended for retroviral and LV culture. The residual serum is to be removed from the medium and/or viral preparations to prevent an immunological response in the patients. Still, retroviral and LV manufacture has been reported to rely on a considerable amount (5%–10% per v/v of animal sera in the culture medium [9,55]. The majority of the latest generation of packaging cell lines, especially the HEK293 (human embryo kidney) derived required high concentrations of serum in the culture medium to support elevated viral productivities for long-term culture [50,56–58]. The need for serum for retroviral and LV production has been mainly associated with the lipidic needs of packaging cell lines.

Glucose is a traditional carbohydrate source in animal cell culture. It is commonly used in the culture media for the production of retroviral and LVs. In combination with glutamine, glucose is the major energy and carbohydrate source in the culture medium. It is one of the main components in the culture media for animal culture, since glucose cellular transporters are present in the majority of mammalian cell types. However, glucose is rapidly consumed and inefficiently metabolized to lactate. This process of high glucose consumption may have adverse effects on productivity. In order to overcome this problem, alternative substitutes for sugar can be used to decrease lactate production. The use of fructose and galactose was shown to improve retroviral production in Te FLY A7, Te FLY Ga 18, PG13, and Tel CeB cell lines [59,60].

6.11.2 Subunit manufacturing

As discussed earlier, the insertion of foreign genetic material(s) into a side cell can be conducted either in "in vivo" or "ex vivo" process. In ex vivo applications, cells are modified in culture and thereafter, prior to re-infusion, can be used for cell expansion and analytical characterization, and they are also examined pretreatment and posttreatment for viability, density, expression level, etc. Often, ex vivo process is proven to be more efficient for transduction.

Varieties of viral vectors are in use, on the basis of suitability as expression systems. This is mainly dependent on many factors that include tissue tropism, desire for integrating or nonintegrating modification, in vivo or ex vivo process, prior immune exposure, whether the target cells are replicating or nonreplicating, and safety.

Commonly, subunit vaccines are derived from living resources such as bacteria and yeasts. Specific microorganisms of interest are grown under prescribed norms as depicted in an approved protocol by international monitoring agencies. This makes them more expensive to produce than chemically synthesized vaccines, such as RNA vaccines. The precise manufacturing method depends on the type of subunit vaccine being produced. Protein subunit vaccines, such as the recombinant hepatitis B vaccine, are made by inserting the genetic code for the antigen into yeast cells, which are relatively easy to grow and capable of synthesizing large amounts of protein. The yeast is grown in turnkey-level fermentation tanks and split open, allowing the antigen to be harvested. This purified protein is then added to other vaccine components, such as preservatives to keep it stable, and adjuvants to boost the immune response. For polysaccharide or conjugate vaccines, the polysaccharide is produced by growing bacteria in industrial bioreactors before splitting them open and harvesting the polysaccharide from their cell walls. In the case of conjugate vaccines, the protein that the polysaccharide is attached to must also be prepared by growing a different type of bacteria in separate bioreactors. Once its proteins are harvested, they are chemically attached to the polysaccharide, and then the remaining vaccine components are added.

References

[1] E. Sasso, A.M. D'Alise, N. Zambrano, E. Scarselli, A. Folgori, A. Nicosia, New viral vectors for infectious diseases and cancer, Semin. Immunol. 50 (2020), 101430, https://doi.org/10.1016/j.smim.2020.101430. PMID 33262065.

[2] D. Shaofeng, L. Hui, C. Pin, L. Yuwan, L. Zhaoyao, F. Shuangqi, W. Keke, L. Xiaowen, C. Wenxian, Q. Yuwei, Y. Lin, C. Jinding, Viral vector vaccine development and application during the COVID-19 pandemic, Microorganisms 10 (7) (2022) 1450.

[3] U.S. Centers for Disease Control and Prevention, Understanding and Explaining Viral Vector COVID-19 Vaccines, 25 February, 2021.

[4] F. Wang, Z. Qin, H. Lu, S. He, J. Luo, C. Jin, X. Song, Clinical translation of gene medicine, J. Gene Med. 21 (7) (2019), e3108, https://doi.org/10.1002/jgm.3108. 31246328.

[5] M.B. Mathews, Interactions between viruses and the cellular machinery for protein synthesis, in: J.W.B. Hershey, M.B. Mathews, N. Sonenberg (Eds.), Translational Control, Cold Spring Harbor Laboratory Press, Plainview, NY, 1996, pp. 505–548.

[6] J.C. Alwine, Modulation of host cell stress responses by human cytomegalovirus, Curr. Top. Microbiol. Immunol. 325 (2008) 263–279.

[7] P. Barozzi, L. Potenza, G. Riva, D. Vallerini, C. Quadrelli, R. Bosco, et al., B cells and herpesviruses: a model of lymphoproliferation, Autoimmun. Rev. 7 (2) (2007) 132–136.

[8] D. Subramanya, P.D. Grivas, HPV and cervical cancer: updates on an established relationship, Postgrad. Med. 120 (4) (2008) 7–13.

[9] J. Sinclair, Human cytomegalovirus: latency and reactivation in the myeloid lineage, J. Clin. Virol. 41 (3) (2008) 180–185.

[10] M.C. Jordan, G.W. Jordan, J.G. Stevens, G. Miller, Latent herpesviruses of humans, Ann. Intern. Med. 100 (6) (1984) 866–880.

[11] J.G. Sissons, M. Bain, M.R. Wills, Latency and reactivation of human cytomegalovirus, J. Infect. 44 (2) (2002) 73–77.

[12] D.H. Crawford, Viruses: A Very Short Introduction, Oxford University Press, US, 2011, p. 16, ISBN: 978-0-19-957485-8.

[13] T. Ura, K. Okuda, M. Shimada, Developments in viral vector-based vaccines, Vaccine 2076-393X, 2 (3) (2014) 624–641, https://doi.org/10.3390/vaccines2030624. PMC 4494222.

[14] J.X. Li, L.H. Hou, F.Y. Meng, S.P. Wu, Y.M. Hu, Q. Liang, et al., Immunity duration of a recombinant adenovirus type-5 vector-based Ebola vaccine and a homologous prime-boost immunisation in healthy adults in China: final report of a randomised, double-blind, placebo-controlled, phase 1 trial, Lancet Glob. Health 5 (3) (2017) e324–e334, https://doi.org/10.1016/S2214-109X(16)30367-9.

[15] T. Travieso, J. Li, S. Mahesh, J. Da Mello, R.E. Fonzeca, M. Blasi, The use of viral vectors in vaccine development, npj Vaccines 2059-0105, 7 (1) (2022) 75, https://doi.org/10.1038/s41541-022-00503-y.

[16] J. Sadoff, G. Gray, A. Vandebosch, V. Cárdenas, G. Shukarev, B. Grinsztejn, P.A. Goepfert, C. Truyers, H. Fennema, B. Spiessens, K. Offergeld, G. Scheper, K.L. Taylor, M.L. Robb, J. Treanor, Safety and efficacy of single-dose Ad26.COV2.S vaccine against Covid-19, N. Engl. J. Med. 0028-4793, 384 (23) (2021) 2187–2201, https://doi.org/10.1056/NEJMoa2101544.

[17] D.A. Jackson, R.H. Symons, P. Berg, Biochemical method for inserting new genetic information into DNA of simian virus 40: circular SV40 DNA molecules containing lambda phage genes and the galactose operon of Escherichia coli, Proc. Natl. Acad. Sci. USA 69 (10) (1972) 2904–2909, https://doi.org/10.1073/pnas.69.10.2904.

[18] H. Fausther-Bovendo, G.P. Kobinger, Pre-existing immunity against Ad vectors: humoral, cellular, and innate response, what's important? Hum. Vaccines Immunother. 10 (10) (2014) 2875–2884, https://doi.org/10.4161/hv.29594.

[19] D.H. Barouch, S.V. Kik, G.J. Weverling, R. Dilan, S.L. King, L.F. Maxfield, et al., International seroepidemiology of adenovirus serotypes 5, 26, 35, and 48 in pediatric and adult populations, Vaccine 29 (32) (2011) 5203–5209, https://doi.org/10.1016/j.vaccine.2011.05.025.

[20] D.D. Pinschewer, Virally vectored vaccine delivery: medical needs, mechanisms, advantages and challenges, Swiss Med. Wkly. 147 (3132) (2017), https://doi.org/10.4414/smw.2017.14465.

[21] World Health Organisation, The Oxford/AstraZeneca (ChAdOx1-S [recombinant] vaccine) COVID-19 vaccine: what you need to know; 2022. http://www.Who.int>vaccine>astrazeneca.

[22] Clinical trial Number NCT04400838 for "Investigating a Vaccine Against COVID-19" at ClinicalTrials.gov.

[23] University of Oxford, A phase 2/3 study to determine the efficacy, safety and immunogenicity of the candidate coronavirus disease (COVID-19) vaccine ChAdOx1 nCoV-19, in: EU Clinical Trials Register, European Union, 21 April 2020. EudraCT 2020-001228-32. Archived from the original on 5 October 2020.

[24] A. Chauhan, A. Agarwal, N. Jaiswal, M. Singh, ChAdOx1 nCoV-19 vaccine for SARS-CoV-2, Lancet 396 (10261) (2020) 1485–1486, https://doi.org/10.1016/S0140-6736(20)32271-6.

[25] J. Corum, Z. Carl, How Gamaleya's vaccine works, The New York Times (2021). 28 January.

[26] Clinical Trial Number NCT04436471 for "An Open Study of the Safety, Tolerability and Immunogenicity of the Drug 'Gam-COVID-Vac' Vaccine Against COVID-19" at ClinicalTrials.gov.

[27] Clinical Trial Number NCT04436276 for "A Study of Ad26.COV2.S in Adults" at ClinicalTrials.gov.

[28] Clinical Trial Number NCT04505722 for "A Study of Ad26.COV2.S for the Prevention of SARS-CoV-2-Mediated COVID-19 in Adult Participants" at ClinicalTrials.gov.

[29] FDA Briefing Document Janssen Ad26.COV2.S Vaccine for the Prevention of COVID-19 (PDF) (Report). U.S. Food and Drug Administration (FDA).

[30] F.C. Zhu, X.H. Guan, Y.H. Li, J.Y. Huang, T. Jiang, L.H. Hou, et al., Immunogenicity and safety of a recombinant adenovirus type-5-vectored COVID-19 vaccine in healthy adults aged 18 years or older: a randomised, double-blind, placebo-controlled, phase 2 trial, Lancet 396 (10249) (2020) 479–488, https://doi.org/10.1016/S0140-6736(20)31605-6.

[31] Clinical Trial Number NCT04566770 for "Phase IIb Clinical Trial of A COVID-19 Vaccine Named Recombinant Novel Coronavirus Vaccine (Adenovirus Type 5 Vector)" at ClinicalTrials.gov.
[32] A. Roberts, E. Kretzschmar, A.S. Perkins, J. Forman, R. Price, L. Buonocore, et al., Vaccination with a recombinant vesicular stomatitis virus expressing an influenza virus hemagglutinin provides complete protection from influenza virus challenge, J. Virol. 72 (6) (1998) 4704–4711, https://doi.org/10.1128/JVI.72.6.4704-4711.1998.
[33] C. Martínez-Romero, A. García-Sastre, Against the clock towards new Ebola virus therapies, Virus Res. 209 (2015) 4–10, https://doi.org/10.1016/j.virusres.2015.05.025.
[34] W.Y. Choi, K.J. Hong, J.E. Hong, W.J. Lee, Progress of vaccine and drug development for Ebola preparedness, Clin. Exp. Vaccine Res. 4 (1) (2015) 11–16, https://doi.org/10.7774/cevr.2015.4.1.11.
[35] J.A. Regules, J.H. Beigel, K.M. Paolino, J. Voell, A.R. Castellano, Z. Hu, et al., A recombinant vesicular stomatitis virus Ebola vaccine, N. Engl. J. Med. 376 (4) (2017) 330–341, https://doi.org/10.1056/NEJMoa1414216.
[36] S.P. Goff, P. Berg, Construction of hybrid viruses containing SV40 and lambda phage DNA segments and their propagation in cultured monkey cells, Cell 9 (4 PT 2) (1976) 695–705, https://doi.org/10.1016/0092-8674(76)90133-1.
[37] R.M. Hoffman, The Beginning of Tissue Culture, Elsevier SciTech Connect, 2016 (Retrieved November 29, 2022).
[38] R.G. Harrison, The Embryo Project Encyclopedia, 1870–1959. embryo.asu.edu.
[39] Alexis Carrel's tissue culture techniques. The Embryo Project Encyclopedia. embryo.asu.edu.
[40] M. Jedrzejczak-Silicka, History of Cell Culture, IntechOpen, 2017, ISBN: 978-953-51-3134-2.
[41] R. Skloot, The Immortal Life of Henrietta Lacks, Crown Publishers, New York, 2010. ISBN:978-1-4000-5217-2. OCLC 326529053.
[42] Y. Genzel, U. Reichl, Vaccine production, in: R. Pörtner (Ed.), Animal Cell Biotechnology: Methods and Protocols, Humana Press, Totowa, NJ, 2007, pp. 457–473.
[43] E.A. Chiocca, Oncolytic viruses, Nat. Rev. Cancer 2 (12) (2002) 938–950.
[44] M.A. Kotterman, T.W. Chalberg, D.V. Schaffer, Viral vectors for gene therapy: translational and clinical outlook, Annu. Rev. Biomed. Eng. 17 (1) (2015) 63–89.
[45] K.S. Sanber, S.B. Knight, S.L. Stephen, R. Bailey, D. Escors, J. Minshull, et al., Construction of stable packaging cell lines for clinical lentiviral vector production, Sci. Rep. 5 (2015) 9021.
[46] L.P. Hensel, M. Emerman, Human immunodeficiency virus infection of cells arrested in the cell cycle, EMBO J. 11 (8) (1992) 30538.
[47] L. Naldini, U. Blomer, P. Gallay, D. Ory, R. Mulligan, F.H. Gage, V.I.M. Trono, In vivo gene delivery and stable transduction of nondividing cells by a lentiviral vector, Science 272 (5259) (1996) 2637.
[48] A.S. Coroadinha-Norton Gama, L. Amaral, A.I. Hauser, H. Alves, P.M.P.E. Cruz, Production of retroviral vectors: review, Curr. Gene Ther. 10 (6) (2010) 45673.
[49] M. Schweizer, O. Merten, W., Large-scale production means for the manufacturing of lentiviral vectors, Curr. Gene Ther. 10 (6) (2010) 47486.
[50] M. Pizzato, O. Merten, W. Blair, E.D. Takeuchi, Development of a suspension packaging cell line for production of high titre, serum-resistant murine leukemia virus vectors, Gene Ther. 8 (10) (2001) 73745.
[51] K. Ghani, S. Cottin, A. Kamen, M. Caruso, Generation of a high-titer packaging cell line for the production of retroviral vectors in suspension and serum-free media, Gene Ther. 14 (24) (2007), 170511.
[52] K. Ghani, A. Garnier, H. Coelho, J. Transfiguracion, P. Trudel, A. Kamen, Retroviral vector production using suspension-adapted 293GPG cells in a 3L acoustic filter-based perfusion bioreactor, Biotechnol. Bioeng. 95 (4) (2006) 65360.
[53] S. Ansorge, S. Lanthier, J. Transfiguracion, Y. Durocher, O. Henry, A. Kamen, Development of a scalable process for high-yield lentiviral vector production by transient transfection of HEK293 suspension cultures, J. Gene Med. 11 (10) (2009) 86876.

[54] S. Broussau, N. Jabbour, G. Lachapelle, Y. Durocher, R. Tom, J. Transfiguracion, R. Gilbert, B. Massie, Inducible packaging cells for large-scale production of lentiviral vectors in serum-free suspension culture, Mol. Ther. 16 (3) (2008) 5007.

[55] T.S. Mc, M. Al-Rubeai, Effects of culture parameters on the production of retroviral vectors by a human packaging cell line, Biotechnol. Prog. 16 (5) (2000) 85965.

[56] L.M. Chan, C. Coutelle, M. Themis, A novel human suspension culture packaging cell line for production of high-titre retroviral vectors, Gene Ther. 8 (9) (2001), 697703.

[57] P.A. Gerin, M.G. Gilligan, P.F. Searle, M. Al-Rubeai, Improved titers of retroviral vectors from the human FLYRD18 packaging cell line in serum- and protein-free medium, Hum. Gene Ther. 10 (12) (1999), 196574.

[58] A.F. Rodrigues, M. Carmo, P.M. Alves, A.S. Coroadinha, Retroviral vector production under serum deprivation: the role of lipids, Biotechnol. Bioeng. 104 (6) (2009), 117181.

[59] M.J.T. Carrondo, O.W. Merten, M. Haury, P.M. Alves, A.S. Coroadinha, Impact of retroviral vector components stoichiometry on packaging cell lines: effects on productivity and vector quality, Hum. Gene Ther. 19 (2) (2008), 199210.

[60] A.S. Coroadinha, J. Ribeiro, A. Roldao, P.E. Cruz, P.M. Alves, O.W. Merten, M.J. Carrondo, Effect of medium sugar source on the production of retroviral vectors for gene therapy, Biotechnol. Bioeng. 94 (1) (2006) 2436.

CHAPTER

Quality control

7

Abbreviations

ADME	absorption distribution metabolism and excretion
AIQ	analytical instrument qualification
BMV	bioanalytical method validation
CV	control valve
CAPAs	corrective and preventive actions
CC	calibration curve
cGMP	cGood manufacturing process
DMRs	device master records
DQ	design qualification
eQMS	enterprise quality management system
GLPs	good laboratory practices
ISO	International Organization for Standardization
IDA	isotope dilution assays
IQ	installation qualification
LT	level transmitter
LC	level controller
LOD	limited of detection
LOQ	limit of quantification
NDA	new drug applications
NAA	neutron activation analysis
OQ	operational qualification
OECD	Organization for Economic Co-operation and Development
PD	pharmacodynamics
PDCA	Plan-Do-Check-Act
PK	pharmacokinetics
PLAs	product license applications
PQ	performance qualification
QC	quality control
TPS	Toyot production system
TQM	total quality management

Techniques for Downstream process for Biologic Drugs and Vaccines. https://doi.org/10.1016/B978-0-443-19157-2.00008-X

7.1 Quality control

The commercial significance of quality control (QC) is to keep a company in a sustainable pattern of services with the assurance of quality product and to increase the trust of consumers. It also ensures that the company stands for factual data and research to maintain continuity of product quality on the basis of day-to-day demand.

Quality is not referred to exclusively for its singular definition. Despite the relative meaning of "value" QC is cumulating represents the entire process through which product/services are assessed to fulfill the norms for quality as prescribed by international QC authority (Fig. 7.1).

So, quality is a relative word, which can be interpreted in variety of ways. Quality process is a variable practice, which may differ in different industries. For instance, biopharmaceutical product control system depends on the nature of the product and its resources.

So, the crucial goals of QC are: (1) to ensure that products are as uniform as possible and (2) to maintain consistency in product quality without any compromise. Both of these crucial goals are mandatory in biopharmaceutical products, as they play critical roles in life-saving processes under constraint.

Generally and particularly for biopharmaceuticals, quality testing is part and parcel of every stage of a manufacturing or business process. Testing a product during various production phases and until it reaches the end user can help in identifying the cause of production problems and making necessary amendments, on the basis of feedback information. Customer service reviews, questionnaires, surveys, inspections, and audits are a few follow-up examples of quality testing procedures.

7.2 Types of quality control

There are four types of QC: (i) process control, (ii) control chart, (iii) acceptance sampling, and (iv) product QC.

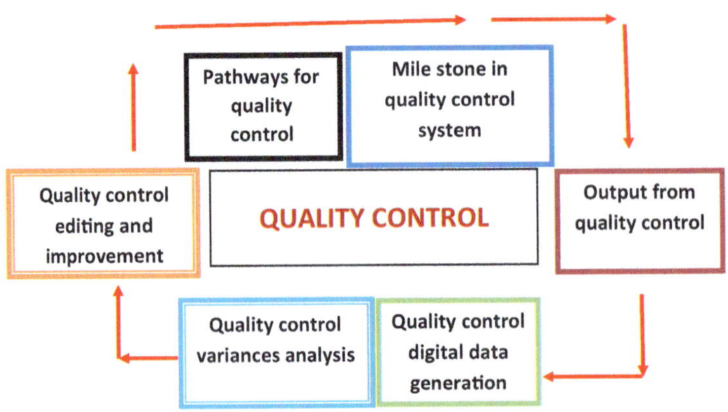

FIG. 7.1

Comprehensive aspects of quality control.

7.2.1 Process control

Process control refers to continuous production processes with consistency in quality production. In the case of biopharmaceutical companies engaged in the production of life-saving drugs and healthcare therapeutic molecules, high-level process care is mandatory.

Due to the complexity of the structure of biologic drugs and their sensitivity to stress and strain, the manufacturing process is more complex and highly expensive. The development of large-scale industrial processes is highly challenging and risky. However, development of complex microchip-based electronic regulatory process has made the process control more autonomous than manual operating process (Fig. 7.2).

For example, a biologic drug is highly temperature-sensitive. The biologic is produced by culturing host cells in a temperature-regulated fermentation system. Besides, the other physical factors like pH, stirring process, oxygen bubbling, etc., necessary for keeping favorable conditions are also regulated through complex auto-regulatory systems.

Overall, process system is composed of three main factors: the manipulated variables, disturbances, and the controlled variables (Fig. 7.3).

Typical manipulated variables are valve position, motor speed, damper position, and blade pitch. The controlled variables are those conditions, such as temperature, level, position, pressure, pH, density, moisture content, weight, and speed that must be maintained at some desired value. For each controlled variable there is an associated manipulated variable. The control system must adjust the manipulated variables, so that the desired value or "set point" of the controlled variable is maintained despite any disturbances.

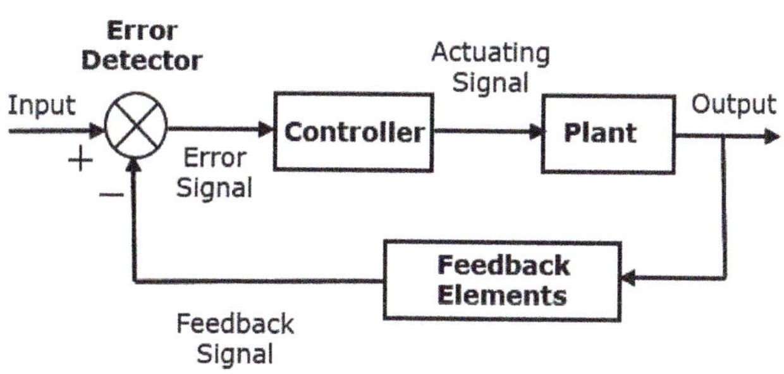

FIG. 7.2

Showing the block diagram of closed-loop system or process control system. The process control system consists of process or plant, feedback elements, error detector, automatic controller, actuator, or control element.

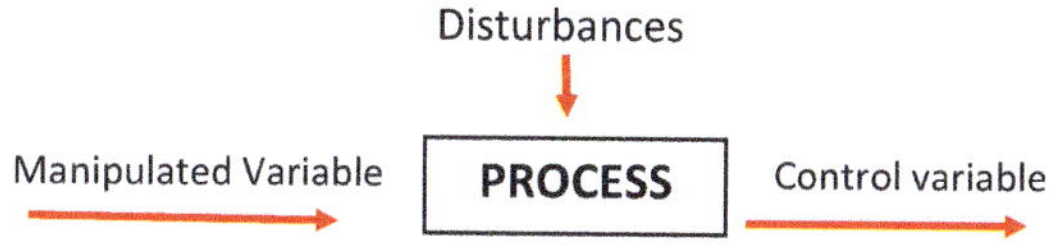

FIG. 7.3

Main factors for process control system.

7.2.2 Element of a process control system

The process control system consists of three elements: (i) level transmitter (LT), (ii) level controller (LC), and (iii) control valve (CV) (Fig. 7.4).

These elements are used to control the liquid level in a process tank. The height of the liquid level (H) is managed by this control system. The level transmitter is a device that measures the fluid level in the tank and converts it into a useful measurement signal, which is sent to a level controller. The level controller evaluates the measurement, compares it with a desired set point (SP), and produces a series of corrective actions that are set to the control valve. The valves control the flow of fluid in the outlet pipe to maintain a level in the tank.

A process control system consists of four essential elements: process, measurement, evaluation, and control (Fig. 7.5). The followings are the elements of process control for quality production.

7.2.2.1 Process

Commonly, process is refers to the functional aspect of assembly equipment for the sequential manufacturing process consisting of the liquid in the tank, the flow of liquid into and out of the tank, and the inlet and outlet piping system. Any given process can involve many dynamic variables, and it may be desirable to control all of them. However, in most cases, controlling only one variable will be sufficient to control the process within acceptable limits. One occasionally encounters a multivariable process in which many variables, some interrelated, require regulation.

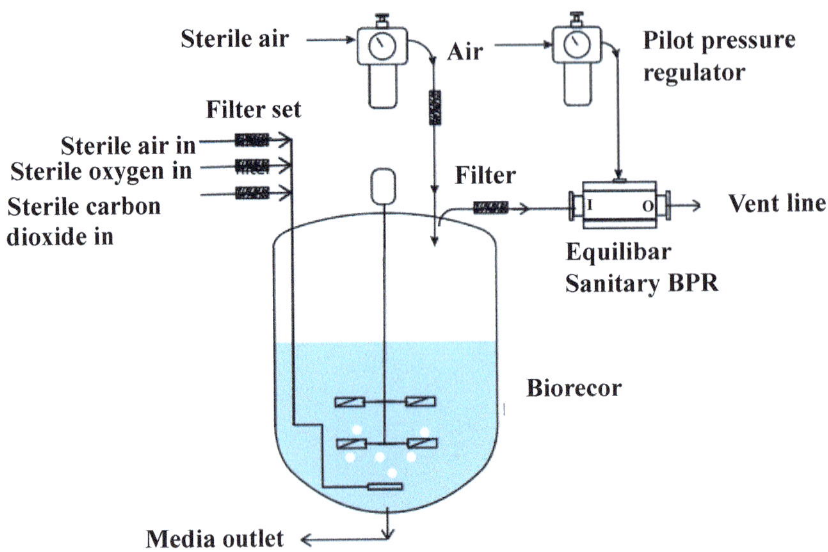

FIG. 7.4

The valve controlling system in a bioreactor.

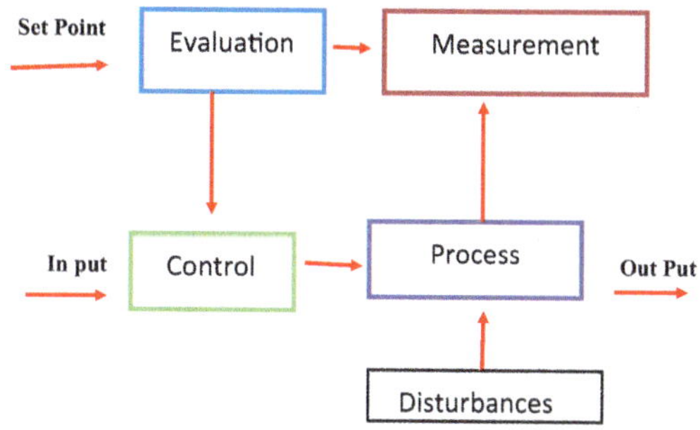

FIG. 7.5

Essential elements for process control.

7.2.2.2 Measurement

Commonly, typical measurements are pressure, level, temperature, flow, position, and speed. The result of any measurement is the conversion of a dynamic variable into some proportional information that is required by the other elements in the process control loop or sequence.

7.2.2.3 Evaluation

In this process, the measurement values of different parameters are examined with reference to a set of desired values or set points, and the amount of corrective action needed to maintain proper control is determined. All the measure values are integrated with the help of controller port to have comparative assessment within different types of variables.

7.2.2.4 Monitor

The process of manufacturing sequence is monitored in a control loop that exerts a direct influence on the process. The central control port receives inputs from the other controller and transforms them into some proportional operation that is performed on the process. In most cases, this final control element will be a control valve that adjusts the flow of fluid in a process. Devices such as electrical motors, pumps, and dampers are also used as control elements.

7.2.3 Control chart

A control chart refers to a graphic expression whether sampled products or processes are meeting their intended specifications rather than the degree of specification. A QC chart that represents a specific character of a product is known as a univariate chart, while a chart expressing variances in several product attributes is called a multivariate chart. Randomly selected products are tested for the given attribute(s) the chart is tracking. The most common form of QC chart is having x-bar (denoted as $\overline{x}$) and y-axis represents the degree to which the variance of the tested attribute is acceptable. The x-axis tracks the samples tested.

7.2.4 Acceptance of sampling

It refers to a statistical measure used in QC. The quality of a batch is determined by random sampling, which has been tested already. This sort of practice helps the company save time and avoid unnecessary expenses. Hence, correctly acceptance sampling is a quality-control measure on which the state of entire quality is expressed. This sort of practice developed during World War II as a quick fix for manufacturing.

Acceptance sampling tests a reprehensive sample of the product for defects. The process involves first, determining the size of a product lot to be tested, then the number of products to be sampled, and finally the number of defects acceptable within the sample batch. Commonly, product sampling test is performed before the final packing is disposed of for shipping.

Several methods are in practice for acceptance sampling. The simplest method involves testing a single unit at random, per x units produced (sometimes called an (n, c) plan). The acceptance is evaluated on the basis of the number of defective units © found in the sample size (n). Other methods involve multiple sampling, which relies on several such (n, c) evaluations. While using multiple samples is more costly, it may be more accurate.

7.2.5 Product quality control

Product quality management ensures a manufacturer's sustainability in the competitive market. This is maintained by a team consisting of skilled workers to create an environment where management and employees strive for perfection. This is done by well-trained personnel, who create benchmarks for product quality and test products to check for statistically.

QC is a process through which a business seeks to ensure that product quality is maintained or improved. The testing units of QC should be well framed with the specific requirements of the final product. The biopharmaceutical industry uses QC methods to ensure customers do not get stuck with their products. QC creates safe measures that can be implemented to make sure deficient or damaged products do not end up with customers.

In a nonmanufacturing business, quality testing can involve customer service evaluations, questionnaires, surveys, inspections, or audits. A business can use any process or method to verify that its end product or service meets the customer's needs and is safe and legal.

7.3 Conceptual development of quality control

The basic concept of QC can be traced back to medieval Europe, when stringent guidelines were framed for tracing defective products. This conceptual development of craftsmanship model with an emphasis on inspections and QC extended through the early years of the Industrial Revolution. In the late 19th century, Frederick Winslow Taylor, a mechanical engineer had drawn the attention of business groups by publishing "The Principles of Scientific Management", which helped in increasing the efficacy of the operation of screening quality product [1].

In 1920s, QC idea was initiated with the introduction of statistical sampling techniques for the analysis of QC methodology, pioneered by engineer Walter A Shewhart—sometimes referred to as the father of statistical QC. During World War II, Walter A Shewhart in collaboration with statistician

William Deming successfully developed methods for inspecting military goods. This enables the armed forces to speed up inspection without compromising product safety or quality. At present, Shewhart's methods (also known as the Shewhart Cycle) serve as the basis for the Plan-Do-Check-Act (PDCA) cycle, which is also effectively applicable as key criteria for quality product screening [2].

At a later stage, the business sector started understanding the drawbacks of ignoring the QC process, and it was realized that an immediate requirement was necessary to develop a more robust, structured, and logical approach to QC. So, in this connection, at the beginning, the business promoters started to inspect the end product, reject defective piece, and take necessary measures to improve the earlier production process. In 1960s, immediately after the devastated World War II, Japanese manufacturers, with the help of Deming and engineer Joseph Juran, started giving intense focus on producing higher quality consumer goods and minimizing raw material waste. Instead of focusing only on defective product screening, Japanese manufacturers adapted a total quality strategy, which pursued the workers for improving operational processes. This practice resulted in Japanese industrialists for managing quality and quantity of products on the basis of market demand [3,4].

Around this same period, Japanese car manufacturer Toyota introduced the Toyot production system (TPS), which was aimed at continually improving the way in which value is delivered to customers. TPS was the precursor to learning manufacturing, which focuses on increasing productivity while minimizing waste.

During 1980s, there was a major setback in the American economy due to its inability to compete with Japan. This resulted in embarrassment for the total quality management (TQM) movement. Consequently, TQM emphasized the need for manufacturers to develop a strategic framework for implementing factual quality production processes across their entire organization and opened a new window of excellence for business in the United States. In 1987, the International Organization for Standardization (ISO) published the ISO 9000 series of quality management standards with the intention of helping companies document and manage the various components of a quality management system so that they could develop trust with customers regarding the safety of the product and its timely availability.

In a later phase, at the end of 1980s, more attention was given to QC development, action was taken to emphasize on quality improvement, and the adoption of new practices, such as just-in-time (JIT). Just-in-time, or JIT, is an inventory management method in which goods are received from suppliers only as they are needed. The main objective of this method is to reduce inventory holding costs and increase inventory turnover. By 1990s, industries around the globe wanted to follow up on QC methodologies to challenge their respectable counterparts in the production of quality products. In 1986, Motorola developed the concept of Six Sigma, which is a set of techniques and tools for process improvement. It was introduced by American engineer Bill Smith while working at Motoral in 1986. The 2000s witnessed the combination of Lean and Six Sigma, plus other continuous improvement methods. The Japanese word for waste is "Muda," which is defined as "uselessness." Lean tools are designed to reduce "Muda" in organizations and improve QC. In other words, Lean tools seek to eliminate processes that are not valuable. Finally, in 2010s, continuous improvement and quality 4.0 were introduced. "Quality 4.0" is a term that refers to the future of quality and organizational excellence within the context of industry 4.0. Quality professionals can play a vital role in leading their organizations to apply proven quality disciplines to new, digital, and disruptive technology.

The last decade has witnessed significant shifts in quality management systems. The 2000 ISO 9000 series of quality management standards were revised to place greater emphasis on customer satisfaction, and again in 2015, the ISO 9001 standard was revised to focus more on risk management. With the modernization of microchip-based information technology the file system management has been replaced with a cloud-based system for ultra-rapid communication system with a wide spectrum of coverage. With cloud-based enterprise quality management system (eQMS) solutions, companies can overcome with communication barriers to quickly resolve issues and drive continuous improvement. Advanced solutions like Arena's product-centric QMS connect product and quality records into a single electronic system. This enables organizations to gain greater control, visibility, and traceability of corrective and preventive actions (CAPAs), device master records (DMRs), standard operating procedures (SOPs), training records, and other quality processes that are essential to meeting regulatory standards, such as FDA, ISO, and EU MDR. Because automated change processes and revision control are applied to this information, companies have greater confidence in passing audits and meeting their commercialization milestones.

7.3.1 Quality control & quality assurance in biopharmaceuticals

QC and quality assurance are two different aspects for cGood manufacturing process (cGMP) (Fig. 7.6).

Quality assurance refers to the implementation of protocols, policies, and procedures during production that help in avoiding defective problems with the finished product. In order to assure the quality of a product, tools, and techniques, such as documentation, checklists, certificates of analysis and validations play important roles.

QC is for testing the processes involved in manufacturing a product, whereas quality assurance involves preventing the defectiveness. Now a days, QC and quality assurance are unavoidable to bring sustainability in manufacturing process. So, all types of industries, including biopharmaceutical strictly follow the QC rules and regulations to bring trust in the consumers for their product.

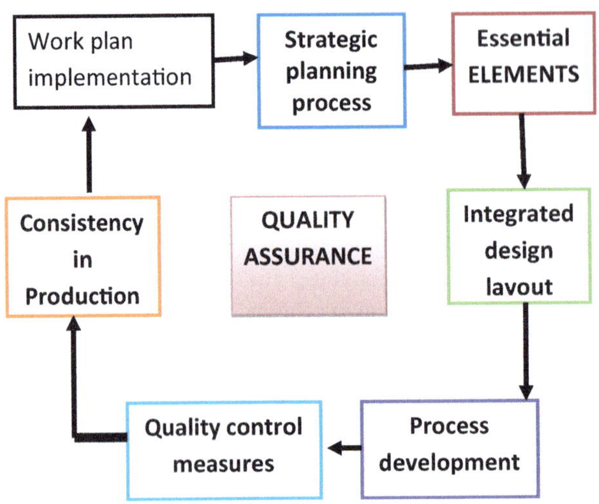

FIG. 7.6

Quality assurance versus quality control.

7.4 How to gain consumer's trust

Consumers appreciate businesses and develop trust in the way they treat other people by holding them to certain standards and building relationships with those they trust. Following are a few tips to gain consumer's trust:

7.4.1 Customer's service

Customer's service and confidence are the basic requirements for sustainable development of goodwill among manufacturers. A good manufacturer pays maximum attention to customer interest, and has a focus on accountability. A company should care enough about its customers to patiently and reliably walk them through any issues they might have with product or service.

7.4.2 Customer's critical review

A manufacturer should have the quality of appreciation in entertaining critical review from consumers on quality of product, and try to bring necessary amendments, accordingly.

7.4.3 Transparency in dealing

The manufacturer should be transparent in QC protocols being followed by internationally approved authority.

7.4.4 Relationship with stakeholders

Top management stakeholders include anyone with responsibility for the strategy and development of a company from c-suite executives to department directors. So, bringing to notice the strategic development of a company, the role of stakeholders is critical and highly essential.

7.4.5 Building relationship with customers

A company must cordially welcome a customer who wants to contact with an issue related to product or service. With the best interest of the company, it is necessary to provide possible information to resolve the problem in marketing the product.

7.5 Tips for identifying & managing project stakeholders

The success of project, solely, depends on the involvement of stakeholders.

It is easy and convenient to manage with a single quality management stakeholder with limited access to company's activities. But, it is rare to have only one stakeholder when a project under operational stage. It is likely to have several quality management stakeholders, both internal and external from CEO to the customer, invested in the outcome of the project.

For operating a hurdle-free project, it is necessary to create a stakeholder management strategy that helps in understanding and managing each stakeholder relationship, so that a project management efficacy is smooth. This could be possible by building a holistic stakeholder management strategy with selective quality management stakeholders and understanding the key differences between them. So, it is necessary to identify and define the internal and external project stakeholder types and offer a few tips on how to manage each one.

7.5.1 Internal quality management stakeholders

7.5.1.1 Top management

Top management stakeholders refers to person(s) responsible for strategy and development of the organization from c-suite executives to department directors. These stakeholders are solely responsible for continuity or discontinuity of a project on performance basis. Following are the few assignments allotted to these stakeholders:

- To design project plans for top management to review and approve.
- To monitor the path ways for time-bound project by eliminating hurdles of different nature.
- To keep a track with other top management stakeholder regarding update project development.
- To distribute assignments to each top stakeholder and instruct them how to prepare notes on update information on digital platform.
- Put a reporting process for the top management in place and stick to a standard delivery schedule.

7.5.1.2 Direct manager

Commonly, boss dictates how a project is to be operated on the basis of availability of the resources. It is desirable to keep transparency in discussion, so that you can manage expectations and build a case for more resources as needed. Following are few tips for influencing the project manager:

- To clarify in detail about his or her expectations as well as how your performance will be measured, on the basis of documented goals and time lines for achievement.
- Set regular meetings to bring the manager's attention about project progress and challenges being encountered while operation.
- Put a reporting process with the manager, directly

7.5.1.3 Internal customers

Anyone within the organization can be an internal customer having the capacity to sign-off a quality management project, after finalization. It is vital important to understand as many details as possible from the internal customer and then negotiate vital deliverables and timelines based on resource availability. Following are few tips for managing internal customers:

- It is foremost necessity to satisfy internal customers by preparing detailed project brief or work sheet, and timely providing them for assessment.
- The written agreement of the project should be circulated among internal customers, after signature.
- Both parties, the project manager and internal customers, should be flexible in bring any changes in the project documented on the basis of quality production.
- The group leader should understand the culture of internal customers and "speak their language" while communicating.

- The project manager should make his or her mind as the main point of contact for the project, but not the team members, to keep transparency in communication

7.5.1.4 Project team

A project can be sustainable in smooth operational conditions, only, with a skilled team on a full- or part-time basis. The structure of project team should be composed of departmental members, borrowed members from other departments, interns, and more. The project manager should have the skill of well presenting the significance of project on the basis of market demand. In addition, the project manager should clearly define the roles and tasks, and to support and motivate them in these efforts on a need-based. It's important to create a culture of collaboration in order to support, learn, and problem-solve together.

Following are few tips for managing the project team:

- The project manager should, always, associate with the team in the planning process to increase project buy-in and encourage members to volunteer for responsibilities that best match their strengths.
- Clearly define and document the responsibilities of each team members, including relevant deadlines.
- The availability of project manager before the project operating team should be ensured, all the time.
- Regular short-term meeting should be kept to resolve the project-related problem, if any, immediately.
- The personal conflict among the members of the project team should be resolved immediately, without delay.

7.5.1.5 Company peers

Company peers include anyone who's at the same level as the lead position in a project, which could include a vast number of people, depending on the structure and function of organization. The specific peers are needed to consider may be assigned to project team or otherwise be invested in the success of the quality management of project.

Here are a few tips for managing company peers:

- It is the responsibility of top management to frame project roles clear to all project team members and emphasize that cooperation is expected from all parties.
- The aim and objectives of the project should be highlighted, clearly.
- Have regular meeting to discuss the progress of each invested peer.
- Act early, if it is clear that a peer is hampering the project in any way, such as being uncooperative or choosing gossip and bad-mouthing over moving the project forward.

7.5.1.6 Internal resource managers

Besides peers, other experts are also available within the organization that can control resources needed for successfully operating project management programs by suitably exploiting the available resources. Fostering goodwill to build a cordial working relationship with any coworkers should be welcomed wholeheartedly.

Followings are few tips for managing internal resource managers:

- On need-based, ask for access to resources for the project operation. This encourages the resource managers to increase their respective moral confidence to plan and adjust schedules for other projects that require same resource.
- Listen actively and show compassion about their workload and challenges to help for goodwill and encourage collaboration.
- Be clear and firm when communicating what is needed for the project, but to remain flexible and willing to compromise as needed.

7.5.2 External quality management stake holders

7.5.2.1 *Government and standards bodies*

In order to meet international market demand, it is mandatory for an industry, in particularly, a biopharmaceutical industry, to follow the QC role and regulation of internal standard. Even if a company is not tightly regulated, there are likely industry best practices that help the company stay certified, and competitive in the market.

Following are few remarkable tips for managing the requirements of government and standard bodies.

- Keep sustainability in QC, especially in case of any hamper in quality management and product safety, while delivering to external customers.
- The documentation should be kept properly for any instant requirements to be projected to customers or regulatory authority.

7.5.2.2 *Suppliers, vendors & contractors*

In case of inadequate internal resources, a company can have the option to hire outside help. On the basis of requirement, this could be anything from consulting with an industry expert to hiring a graphic designer.

Here are tips for managing project suppliers, vendors, and contractors:

- Never try to deliver any goods with spoken promises or goodwill.
- Schedule frequent meetings to check on their progress
- Be aware of any roadblocks or such issues, and keep immediate action for safety of Goods, especially biopharmaceuticals and foods.
- Refer also to the tips for managing peers and internal resource manager above.

7.5.2.3 *External customers*

External customers refer to those people who are the ultimate consumer of the product or availing the service of a company. Indirectly, external consumers are also important stakeholders of a company. The income of revenue, also, depends on avoiding bad press, creating positive experiences through safe and quality products, increasing customer's retention, and gaining new customers are all the examples of success that drives revenue.

Following are few important tips to keep in touch with external customers:

- While bringing amendment in quality, it is advisable to take consents of customers to improve quality of the product.
- The internal stakeholder should be aware of about customer's needs when planning and negotiating projcct dctails.
- In order to bring sustainability in quality management projects and all internal stakeholders is what help you keep brand promises to company's customers.

7.6 Benefits of quality control and quality assurance

In order to bring sustainability in a company, especially with reference to biopharmaceutical company, it is necessary to avail QC and quality assurance practices to bring a company to limelight. QC practice develops and encourages quality consciousness among the workers in the factory, which is greatly helpful in achieving desired level of quality in the product.

Quality products increase the trust of customers and bring the company into the limelight. In order to compete with the competitive contemporary market, a company has to manage manufacturing process control systems rigorously under the stringent protocols as prescribed by regulatory authority. Moreover, QC is necessary because it ensures that a company will look at evidence-based data and research, not just anecdotal observations, to ensure that products are living up to their standard. The most essential requirement of QC is consistency in the routinely work process and product development.

A company's reputation, reliability, and efficiency are all at risk, if QC is ignored or neglected. A product's testing can play a role in marketing and sales as well, since consumers may trust it more.

Receiving best quality product increases satisfaction. This could be possible by integrating operational steps and saving time and unnecessary expenses. QC system can most effectively utilize available resources by minimizing wastage and inefficiency in product output. Practice of quality production satisfy the consumers and indirectly help in increasing morale of employees and induce the feeling that the company where they serve never compromise with quality production. By this, a cordial relationship is established between company and employees. Adaptation with emerging tools and technology, a company can compete with the counter partners with confidence. Effective advertisement on transparency in manufacturing quality product may helpful in winning the public confidence. By introducing QC measures, uniform products of same quality are produced. This greatly facilitates the problem of price fixation. Guarantee in quality output would be immensely helpful in bringing sustainability in market demand for the product. So, QC can be a powerful instrument with the help of which market, both at home and abroad, can be expanded.

7.7 Launching biopharmaceuticals in the market

Biologic drugs are heterogeneous biomolecules derived from the cells of wide variety resources like human, animals, plants, insects, and microorganisms by using biotechnology. Biologic drugs are clinically used for variety of diseases, which are not under the control of traditional pharmaceuticals.

Since the last two decades the market demand of biologic drugs are in increased magnitude due to their high efficacy in curing diseases. So, biological drugs approval by FDA-like regulatory body is in progress with updated data. Table 7.1 depicts the newly FDA-approved biologic drugs for the yeast 2021.

Biopharmaceuticals are highly expensive and consume long time from discovery to marketing. Lots of differentiations are noticed between biopharmaceutical and traditional pharmaceuticals, both in structure and function (Table 7.2).

Table 7.1 Biological drugs approved FDA regulatory body for the year 2021.

No.	Trade name[a]	Active ingredient[a] class indication	Approval date	FDA-approved use on approval date
50.	Adbry	Tralokinumab-ldrm	12/27/2021	To treat moderate-to-severe atopic dermatitis
49.	Leqvio	Inclisiran	12/22/2021	To treat heterozygous familial hypercholesterolemia or clinical atherosclerotic cardiovascular disease as an add-on therapy
48.	Vyvgart	Efgartigimod alfa-fcab	12/17/2021	To treat generalized myasthenia gravis Press Release
47.	Tezspire	Tezepelumab-ekko	12/17/2021	To treat severe asthma as an add-on maintenance therapy
46.	Cytalux	Pafolacianine	11/29/2021	To help identify ovarian cancer lesions Press Release
45.	Livtencity	Maribavir	11/23/2021	To treat post-transplant cytomegalovirus (CMV) infection/disease that does not respond (with or without genetic mutations that cause resistance) to available antiviral treatment for CMV Press Release
44.	Voxzogo	Vosoritide	11/19/2021	To improve growth in children five years of age and older with achondroplasia and open epiphyses Press Release
43	Besremi	Ropeginterferon alfa-2b-njft	11/12/2021	To treat polycythemia vera, a blood disease that causes the overproduction of red blood cells Press Release
42.	Scemblix	Asciminib	10/29/2021	To treat Philadelphia chromosome-positive chronic myeloid leukemia with disease that meets certain criteria
41.	Tavneos	Avacopan	10/7/2021	To treat severe active anti-neutrophil cytoplasmic autoantibody-associated vasculitis (granulomatosis with polyangiitis and microscopic polyangiitis) in combination with standard therapy, including glucocorticoids

Table 7.1 Biological drugs approved FDA regulatory body for the year 2021—cont'd

No.	Trade name	Active ingredient class indication	Approval date	FDA-approved use on approval date
40.	Livmarli	Maralixibat	9/29/2021	To treat cholestatic pruritus associated with Alagille syndrome
39.	Qulipta	Atogepant	9/28/2021	To prevent episodic migraines
38.	Tivdak	Tisotumab vedotin-tftv	9/20/2021	To treat recurrent or metastatic cervical cancer with disease progression on or after chemotherapy
37.	Exkivity	Mobocertinib	9/15/2021	To treat locally advanced or metastatic non-small cell lung cancer with epidermal growth factor receptor exon 20 insertion mutations
36.	Skytrofa	Lonapegsomatropin-tcgd	8/25/2021	To treat short stature due to inadequate secretion of endogenous growth hormone
35.	Korsuva	Difelikefalin	8/23/2021	To treat moderate-to-severe pruritus associated with chronic kidney disease in certain populations
34.	Welireg	Belzutifan	8/13/2021	To treat von Hippel-Lindau disease under certain conditions
33.	Nexviazyme	Avalglucosidase alfa-ngpt	8/6/2021	To treat late-onset Pompe disease Press Release
32.	Saphnelo	Anifrolumab-fnia	7/30/2021	To treat moderate-to severe systemic lupus erythematousus along with standard therapy
31.	Bylvay	Odevixibat	7/20/2021	To treat pruritus
30.	Rezurock	Belumosudil	7/16/2021	To treat chronic graft-versus-host disease after failure of at least two prior lines of systemic therapy
29.	Fexinidazole	Fexinidazole	7/16/2021	To treat human African trypanosomiasis caused by the parasite *Trypanosoma brucei gambiense*
28.	Kerendia	Finerenone	7/9/2021	To reduce the risk of kidney and heart complications in chronic kidney disease associated with type 2 diabetes
27.	Rylaze	Asparaginase erwinia chrysanthemi (recombinant)-rywn	6/30/2021	To treat acute lymphoblastic leukemia and lymphoblastic lymphoma in patients who are allergic to *E. coli*-derived asparaginase products, as a component of a chemotherapy regimen Press Release
26.	Aduhelm	Aducanumab-avwa	6/7/2021	To treat Alzheimer's disease Press Release
25.	Brexafemme	Ibrexafungerp	6/1/2021	To treat vulvovaginal candidiasis
24.	Lybalvi	Olanzapine and samidorphan	5/28/2021	To treat schizophrenia and certain aspects of bipolar I disorder
23.	Truseltiq	Infigratinib	5/28/2021	To treat cholangiocarcinoma whose disease meets certain criteria

Continued

Table 7.1 Biological drugs approved FDA regulatory body for the year 2021—cont'd

No.	Trade name	Active ingredient class indication	Approval date	FDA-approved use on approval date
22.	Lumakras	Sotorasib	5/28/2021	To treat types of non-small cell lung cancer Press Release
21.	Pylarify	Piflufolastat F 18	5/26/2021	To identify prostate-specific membrane antigen-positive lesions in prostate cancer
20.	Rybrevant	Amivantamab-vmjw	5/21/2021	To treat a subset of non-small cell lung cancer Press Release
19.	Empaveli	Pegcetacoplan	5/14/2021	To treat paroxysmal nocturnal hemoglobinuria
18.	Zynlonta	Loncastuximab tesirine-lpyl	4/23/2021	To treat certain types of relapsed or refractory large B-cell lymphoma
17.	Jemperli	Dostarlimab-gxly	4/22/2021	To treat endometrial cancer Press Release
16.	Nextstellis	Drospirenone and estetrol	4/15/2021	To prevent pregnancy
15.	Qelbree	Viloxazine	4/2/2021	To treat attention deficit hyperactivity disorder
14.	Zegalogue	Dasiglucagon	3/22/2021	To treat severe hypoglycemia
13.	Ponvory	ponesimod	3/18/2021	To treat relapsing forms of multiple sclerosis
12.	Fotivda	Tivozanib	3/10/2021	To treat renal cell carcinoma
11	Azstarys	Serdexmethylphenidate and dexmethylphenidate	3/2/2021	To treat attention deficit hyperactivity disorder
10.	Pepaxto	Melphalan flufenamide	2/26/2021	To treat relapsed or refractory multiple myeloma
9.	Nulibry	Fosdenopterin	2/26/2021	To reduce the risk of mortality in molybdenum cofactor deficiency Type A Press Release
8.	Amondys 45	Casimersen	2/25/2021	To treat Duchenne muscular dystrophy Press Release
7.	Cosela	Trilacicilib	2/12/2021	To mitigate chemotherapy-induced myelosuppression in small cell lung cancer Press Release
6.	Evkeeza	Evinacumab-dgnb	2/11/2021	To treat homozygous familial hypercholesterolemia
5.	Ukoniq	Umbralisib	2/5/2021	To treat marginal zone lymphoma and follicular lymphoma
4.	Tepmetko	Tepotinib	2/3/2021	To treat nonsmall cell lung cancer
3.	Lupkynis	Voclosporin	1/22/2021	To treat lupus nephritis Drug Trials Snapshot
2.	Cabenuva	Cabotegravir and rilpivirine (copackaged)	1/21/2021	To treat HIV Press Release Drug Trials Snapshot
1.	Verquvo	Vericiguat	1/19/2021	

[a]*Trade name used in the USA.*

Table 7.2 Different between biologic drugs and traditional pharmaceuticals.

Chemical	Biological
Produced by chemical synthesis	Produced by living cell cultures
Low molecular weight	High molecular weight
Well-defined structure	Complex, heterogeneous structure
Mostly process-independent	Strongly process-dependent
Completely characterized	Impossible to fully characterize the molecular composition and heterogeneity
Stable	Unstable, sensitive to external conditions
Mostly non-immunogenic	Immunogenic

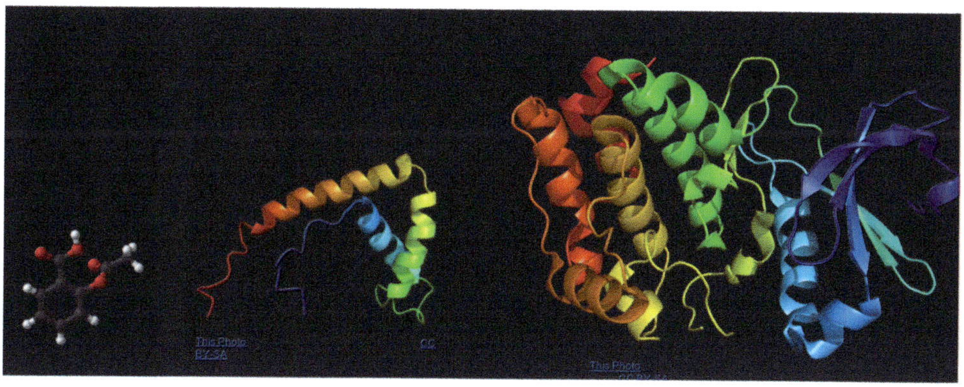

Aspirin 180 Da **Peptide -40aa- 4,400Da** **Antibody 150,000 Da**

FIG. 7.7

Structural complexity of biological drugs (Peptide, Antibody) as compared to a traditional pharmaceutical drug (Aspirin).

Basically, biopharmaceuticals are high molecular complex compounds as compared to traditional generic drugs (Fig. 7.7). The best example to understand biopharmaceutical structure complexity is to compare it with a fighter plane having thousands of parts with manifold functions. But in case of traditional generic drug is like a plastic plane with limited function and structural simplicity.

Due to structural complexity and high-temperature sensitivity, it is necessary to follow protocol as prescribed by some international recognized regulatory bodies. Following are some phases of development to be followed by biopharmaceutical products before reaching to competitive market: biologic drug design; production strategy development for preclinical trials; large-scale production planning for clinical trials, and selection of volunteers for human trials; bringing amendment in structure and function of biologic drugs, based on the feedback from market; and product transfer to customers under safe and secure condition (Table 7.3).

Table 7.3 Product transfer to customers, under safe and secure condition.

Preclinical drug development	An overview of biologic	Product manufacturing	Downstream development
The preclinical phase of drug development refers to the testing of activity and toxicity of a drug candidate in in vitro and in vivo studies before testing in patients can be performed. The preclinical development stage begins when a viable candidate for formal preclinical development has been chosen	Biology is a branch of science that deals with living organisms and their vital processes. Biology encompasses diverse fields, including botany, conservation, ecology, evolution, genetics, marine biology, medicine, microbiology, molecular biology, physiology, and zoology		Downstream refers to the processes applied after extraction through to it being delivered to the customer in whatever format required
Analytical method development	**Drug discovery**	**QA/QC testing**	**Assay development**
Analytical method development is the creation of a set of experimental conditions to perform analytical procedures in chemical samples. Developed analytical methods can be used to identify, separate, quantify, and learn more about the chemical components in drug products intended for commercial manufacturing	In the fields of medicine, biotechnology and pharmacology, drug discovery is the process by which new candidate medications are discovered	While QA testing focuses on providing assurance that quality requested will be achieved, QC testing focuses on fulfilling the quality requested. QA focuses on preventing defect while QC focuses on identifying the defect	Assay development is a procedure in molecular biology for testing or measuring the activity of a drug or biochemical in an organism or organic sample. A quantitative assay may also measure the amount of a substance in a sample. Bioassays and immunoassays are among the many varieties of specialized biochemical assays
Target discovery	**Upstream development**		
Target discovery, which involves the identification and early validation of disease-modifying targets, is an essential first step in the drug discovery pipeline. Indeed, the drive to determine protein function has been stimulated, both in industry and academia, by the completion of the human genome project	Simply put, upstream works include the exploration and production of crude biologic drugs		

7.7.1 Preclinical drug development

Preclinical drug development is a phase of research that begins with animals, before clinical trial on humans to understand feasibility, iterative testing, and drug safety, typically in laboratory condition (Fig. 7.8).

The target of preclinical development to determine favorable starting dose for trial on limited number of human volunteers compare the results obtained from animal system. Only one out of every 5000 compound that enters drug discovery to the stage of preclinical development becomes an approved drug.

In preclinical trials, a specific biologic drug is tested for pharmacodynamics (reaction of drug to the body) (PD); pharmacokinetics (how the body accept the drug) (PK); ADME ("absorption, distribution, metabolism, and excretion"); and toxicology testing. Integrated result obtained from the above data is estimated allometrically to assess the safety of starting dose of the drug for clinical trials in humans.

A satisfied result obtained after an allometric test, medical devices that do not have drugs attached will not undergo these additional tests and may be recommended for direct good laboratory practice (GLP) testing for safety of the device and its components. All types of clinical studies abide to GLP in ICH guidelines to be acceptable for submission to regulatory agencies, such as the Food & Drug Administration in the United States.

For animal testing, typically in drug development studies, murine and canine are used, although primate and porcine are acceptable. The choice of animal species is based on which will give the best correlation to human trials. Most studies are performed on large species, such as dogs, pigs, and sheep, which allow for testing on a similar sized model as that a human. In addition, some selective species are used for similarity in specific organs or organ system physiology (swine for dermatological and coronary stent studies; goats for mammary implant studies; dogs for gastric and cancer studies; etc.). It is mandatory, as per regulatory guidelines of FDA, EMA, and other similar international and regional

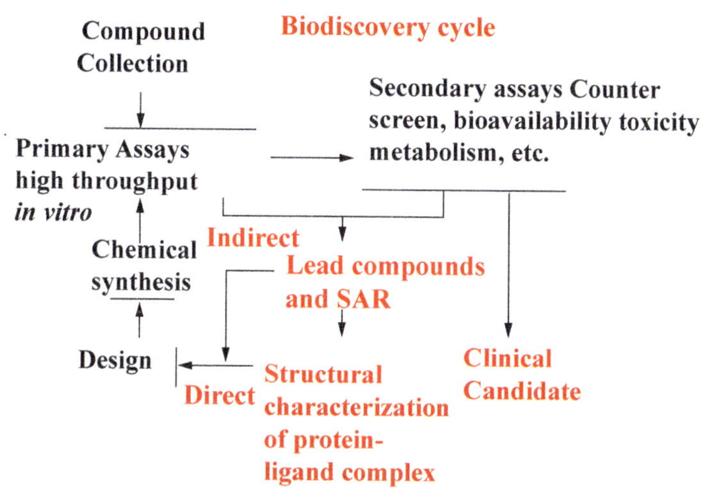

FIG. 7.8

Line diagram showing preclinical drug development.

authorities that safety testing to be conducted in at least two mammalian species, including one non-rodent species, prior to human trials authorization [5].

7.7.2 Analytical methods for preclinical trials

Mostly, the analytical methods for preclinical trials are based on the following categories:

(i) Basic regulatory aspects: Bioanalytical method validation (BMV), as per the guidelines of FDA, include: specificity, selectivity, sensitivity, linearity, accuracy, precision, stability, matrix effects, carry over, and contamination.
(ii) Default batch: It increases the possibility of reliability. Bioanalytical method gives 80% accuracy, excluding execution errors.
(iii) Metabolic stability: Stabilization in metabolic activity is governed by adjusting pH or addition of enzymatic inhibitors.
(iv) Strategic development: This is managed by three important inter-related parts: Sample preparation, chromatography, mass spectroscopy (MS) detection.
(v) Analytical issues: Loss of analyte while collecting samples preparation of QC, storage and analysis due to adsorption to the container should be taken care.
(vi) Automation strategy: Automation brings consistency in results and quickens the overall process.
(vii) Method specificity: It is based on interaction of antibodies with the drug molecules or opposed to interact with nondrug molecules. The evaluation of antibody specificity is referred to cross reactivity of antibody with nonspecific drug molecules including drug metabolites and structurally related drug.
(viii) Screening of methods: Commonly, drug method development is performed based on a minimum of 10 individual samples from 10 normal animals that are assayed attached or detached with a drug at lower limit concentration of calibration curve (CC) [6].
(ix) Minimum dilution point: It is referred to the dilution point at which matrix interference is minimized to an acceptable level.
(x) Biological matrix stability: Commonly, proteolytic enzyme has effect on drug in biological matrix. For instance, the stability of an analyte in rabbit or monkey serum may vary and be highly unstable in rat or mouse serum. In such cases, it is necessary to study in the presence of a cocktail of proteolytic enzyme inhibitors and/or keep samples on ice during analysis [7].

7.7.3 Preclinical analytical methods

The reliability of drug development from the initial stage of synthesis to manufacture is based on the development of preclinical method development. Data on procedural develop of preclinical methods to meet the requirements of new drug applications (NDAs), biologics license applications (BLAs), or product license applications (PLAs), before submitting to FDA to fulfill the norms as prescribed in GLP and ICH guidelines of accuracy and reliability and informational method validation [8]. The active stable pharmaceutical/biopharmaceuticals ingredient with long shelf-life depicts:

- Synthetic contaminants, degradation products-manufacturing, stability, shelf-life forecast.
- Pharmaceutical excipients
- Drug and metabolites in biological fluid and tissues

7.7.4 Analytical method development

(i) Sample handling: The most significant factors for sample handling include [9]:
- Appropriate sample collection
- Preparation of sample for analysis
- Analysis using appropriate method

(ii) Sample preparation: It is the most critical step in the analytical method, in particularly when biological samples are involved. It is necessary to select the most suitable medium for keeping sample stability and its compatibility with the chromatographic system [10].

(iii) Purpose: To make the sample contamination free and dissolve in a suitable solvent for subsequent separation and detection [11].

7.7.5 Analytical techniques for method development

Following techniques are used for analysis of sample:

Nonseparation methods [12,13]

- Absorption and emission spectroscopy, isotope assays, isotope dilution assays (IDA), neutron activation analysis (NAA).
- Separation methods—chromatography (TLC, GC, HPLC), chiral method.

7.7.6 Analytical method validation for clinical analysis

For successful method validation, a qualified instrument is required. Hence, "Analytical Instrument Qualification (AIQ)" should be used instead "Analytical Instrument Validation."

AIQ is divided into four main phases [14]: (i) design qualification (DQ) for testing functional and performance specification; (ii) installation qualification (IQ) for performing documenting the installation in the selected user environment; (iii) operational qualification (OQ) for testing the equipment in the selected user environment to ensure that it meets the previously defined functional and performance specifications; (iv) performance qualification (PQ) for testing that the system consistently performs as intended for the selected application.

Guidelines: ICH, FDA, AOAC, USP, ISO 9000, and 17025

It includes followings:

Q1A: (R2): Stability testing of new drug substances and products [15].
Q1B: Photo stability testing of new drug substances and product [16].
Q1C: Stability testing for new dosage forms [17].
Q1D: Bracketing and matrixing designs for stability testing of drug substances and drug products [18].
Q1E: Evaluation of stability data [8].
Q1F: Stability data package for registration applications in climatic zones III and IV [19].
Q2A: Text on validation of analytical procedures
Q2B Validation of analytical procedure-methodology
Q3AR Impurities in new drug substances
Q3B R: Impurities in new drug product
Q3C: Impurities-guideline for residual solvent [20].
Q9: Quality risk management [21].

7.8 Validation parameters

The efficacy of QC depends on the accuracy of analytical method validation performance [22–24]. It is a practice of performing numerous assessment designed to verify for giving guarantee for beneficial and legit analytical data [25–28]. Process of validation analyze the testing multiple attributes of a method to determine that it may provide useful and valid facts [29–31]. The accuracy of method parameters includes normal test conditions on excipients [32–35]. So, a method validation represents product-specific. Following are 10 important criteria:

(1) Specificity: It is the ability to assess clearly the analyte in the presence of components that may be expected to be present.
(2) Accuracy and precision: It represents between measured and theoretical values. Five samples per concentration should be used to validate.
(3) Limited of detection (LOD): There should be 3:1 signal to noise ratio or three times the standard deviation of the response divided by slope of the calibration curve, while detecting lowest amount of analytes.
(4) Limit of quantification (LOQ): With sufficient accuracy, the lowest and highest concentration should be quantified.
(5) Linearity and range: There is direct variation relationship between response and analyte concentration. Commonly, 5–8 concentrations are used for standard curve. Range refers to concentration between the low and high limit of quantification.
(6) Stability: Stress studies performed by the analytes to acid, base, heat, UV light to assess their stability.
(7) Ruggedness: Refers to reproducibility under normal but variable conditions (different instruments, operators, laboratories, and reagents).
(8) Robustness: Refers to analytical methods ability to remain unaffected by small changes in operational parameters and in used to define acceptable tolerances.
(9) System suitability: It is determined by checking a system to ensure system performance before and during.
(10) Transferability and revalidation: Transfer of validated methods to others should be followed by after the method validation.

7.9 Good laboratory practice

Good laboratory practice (GLP) referred to a quality system of management control for research laboratories and organization to ensure the uniformity, consistency, reliability, reproducibility, quality, and integrity of products to be utilized by human beings and animals for good health.

In 1972, due to industrial biotest laboratories scandal New Zealand and Denmark introduced GLP, and at later stage, it was followed in the US research laboratories. In 1992, Organization for Economic Co-operation and Development (OECD) followed the basic principles of GLP and helped promulgate GLP to many countries.

The basic principles of GLP are, also, applicable for assessment of the safety or efficacy of products in development for human beings, animals, and the environment. GLP is not limited to chemicals and also applicable to medical devices, food additives, food packaging, color additives, animal food additives, and other nonpharmaceutical products or ingredients, biological products, and electronic products. Followings are the main basic principles govern by GLP rules:

1. Structural getup of organization
   - Management—tasks
   - Nature of sponsorship
   - Study direct-responsibility
   - Principal investigator responsibilities
   - Study personnel-responsibilities
2. Quality assurance program
   - Quality assurance personnel
3. Facilities
   - Test system facilities
   - Facilities for test references items
4. Equipment, reagents and materials
5. Test system
   - Physical/Chemical
   - Biological
6. Test and reference items
7. Standard operating procedures
8. Performance of study
   - Study plan
   - Conduct of study
9. Reporting of results
10. Archival-Storage of records and reports

7.9.1 Guidelines for GLP

OECD guidelines: OECD publishes guidelines for the testing of chemicals, which are to be followed for GLP compliance being followed by agencies doing risk assessment of chemicals.

The US FDA: The United States FDA has prescribed rules for GLP. Preclinical trials on animal in the United States of America use these rules prior to clinical research in humans.

European Union: Since 1987, the European Council had adopted two basic directives and a decision relating to the application of the GLP principles.

References

[1] https://www.qad.com/blog/2018/04/frederick-winslow-taylor-scientific-management.
[2] https://www.leansixsigmadefinition.com/glossary/walter-shewhart/.
[3] https://www.bl.uk/people/w-edwards-deming.
[4] https://www.juran.com/blog/the-history-of-quality/.
[5] A.G. Atanasov, B. Waltenberger, E.M. Pferschy-Wenzig, T. Linder, C. Wawrosch, P. Uhrin, et al., Discovery and resupply of pharmacologically active plant-derived natural products: a review, Biotechnol. Adv. 33 (8) (December 2015) 1582–1614, https://doi.org/10.1016/j.biotechadv.2015.08.001.

[6] E. Brewer, J. Henion, Atmospheric pressure ionization LC/MS/MS techniques for drug disposition studies, J. Pharm. Sci. 87 (1998) 395–402.
[7] M. Jemal, Y.-Q. Xia, LC-MS development strategies for quantitative bioanalysis, Curr. Drug Metab. 7 (2006) 491–502.
[8] J.N. Kyranos, H. Cai, D. Wei, W.K. Goetzinger, et al., High-throughput high-performance liquid chromatography/mass spectrometry for modern drug discovery, Curr. Opin. Biotechnol. 12 (2001) 105–111.
[9] V.P. Shah, K.K. Midha, S.V. Dighe, et al., Analytical method validation assessment, J. Pharm. Sci. 81 (1992) 309–312.
[10] G. Vas, K. Vekey, Solid-phase micro extraction: a powerful sample preparation tool prior to mass spectrometric analysis, J. Mass Spectrum. 39 (3) (2004) 233–254.
[11] Y. Ohkura, M. Kai, H. Nohta, et al., Fluorogenic reactions for biomedical chromatography, J. Chromatogr. B Biomed. Appl. 659 (1–2) (1994) 85–107.
[12] J.M. Halket, V.V. Zaikin, Derivatization in mass spectrometry, Eur. J. Mass Spectr. (Chichester) 10 (1) (2004) 1–19.
[13] S.K. Bansal, T. Layoff, E.D. Bush, Qualification of analytical instruments for use in the pharmaceutical industry: a scientific approach, AAPS PharmSciTech 5 (1) (2004) 22–25.
[14] V.P. Shah, K.K. Midha, J.W. Findlay, et al., Bioanalytical method validation—a revisit with a decade of progress, Pharm. Res. 17 (12) (2000) 1551–1557.
[15] ICH Q1A (R2), Stability testing of new drug substances and products (second revision), Fed. Regist. 68 (225) (2003) 2844–2945.
[16] ICH Q1B, Photo stability testing of new drug substances and products, Fed. Regist. 62 (95) (1997) 27115–27122.
[17] ICHQ1C, Stability testing for new dosage forms, Fed. Regist. 62 (90) (1997) 25634–25635.
[18] ICH Q1D, Bracketing and matrixing designs for stability testing of drug substances and drug products, Fed. Regist. 68 (11) (2003) 2339–2340.
[19] ICH Q1E, Evaluation of stability data, Fed. Regist. 69 (110) (2004) 32010–32011.
[20] ICH Q1F, Stability data package for registration applications in climatic zones III and IV, Fed. Regist. 68 (225) (2003) 65717–65718.
[21] ICH Q3C, Impurities: guideline for residual solvents, Fed. Regist. 62 (247) (1997) 67377–67379.
[22] Validation of analytical procedure: methodology Q2B, in: ICH Harmonized Tripartite Guidelines, 1996, pp. 1–8. Geneva, Switzerland.
[23] CIPAC Document No. 3807—Guidelines on method Validation to be performed in support of analytical methods for agrochemical Formulations, cipac.org on 28 July 2003.
[24] International Standard ISO 5725, Precision of Test Methods—Repeatability and Reproducibility, 1986.
[25] ISO Standard 11095, Linear Calibration Using Reference Material, 1996.
[26] W. Horwitz, Evaluation of analytical methods used for regulation of foods and drugs, Anal. Chem. 54 (1) (1982) 67A–76A, https://doi.org/10.1021/AC00238A765.
[27] M. Thompson, S.L.R. Ellison, R. Wood, Harmonised guidelines for single laboratory validation of method of analysis, Pure Appl. Chem. 74 (5) (2008) 835–855.
[28] European Commission, Annex 15. EU Guide to Good Manufacturing Practice: Qualification and Validation, vol. 4, 2010, pp. 1–10.
[29] V. Ravichandran, S. Shalini, K.M. Sundram, H. Rajak, Validation of analytical methods—strategies & importance, Int. J. Pharm. Pharm. Sci. 2 (3) (2010) 340–345.
[30] P. Tangri, P.S. Rawat, V. Jakhmola, Validation: a critical parameter for quality control of pharmaceuticals, J. Drug Deliv. Therap. 2 (3) (2012) 34–40.
[31] A. Sharma, R. Sharma, Validation of analytical procedures: a comparison of ICH vs Pharmacopoiea (USP) vs FDA, Int. Res. J. Pharm. 3 (6) (2012) 39–42.

[32] J. Lambert, Validation guidelines for pharmaceutical dosage forms, Health Canada Health Prod. Food Branch Inspect. 26 (2004) 7–15.
[33] S.S. Prabh, S. Gagan, Analytical method development and validation, J. Pharm. Res. 4 (7) (2011) 2330–2332.
[34] M. Ramamurthy, K. Sarvanakumar, Pharmaceutical validation, East. Pharm. 476 (1997) 45–47.
[35] J. Agalloco, Validation: an unconventional review and reinvention, PDA J. Pharm. Sci. Technol. 49 (1995) 175–179.

CHAPTER

Quality by design

8

Abbreviations

CCP	critical process parameters
cGMP	current good manufacturing practice
CMA	critical material attributes
CMC	chemistry manufacturing and control
CQAs	critical quality attributes
DoE	design of experiments
EMA	European Medicines Agency
FAD	Food and Drug Administration
ICH	International Conference on Harmonization
IPC	in-process-control
QbD	quality by design
PAT	process analytical technology
PCA	principal component analysis
PLSR	partial least squares regression
PPIs	process performance indicators
QTPP	quality target product profile
USFDA	United States Food and Drug Administration
WHO	World Health Organization

8.1 Qualities by design

Quality by design (QbD) refers to a practice for developing a material from an animate or inanimate entity by giving maximum emphasis on product development or design based on well-defined regulatory act like ICHQ8 (R2). The ICH represents International Conference on Harmonization of Technical Requirements for Registration of Pharmaceuticals for human use considerations guideline Q8 (R2) on pharmaceutical development. So, biologic drugs should be designed on the basis of QbD during the manufacturing process. A defaulted drug is unsafe for healthcare, no matter, how many tests or analyses have been done to verify its quality [1–5].

The QbD is not referred to as simply testing the product but rather as building quality into the manufacturing process by design. This is based on control strategy by establishing close relationship between product and process understanding. In biopharmaceutical terminology QbD represents a

Techniques for Downstream process for Biologic Drugs and Vaccines. https://doi.org/10.1016/B978-0-443-19157-2.00009-1

systematic procedural approach for manufacturing biologic drugs by giving special emphasis on clear process understanding followed by quality production as per internationally recognized regulatory acts with quality risk management.

8.2 Basic of QbD

Quality-by-design can be developed by phases. In the first phase, correlation is established between product understanding and process understanding to develop an overall control strategy for QbD (Fig. 8.1).

Within this workflow system, there are multiple risk assessment and deliverables that should be met to develop a well-authentic process and analytical control strategy (Fig. 8.2).

The deliverables and risk assessments include:

- Establish a basic design format;
- Bring conformation on critical quality attributes (CQAs) and process performance indicators (PPIs);
- Specify process parameters, material attributes, and raw materials with potential impact on CQAs and PPIs; and
- Resulted in classifying process parameters, material attributes, and raw materials, as critical or noncritical.

Based on these facts, a comprehensive protocol strategy for the life cycle of product is developed. It assures perfect process understanding and consistency in QbDs.

8.3 Benefits of QbD

Strategically, data-driven process by QbD helps to serve the progressive manufacturing environment, as they give drug developers a better understanding of manufacturing processes, decreases the amount of batch failures, give more effective control of change, and also, provide a higher return on investment. Following are the few important benefits; a company gets by implementing QbD.

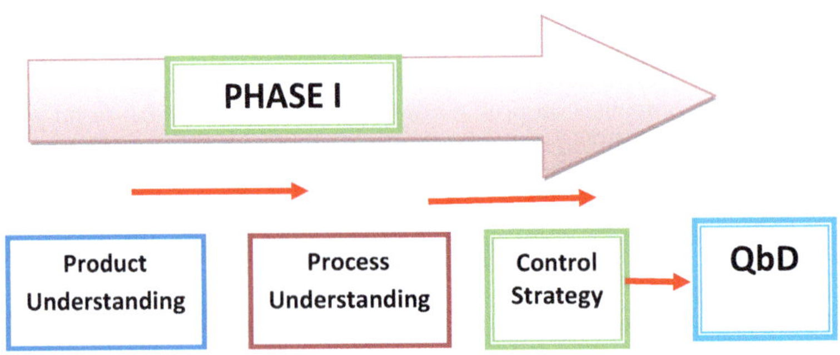

FIG. 8.1

Conceptual development of QbD.

Product Profile--- Quality Target Product Profile (QTPP)

- **CQAs**----Determine "Potential Critical Quality" attributes (CQAs)
- **Risk Assessment**- Link raw material attributes & Process Parameters to CQAs and perform risk assessment
- **Design Space**---- Develop a design space (optional and not required)
- **Control Strategy**--- Design implement a Control strategy
- **Continual Improvement**--- Manage product life cycle, including continual improvement

FIG. 8.2

Q8 (R2) Example of QbD approach.

8.3.1 Sustainable pattern of improvement

It is the core focus of QbD. This results in a better understanding of product, as it moves through its lifecycle, making it much easier to spot areas of concern.

Once the final flow toward QbD is finished, the overall technique can be adapted to create a safer product and efficient manufacturing process. The QbD approach allows for ongoing fine-tuning of operations, even when sealing up from the laboratory to commercial manufacturing.

8.3.2 Consistency

Consistency in batch-to-batch production can be maintained by using QbD in manufacturing process. This practice can be resulted in the most desirable outcomes not only from regulatory bodies but also from consumers.

More moral confidence will develop among the workers and consumers in the process of QbD-based product production. Consistency also encourages a company to maintain a high-level of biologic drug manufacturing operational readiness.

8.3.3 Reduced controls

Manufacturing process by design based on QbD is helpful in reducing the need for controls. Because of this, developers have a good idea of quality even prior to testing. In the end, this can save a company's time and money.

8.3.4 Reduced failure

The QbD practice brings the value of working together with clear and comprehensive understanding of the parameters involved in the development of sustainable product. This deep understanding helps teams assess risk and act accordingly by significantly reducing the likelihood of failure.

8.4 Significance of QbD

Life is the most delicate entity. Its care and sustainable survival need a proper healthcare system with safe and secure therapeutic biomolecules. Quality by design is one of the most important factors when designing biologics QbD tools and studies include prior knowledge, risk assessment, mechanistic models, design of experiments (DoE), data analysis, and process analytical technology (PAT). In brief, quality means fitness for intended use. So, biologics should be free from any foreign contamination which may threaten the structural and functional stability of the product. Both in vivo and in vitro tests are conducted for assuring quality, before marketing. QbD is emerging to enhance safe, effective drug supply to the consumer, and also offers promise to significantly improve manufacturing quality performance.

8.5 Conceptual development on QbD

In the year 1992, Dr. Joseph M. Juran took first credit for developing the concept of QbD [6] for product development [7]. This was mainly based on a model system known as quality trilogy (Fig. 8.3). Later, he developed the concept that quality can be built into the product rather than testing periodically.

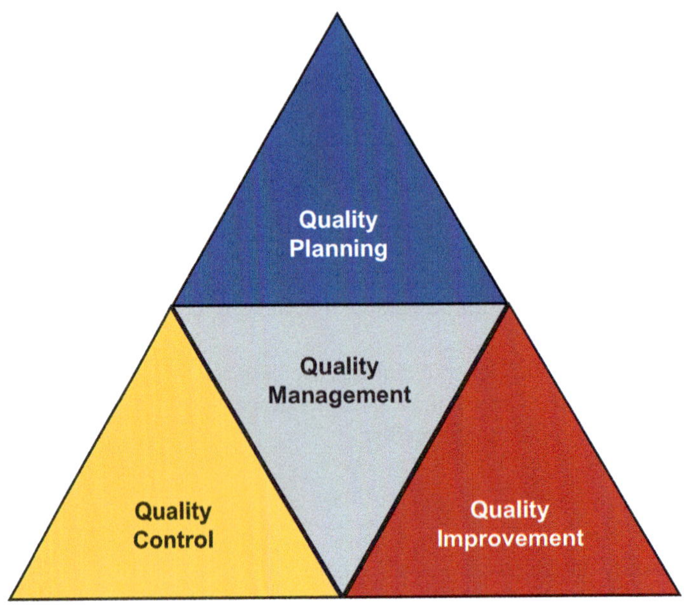

FIG. 8.3

Outline of model system on quality trilogy.

Hence, QbD is also referred to as "Quality by planning, but not by chance." The applicability of QbD concept has percolated in diverse manufacturing industries including the biopharmaceutical industries [8].

In his view, the format for quality should be framed on the basis of the product. Generally, quality crises and problems relate to the way in which a product is designed in the first place. A high-quality product, as defined by Woodcock is free of contamination and reliably delivers the therapeutic benefit promised on the label to the consumer [9].

The QbD development process involves a series of processes based on certain basic principles, as prescribed by regulatory bodies, to adapt by biopharmaceutical industries for highly authentic product manufacturing (Fig. 8.4).

The United States Food and Drug Administration (USFDA) adapted QbD for the discovery, development and manufacture of biologic drugs [10]. FDA's emphasis on QbD began with the recognition that increased testing does not necessarily improve product quality. Quality must be built into the product. At the same time, the FDA issued another guideline on "Process Analytical Technology" (PAT) to guide the generic industry about the advantages of PAT in real time release. This was the first significant step taken by the FDA to lay out the pathway to implement QbD. The main reason behind it was to boost process and product understanding with the help of risk assessment, identifying critical quality attributes, and critical process parameters (CPPs), so that end users benefit from getting quality product, consistently FDA recommended the implementation since 2013.

The ICH Q8 guideline for QbD states: "quality cannot be tested into products, i.e., quality should be built in by design." The ICH Q8 defined QbD as: "a systematic approach to development that begins with predefined objectives and emphasizes product and process understanding and process control, based on sound science and quality risk management."

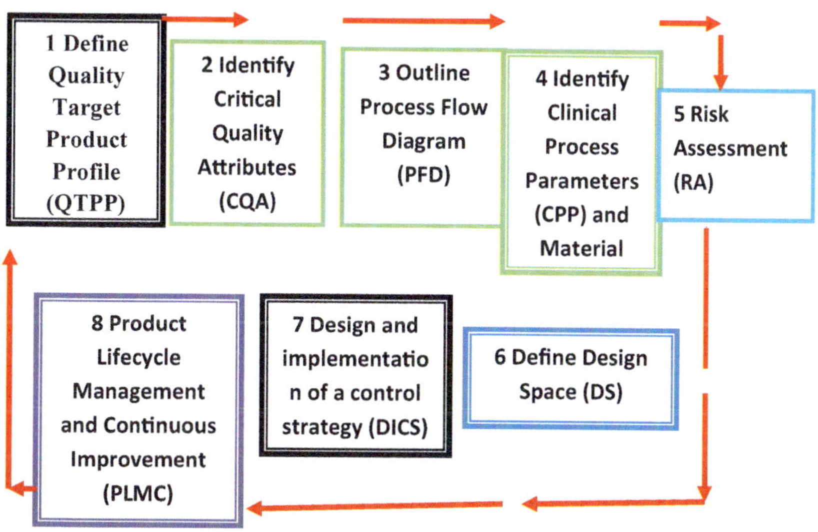

FIG. 8.4

Process for QbD development.

QbD was developed few years ago, only to ensure manufacture of quality products. It is a part of biologic drug control process. Under the banner of QbD, designing and development of a product, it is essential to define a target product performance profile or target product profile (TPP), or target product quality profile (TPQP), and identify critical quality attributed (CQA). On the basis of this, we can design the product formulation and process to meet the product attributes. This is resulted in how to understand the impact of raw materials (critical material attributes [CMA]), critical parameters (CCP) on the CQAs, and identification and control sources of variability. By following QbD, the biopharmaceutical industries can strengthen self-regulated flexibility while maintain high quality standards and real-time release of the drug product.

8.5.1 ICH guidelines for QbD

In the year 2006, ICH introduced the Q8 (Pharmaceutical Development) guideline, which was supplemented between 2009 and 2012 by Q9 (Quality Risk Management [11]), Q10 (Pharmaceutical Quality System [12]), and Q11 (Development and Manufacture of Drug Substances [13]). Besides, the ICH Q8 (R2) guideline published in 2009 describes the principles of QbD which are Quality Target Product Profile (QTPP), Critical Quality Attributes (CQA), Quality Risk Management, Design Space, and Control Strategies [14].

QbD practice in biopharmaceutical industries supports increasing process understanding, better control over product manufacturing and desired quality built into the development of the product. But, at initial stage of QbD implementation, a company has to bear high costs.

The FDA has provided more detailed guidance on QbD for pharmaceutical product design, process understanding, and lifecycle management. A major focus is given to performing in-process testing in order to adjust with variation of price rise and fall, and product quality and process changes. Similarly, the European Medicines Agency (EMA) promotes QbD as an approach that aims to ensure the quality of medicines by employing statistical, analytical, and risk-management methodologies in the design, development, and manufacturing of medicines. The regulatory bodies convey to the customers that quality is developed into biologic drugs from the beginning based on knowledge of their characteristics and perfect understanding of the process by which the product is manufactured.

8.6 Toward QbD

In 2002, the FDA introduced current good manufacturing practice (cGMP) to resolve the problem of limitation of GMP [15,16]. The main target for bringing such amendment is to assure software-based management system in manufacturing process, monitoring the technical education, and updating the training of manpower governing the entire process of manufacturing, quality control, quality assurance, and technical manpower management, during the entire process of production and management.

Despite such amendments and implementation of cGMP, there is still some lacuna in the efficacy of quality management in biologics production, due to strenuous and rigid nature of manufacturing process. Besides this, it was realized that during the approval process of documents for new product to be handed over to the FDA, a company can only produce a limited number of information on the authenticity of the manufacturing process, which may, often, lead to failure to meet the rigid specifications. In order to overcome such a problem, in 2002, the FDA brought certain new amendments in the pharmaceutical cGMP (good manufacturing practice). The contends were incorporated as PAT, which is a system for designing, analyzing, and controlling manufacturing process, which can be helpful in critical quality and performance attributes of raw and processed materials to ensure final product quality. Besides this, it was also ensured by increasing the efficacy of process to reduce waste and over-processing. In 2003, in order to develop a pharmaceutical quality system on the basis of risk management and science, FDA and the International Conference on Harmonization of Technical Requirements for Registration of Pharmaceuticals for Human Use (ICH) proposed regulatory act on QbD. The basic aim of such regulatory acts was to have systematic approach to drug development having well-planed protocol for process development to procure quality product with safe and secure. On the basis of QbD, it is essential to define target product performance profile (TPP), TPQP, and identify critical quality attributed (CQA). The frame work of QbD is based on the ICH Guidelines Q8 for pharmaceutical development, Q9 for quality risk management, and Q10 for pharmaceutical quality systems.

Since 2004, a number of biopharmaceutical companies have been showing interest in incorporating elements of risk and QbD in manufacturing process [17]. However, implementation of QbD has been slow, till 2014. But the FDA has now "strongly suggested" QbD elements, to be implemented in generic-drug applications. The current challenges for QbD implementation are numerous. In order to have more systematic, in 2004, USFDA ordered to pharmaceutical company to submit their chemistry manufacturing and control (CMC) in QbD format [18].

8.7 Elements of QbD

In biopharmaceutical QbD is targeted to product development based on the characteristics that are critical to the quality required by patients, translates them into the drug product with critical quality attributes (CQAs). QbD is a comprehensive process package, which and establishes the relationship between formulation/manufacturing variables and CQAs to deliver biologic drug to patients through supply chain system well-regulated by internationally recognized bodies. QbD consist of the following elements:

8.7.1 The target product profile (TPP)

A target product profile (TPP) refers to "profile" or characteristics of a product that is aimed at a particular disease. TPPs state the intended use, target populations and other desired attributes of products, including safety and efficacy-related characteristics. Such profiles can guide product research and development (R&D) [19,20].

In industry, in-house target product profiles (TPPs) are used as planning tools that guide development toward desired characteristic. With reference to regulatory context, TPPs are considered as tools to frame development for submitting product dossiers [21].

So, TPP is defined as a dosage form, delivery systems, or dosage potential. It represents as a whole, the quality of drug that carries all the prescribed characters (e.g., dissolution, aerodynamic performance) and drug product quality criteria (e.g., sterility, purity, stability, and drug release) that helpful in curing specific dieses.

Since the last decade, the internally recognized funding agencies have framed their own TPPs. It is difficult to say how these TPPs relate to WHO's preferences, creating the possibility that not-for-profile product development programs are not always well-aligned with WHO's identified needs. However, WHO TPPs on priority basis for public health helps in supporting the development of missing health products. The WHO TPP documents intend to bring product developers, regulatory agencies, and funders together on R&D to provide priorities for public health. In this connection WHO wanted to ensure minimally acceptable profiles for vaccines, therapeutics, diagnostic, or medical devices criteria. In addition WHO provides information for funders and developers on the performance and operational characteristics expected of products, if they meet WHO's criteria.

8.7.2 Quality target product profile

QTPP referred to as a key element of QbD and has a significant role in developing biologic drugs. A perfect QTPP contains properties of the drug substance, functional traits, and clinical perspectives of the target product for the intended patients [10]. Holistic implantation of QTPP leads to success in product development [22,23].

8.7.3 Critical quality attributes

It exclusively means that the physical, chemical, biological, or microbiological properties or characteristics of a biologic are in the finished stage (according to ICH q8). Potential drug products (CQAs) derived from the QTPP and/or with prior knowledge on product and process development, within an appropriate limit, range, or distribution to ensure the desired product quality. CQA has been used by someone to describe elements of the TPQP, while others have used it to describe mechanistic factors that determine product performance. The CQA is used to describe both aspects of product performance and determinants of product performance.

8.7.4 Risk assessment

The main target of risk assessment in biologic drugs manufacturing process is to identify the nature of material attributes and process parameters that affect the drug product CQAs, through which one can understand and predict sources of variability in the manufacturing process, so that an appropriate control strategy can be implemented to ensure that the CQAs are within the desired requirements [24,25].

To identify the CPPs and critical matter attributes as a part of product development process. In the beginning, it is necessary to have thorough understanding of the process product to ensure zero default in the final phase of development. Hence, the risks identified at the initial phases are perceived risks; as further process/product understanding is gained, the actual risk become clearer and a control strategy can be better defined. The risk assessment tools used in earlier phases of development, therefore, tend to be more qualitative and serve as a means to prioritize the experimentation.

8.7.5 Design space

Design space is a scientific concept used in the biopharmaceutical/pharmaceutical industries to ensure product quality (Fig. 8.5). The cumulative digital information during product development provides the foundation for the design space.

The International Conference on Harmonization of Technical Requirement for Registration of Pharmaceuticals for Human use (ICH) defines design space as "the multidimensional combination

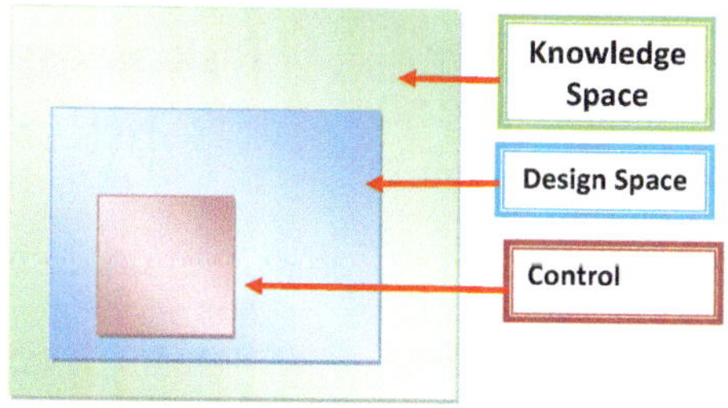

FIG. 8.5

Graphic presentation on concept of design space.

Adapted in biopharmaceutical industry.

and interaction of input variables (e.g., material attributes) and process parameters that have been demonstrated to provide assurance of quality." However, relatively little has been put forth to date on how to frame a design space from data composed of such variable. Design space determined on the laboratory scale may not be relevant to the process at the commercial scale. So, design space verification at the commercial scale becomes essential unless it is confirmed that the design space is scale-independent. Currently, generic drug sponsors obtain information about acceptable ranges for individual CPPs and CMAs at laboratory or pilot scales.

8.7.6 Critical material attributes (CMA)

Besides possessing critical quality attributes (CQAs) characteristics, CMA-based product should be within an appropriate limit, range, or distribution to ensure the desired quality of that drug substance, excipient, or in-process material.

8.7.7 Critical process parameters

CPPs practice is mainly for understanding the appearance, impurity, and yield of product before or in process of manufacturing stages.

8.8 How does QbD works?

The concept of QbD can be well-developed by understanding in depth on stepwise manufacturing process with safety and security. Sometimes, consistency in quality maintenance is mainly dependent on well-trained QbD manpower. Knowledge of risk involved and how to mitigate them is also critical. To achieve this, several detailed steps make up QbD approach (Fig. 8.6).

Basically, the process begins with a target product profile (TPP) that describes the intended use, safety, and efficacy parameters of the product. This, along with the TPQP, is used by formulators and process engineers during product development. Critical quality attributes (CQAs), such as sterility or dose accuracy requirements, are then identified based on prior product knowledge or experience with similar compounds. CQAs are those things that must be controlled to achieve the TPP. A manufacturing process is designed to produce a final product with these CQAs. A control strategy for the entire

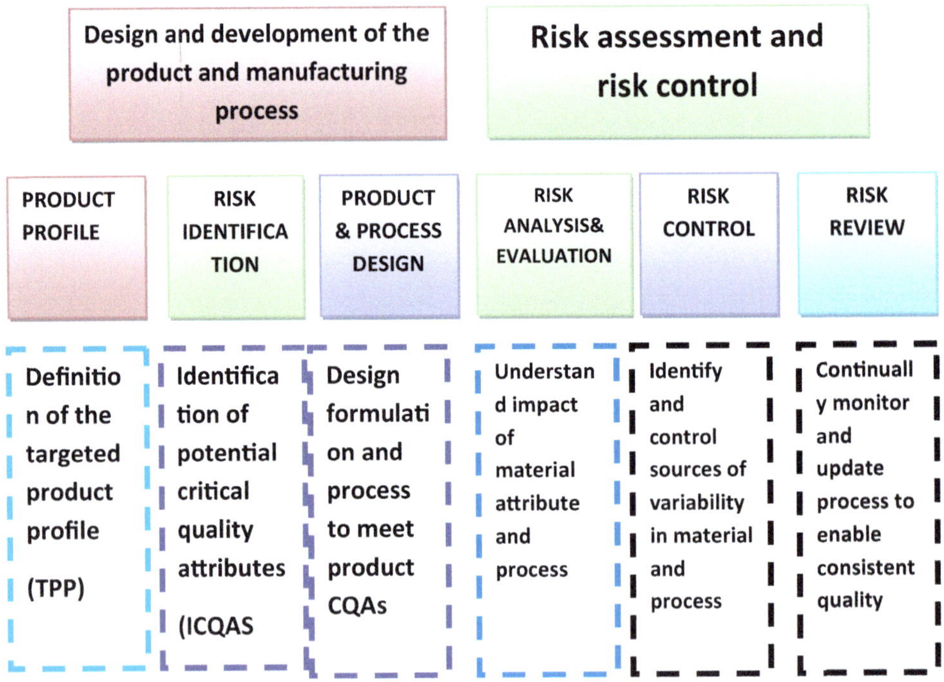

FIG. 8.6

Works detail steps make up QbD approach.

process is established, accounting for variability in material and process as well as expected changes in scale etc. Using such a strategy and various in-process control (IPC) and PAT tools, the process is monitored and quality assessed in real time.

8.9 Control strategy

ICHQ10 stands for a compressive model for an effective pharmaceutical quality system based on International Organization for Standardization (ISO) quality control and incorporating regional GMP regulation. Along with ICH Q8, and ICHQ9, ICHQ10 forms a tripartite set of guidelines that all pharmaceutical companies need to implement to show their end-users and regulatory agencies that their products are safe and efficacious. It is difficult for organizations to continue to use manual, paper-based QMS systems because of the many inherent flaws and drawbacks.

ICH Q10 is framed to have control over the product quality of a drug substance, mostly biology origin, with assured process commitment. The control parameters include drug substances and drug product materials and components, specific finished products and associated methods, and frequency of monitoring and control. A control strategy normally include: input material controls, well-defined design space around individual or multiple unit operations, and/or final product specifications used to ensure consistent quality. The finished drug products are tested for quality by assessing, if they meet specification. In addition, manufacturers are usually expected to conduct extensive in process tests, such as blend uniformity or tablet hardness. So, pharmaceutical and nutraceutical production personnel often struggle to achieve the same tablet hardness as that achieved during R&D and sometimes fail to reach even the specified minimum hardness. While the tablets may be fine in all other aspects, such as weight, thickness, and disintegration time, they are softer than those produced during R&D. Insufficient tablet hardness can lead to defects such as lamination and erosion of some tablets during coating. Often, the problem is related to blending, and fixing the problem begins with a new understanding of the blending unit operation.

A QbD-based control strategy for blending process is shown in Fig. 8.7. Pharmaceutical quality is assured by understanding and controlling formulation and manufacturing variables to assure the quality of the finished product. The end-product testing only confirms the quality of the product.

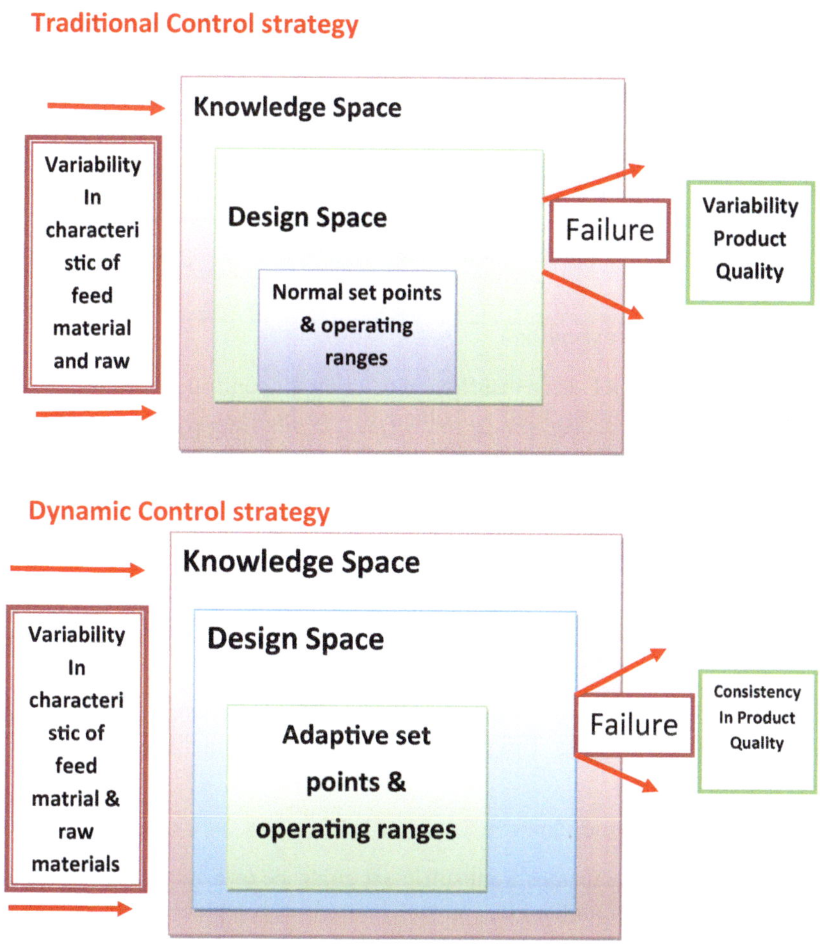

FIG. 8.7

Control strategy for QbD.

8.10 Challenges

To challenge this the FDA released a document on "Pharmaceutical for the 21st century: A risk-based approach." The main target behind this attempt was to bring awareness among pharmaceutical manufacturing companies about innovative methodologies to ensure drug QbD, a risk-based approach that begins with predefined objective for meeting desired clinical performance and emphasizes product and process understanding and process control.

The last decade has witnessed the appreciation of QbD for quality drug manufacturing. In due time, the biotechnology-based industries started understanding and implementing the basic concept of QbD to bring sustainability in their product in the market. Following are few important challenges as FDA confront while promoting QbD implementation in industries:

(a) Internal misalignment

Currently, one of the most unavoidable factors for successfully manufacturing quality products with maximum safety is QbD, as an operational model. But, some of the scientists and promoters hesitate to redesign/or amend certain manufacturing protocols because the initiative is not required and the benefits are not guaranteed.

(b) Lack of confidence in a business deal

Although, it is believed that QbD gives a strong boost in business achievement but sceptics felt that QdD would take maximum time for clinical trials for drug substances approval, and ultimately result in business failure.

(c) Inadequate technology for execution

Some of the critics claim that QbD is an insufficient solution for controlling the variability of raw materials as companies develop new skill sets and obtain new technology that are viewed as out of reach, and would be difficult to gather, manage, and analyze all of the product data generated during development.

(d) Alignment with third parties

Industries feel as challenging task to link QbD with supply chain management system and contract manufacturers to upgrade the efficacy of marketing process and gain trust of the customers.

(e) Contradictory attitude of FDA toward QbD

The pharmaceutical industries have perceived the idea that FDA understood QbD requirements differently both within and between offices. This discrepancy leads to the belief that the FDA may not review filings in a consistent manner.

(f) Nonavailability of proper guide lines

Most of the newly emerging companies want sufficient guidance on how to actually implement QbD. Companies want clarification from the FDA on matters such as acceptable methods, criteria to select, and discard critical quality analytical method substitution.

(g) Unwillingness of handling QbD

In spite of the continuous attempts of the FDA to support and bring remediation to the manufacturing process by implementing guidelines on QbD, but the regulators decline to show their willingness to follow such procedural amendment in the product development process and increase turnover rate.

(h) Lack of trust in regulatory benefit

The companies do not like to avail the challenge of QbD due to the lack of confidence on real regulatory benefits from FDA. Without clear benefits, proponents of QbD in industry have expressed difficulty in promoting the idea within their companies.

(i) Acceptability of other regulatory bodies

Companies have no clearance on acceptability of other regulatory bodies on their QbD-based manufacturing process.

(j) Communication gap

Even in the past, the pharmaceutical manufacturers have any information on comfort discussion with the FDA on success of QbD in increasing manufacturing process; although companies are keen to open reduce the communication gap on the various issues on the possibility of QbD implementation.

8.11 Regulatory aspects to QbD

8.11.1 FDA perspective

Biopharmaceutical companies most often deal with life-saving drugs. So, QbD plays the most critical role in maintaining the consistency in product quality throughout its manufacturing process and pattern of distribution through a well-managed supply chain system and temperature-controlled logistic system. In addition, design space and real-time release risk assessment are other parameters for implementation of QbD. Ultimately, QbD helps to implement Q8 and Q9. FDA's view of QbD is "QbD is a systematic approach to product and process design and development." This concept was accepted by the FDA in 2004 and detailed description was given in "pharmaceutical cGMP for 21st century—a risk-based approach."

8.11.2 ICH guideline and QbD

The International Council for Harmonization of Technical Requirements for Pharmaceutical for Human use (ICH) coordinates in bringing together the regulatory authorities and pharmaceutical industries to have an exchange of idea on scientific and technical aspects of therapeutic molecules for human health. The underlying principles of QbD, i.e., science- and risk-based product development, risk assessment, lifecycle approach, and method design are available in the quality guidelines of the International Conference on Harmonization, i.e., ICH, as already mentioned, earlier.

8.11.3 Regulatory challenges and inspection

Generally, FDA system-based approach is followed while inspecting a company's entire operation and production process in accordance with CDER's Compliance Program "Inspection of Licensed Biological Therapeutic Drug Product." In a QbD mandated company, FDA inspection team will assess the implementation and effectiveness of the process design as described successfully from development to manufacturing. The main target for such inspection is to evaluate the quality system and its efficiency inconsistent production quality, change in control procedure, process improvements, deviation management, and knowledge and risk management during the product cycle. Inspection of manufacturing facilities, equipment standard and maintenance, and raw material screening and supplier management remain same as per the previous performance.

8.12 Process analytical technology

The USFDA has given emphasis on PAT as a noble practice in developing innovative ideas and mechanisms to design, analyze, and monitor the biopharmaceutical manufacturing process to produce biologics under CPPs, which influence critical quality assurance (CQA). The FDA believes that by strict implementation of PAT will develop consumer's confidence on quality biologics with safe and security feeling. On the whole, the PAT is to generate product quality information in real-time. The process of implementation of PAT in biopharmaceutical companies is just as applicable to pharmaceutical industries. Therefore, it is still worthwhile for biopharmaceutical manufacturers to evaluate their need and ability to implement new technologies. It is also possible that the final version of the PAT guidance will apply to biopharmaceutical manufacturing.

8.12.1 Advantages of PAT

Biopharmaceutical manufacturing process is more challenging then the pharmaceutical companies engaged in developing drugs on chemical basis. Biotechnology process presents technical challenges different from those in the pharmaceutical industry. The main differences between biotechnology and pharmaceutical industry are aqueous manufacturing process, and complexity in therapeutic protein development process as compared with generic medicines. The PAT system is not only for testing quality but also assists in online monitoring of product intermediate and control quality in the manufacturing process, earlier and more precisely. Besides this, PAT is to facilitate faster release and elevate manufacturing safety and reliability to higher levels. Since long, in spite of many challenging issues, the PAT applications have been well in use in the biopharmaceutical companies and food processing industries for quality product in consistency manner [26–28] .

Presently, PAT tools are in widely used for routine manufacturing process in order to develop a greater understanding of the process design space under a Quality-by-design frame work. In the modern biopharmaceutical companies PAT has become exclusive tool for the successful development of innovative products with utmost confidence as "quality cannot be tested into a product, it should be built-in or should be by design" [29].

In addition, PAT tools controlling critical quality attribute (CQA) can be implemented commercially for process control. However, there are business drivers and regulatory aspects that will contribute to a final control strategy. While the process monitoring in pharmaceutical industry involved temperature, pressure, flows, pH, and other physical parameters, whereas the biopharmaceutical company use in-line testing using near infrared, Raman, or other physiochemical techniques to maintain quality control and quality assurance practices on the basis of PAT guideline.

Presently, both on-line and in-line testing are in practice in biopharmaceutical and biotechnology companies. In on-line testing samples are drawn or monitor periodically, where as in case of in-line testing, which places probes in constant contact during manufacturing process. Beyond data such as blending, or drying, the FDA has proposed creating on/at-line, and does not bring any structural modification in the sample, and as a result the quality of product remains unaltered. Many other techniques like Raman, mid-IR, acoustic emission signals, and other imaging techniques are also gaining popularity.

Dissolution is the current primary method for evaluating the consistency and similarity of solid oral dosage form. Using PAT, process would be under such high control that the dissolution results could be accurately predicted well before the product is analyzed. Research on the correlation between dissolution results and measured process parameters would be performed, so that the impact of processes, raw materials, and measured process parameters could be continuously monitored and adjustments made to ensure that the finished product variables would be understood. The manufacturing process could be continuously monitored and adjustments made to ensure that the finished product would meet the desired specifications. Measurements from these techniques have already been used successfully to give predictive values for dissolution, content uniformity, assay, moisture, and hardness. The data produced by these devices are rich with highly complex information. These correlations must be performed with the use of chemometrics.

8.12.2 Chemometrics

It is an analysis method by means of which informative data on a specific matter, irrespective of its nature, can be analyzed on the basis of computer analysis, statistical analysis like multivariate analysis and applied mathematics to assess the physiochemical nature of an entity, and to understand problems in chemistry, biochemistry, medicine, life sciences, and biotechnology.

Earlier, manufacturing processes have been analyzed in an univariate manner, with the help of a standard single parameters stated control chart. But, in reality the physiochemical process are multivariate and interlinked. Due to such complexity nature of data computer software model is utmost necessary to derive substantial conclusion. Generally, two chemometric techniques that have been found to be useful are principal component analysis (PCA) and partial least squares regression (PLSR). These techniques are recognized for their ability to eliminate noise, identify latent variables, and extrapolate missing data.

The PCA is a technique of creating data models of previously produced and tested batches to verify similarity to newly created batches. One advantage of this technique has over the commonly used f2 metric is that the batches are compared to an already set of compiled batches, included in a validated model. Trends could potentially be identified earlier than with f2 comparison. This could help improvement of process consistency after scale up and post approval changes. Another advantage of PCA is that it can handle the large amount of data produced by dissolution fiber optic (Dis-FO) techniques without the need to reduce data points. PLSR is used to correct correlate data, such as finished product dissolution results, to raw material, process parameters, and in-line readings. Variables, which affect the dissolution rate can be better understood and monitored. The effect of scale ups and postapproval changes can be quantified. Critical parameters can be controlled, thereby creating high quality drug product, less/stage-2 testing, and minimal product failure. When out of specification results do occur, drug products can be better investigated through the use of PLSR to determine, which underlying variables contributed to the failing drug product.

8.12.3 Pat tool boxes

Presently, many types of on-line tools are being used to monitor process technology in biopharmaceutical and pharmaceutical companies to produce quality products. However, PAT-specific tools are more acceptable due to its well-integrated nature. Spectroscopic tools near infrared [30–32]; mid-infrared, Raman [33–35] are utilized due to the specificity gained by the presence of functional groups routinely found in raw materials, intermediates, and drug substances that are often not a part of the matrix (e.g., solvents, excipients).

More success in the production of quality biologics depends on the development of more efficient tools for quick and accurate identification of intermediates during manufacturing process. Now efforts are in progress to understand the nature of biological-origin raw materials by NIR [36] for immediate use in the manufacturing process. Even NRI and Raman spectroscopy are in process trial to monitor bioprocesses. A recent study of NIR for demining amino acid concentration in chromatography pools [37] indicates that NIR may be used to detect moisture content of Lyophilized therapeutic protein. Lyophilization is a common process for protein and moisture content in the final product, which is typically part of certificate of analysis testing for lyophilized products. Chemical titration methods for moisture detection, such as Karl Fischer, are a tedious process. The benefit of NIR would be that a large number of final product vials can be tested using a relatively simple procedure. In addition, the testing can be done in real-time, which would decrease lot of release time.

Fluorescence-based sensor for biomass measurement [38]: biosensors for monitoring of specific nutrients for metabolic products [39]; dielectric spectroscopy for biomass measurement [40,41]; radio frequency impedance, probes to measure viable cell concentration of suspended yeast [42,43]; on-line high-pressure liquid chromatography in mammalian cell culture are some of the important on-line/in-line tools to monitor host cells culture at commercial level. The cost of implementation and risk of regulatory impact, discourage the company to incorporate in product analysis. Further, it is difficult to justify the benefit of such technologies when companies are currently manufacturing biologics without such technical provision. New analytical technologies have potential to improve the process monitoring capability, but companies have their own reservation, due to the cost of analytical instruments and their maintenance.

References

[1] J. Woodcock, The concept of pharmaceutical quality, Am. Pharm. Rev. 7 (6) (2004) 10–15.
[2] Q9: Quality Risk Management. ICH Harmonized Tripartite Guidelines International Conference on Harmonization of Technical Requirements for Registration of Pharmaceuticals for Human Use, 2006.
[3] Q10: Pharmaceutical Quality System, ICH Tripartite Guidelines. International Conference on Harmonization of Technical Requirements for Registration of Pharmaceuticals for Human Use, 2007.

[4] R.A. Lionberger, L.S. Lee, L. Lee, A. Raw, L.X. Yu, Quality by design: concepts for ANDAs, AAPS J. 10 (2008) 268–276.
[5] FDA. Guidance for Industry and Review Staff: Target Product Profile—A Strategic Development Process Tool (Draft Guidance).
[6] J.M. Juran, Juran on Quality by Design: The New Steps for Planning Quality Into Goods and Services, The Free Press, New York, 1992.
[7] J.M. Juran, Juran on Quality by Design: The New Steps for Planning Quality Into Goods and Services, Free Press, 1992.
[8] J.M. Juran, The quality trilogy: a universal approach to managing for quality, Qual. Prog. (1986) 19–24.
[9] J. Woodcock, The concept of pharmaceutical quality, Am. Pharm. Rev. (2004) 1–3.
[10] L.X. Yu, Pharmaceutical quality by design: product and process development, understanding, and control, Pharm. Res. 25 (2008) 781–791.
[11] ICH Q9, Quality Risk Management. ICH Harmonized Tripartite Guidelines. International Conference on Harmonization of Technical Requirements for Registration of Pharmaceuticals for Human Use, 2006.
[12] ICH Q10, Pharmaceutical Quality System, ICH Tripartite Guidelines. International Conference on Harmonization of Technical Requirements for Registration of Pharmaceuticals for Human Use, 2007.
[13] ICH Q11, Development and Manufacture of Drug Substances, ICH Tripartite Guidelines. International Conference on Harmonization of Technical Requirements for Registration of Pharmaceuticals for Human Use, 2012.
[14] ICH Q8 (R2), Pharmaceutical development, ICH Harmonized Tripartite Guidelines. International Conference on Harmonization of Technical Requirements for Registration of Pharmaceuticals for Human Use, 2009.
[15] Food and Drug Administration, Pharmaceutical Current Good Manufacturing Practices (cGMP) for the 21st Century—A Risk-Based Approach, 2002.
[16] Food and Drug Administration, Pharmaceutical cGMP s for the 21st Century. A Risk Based Approach: Second Progress Report and Implementation Plan, 2003.
[17] US Food and Drug Administration, Guidance for Industry: Q8 (2) Pharmaceutical Development, 2009.
[18] L.X. Yu, G. Amiddon, M.A. Khan, et al., Understanding pharmaceutical quality by design, AAPS J. 16 (4) (2014) 771–783.
[19] Food and Drug Administration: fda.gov/ohrms/dockets/ac/06/minutes/2006–4228m1.pdf, 2006.
[20] US Food and Drug Administration, Guidance for Industry: Q10 Pharmaceutical Quality Systems, 2009.
[21] R.W. Miller, Process analytical technologies (PAT)—part-2, Am. Pharm. Rev. 6 (2001) 1.
[22] S. Beg, M. Rahman, S.S. Panda, Pharmaceutical QbD: omnipresence in the product development lifecycle, Eur. Pharm. Rev. 22 (1) (2017) 58–64.
[23] S. Beg, M. Rahman, K. Kohli, Quality-by-design approach as a systematic tool for the development of nanopharmaceutical products, Drug Discov. Today 24 (3) (2019) 717–725.
[24] WHO, Guideline on Quality Risk Management a Draft Guidance, August, 2010.
[25] Implementation of ICH Q9 in the pharmaceutical field an example of methodology from PIC/S, 2010.
[26] R.W. Miller, Process analytical technologies (PAT)—part 1, Am. Pharm. Rev. 5 (2) (2002) 25–29.
[27] S.M. Arrivo, The role of PAT in pharmaceutical research and development, Int. J. Pharm. Sci. Rev. Res. 23 (2) (2013) 212–223.
[28] Food and Drug Administration, Guidance for Industry: PAT—A Framework for Innovative Pharmaceutical Development, Manufacturing, and Quality Assurance, Center for Veterinary Medicine, Office of Regulatory Affairs, Center for Drug Evaluation and Research, Rockville, MD, 2004. Docket Number: FDA-2003-D-0032.
[29] B. Diehl, B. Grout, NIR spectroscopy-just one of many analytical tools for PAT, Am. Pharm. Rev. 3 (2011) 70–74.

[30] W. Howard, I.I. Ward, S.S. Sekulic, M.J. Wheeler, G. Taber, F.J. Urbanski, F.E. Sistare, T. Norris, P.K. Aldridge, On-line determination of reaction completion in a closed-loop hydrogenator using NIR spectroscopy, Appl. Spectrosc. 52 (1) (1998) 17–21.
[31] M. Santangelo, B. Maranzano, K. Norris, T. McDermott, Near infrared (NIR) determination of uniformity for a drug product powder for oral suspension (POS) in amber glass bottles, Am. Pharm. Rev. (July/August) (2010) 68–72.
[32] M. Bell, NIR as an in-process tool, Am. Pharm. Rev. 11 (3) (2010) 75–79 (2008).
[33] F.T. Mattrey, S. Dolman, J. Nyrop, P.J. Skrdla, On-line FTIR monitoring and simultaneous optimization of a Strecker reaction performed in a laboratory scale flow-through reactor, Am. Pharm. Rev. 14 (7) (2011). http://www.americanpharmaceuticalreview.com/Featured-Articles/37512-On-line-FTIR-Monitoring-and-Simultaneous-Optimization-of-a-Strecker-Reaction-Performed-in-a-Laboratory-Scale-Flow-Through-Reactor/.
[34] A. Sparen, J. Johansson, O. Svensson, S. Folestad, M. Claybourn, Transmission Raman spectroscopy for quantitative analysis of pharmaceutical solids, Amer. Pharm. Rev. (January/February) (2009) 62–73.
[35] C.L. Robbe, et al., Exploring pharmaceutical applications of near-infrared technology, Am. Pharm. Rev. 6 (3) (2003) 62–73.
[36] J. Goode, The on-line analysis of aqueous based products from a separation column using near infrared spectroscopy, 2004. IFPAC 2004 Meeting.
[37] E. Skibsted, C. Lindemann, C. Roca, L. Olsson, Online bioprocess monitoring with a multi-wavelength fluorescence sensor using multivariate calibration, J. Biotech. 88 (2001) 47–57.
[38] X. Ge, L. Tolosa, J. Simpson, G. Rao, Genetically engineered binding proteins as biosensors for fermentation and cell culture, Biotechnol. Bioeng. 84 (2003) 723–731.
[39] K.R. Foster, H.P. Schwan, Dielectric properties of tissues, in: C. Polk, E. Postorv (Eds.), CRC Handbook of Biological Effects of Electromagnetic Fields, CRC Press, Boca Raton, FL, 1986, pp. 27–95.
[40] C.L. Davey, D.B. Kell, The low-frequency dielectric properties of biological cells, in: D. Walz, H. Berg, G. Milazzo (Eds.), Bioelectrochemistry of Cells and Tissues, Birkhauser Verlag, Basel, Switzerland, 1995, pp. 159–207.
[41] H.W. Maca, et al., The use of radio frequency capacitance for the measurement of yeast viable 14 biomass and its use in the automatic pitching of fermentations, MBAA Techn. Quart. 31 (1994) (1994) 146–148.
[42] J. Carvell, J.E. Dowd, On-line measurements and control of viable cell density in cell culture manufacturing processes using radio-frequency impedance, Cytotechnology 50 (1–3) (2006) 35–48.
[43] T. Larson, et al., Chemometric evaluation of on-line high pressure liquid chromatography in mammalian cell cultures: analysis of amino acids and glucose, Biotechnol. Bioeng. 77 (2002) 553–563.

CHAPTER

Biologic drugs: New frontier

9

Abbreviation

API	active pharmaceutical ingredient
ACE	artificial chromosome expression
ANDA	abbreviated new drug application
AI	artificial intelligence
ABL	abelson murine leukemia viral
BLA	biologic license application
BCR	break point cluster region
cGMP	current good manufacture process
CHO	Chinese hamster ovary
DHFR	dihydrofolate reductase
HPAPIs	high-potent APIs
HCMV	human cytomegalo virus
IND	investigational new drug
mRNA	messenger RNA
MTX	methotrexate
MSX	methionine sulfoxine
MAR	matrix attachment region
ML	machine learning
NDA	new drug application
OTC	over the-counter drugs
PTMs	posttranslational modifications
PCR	polymerase chain reaction
PDUFA	prescription drug user fee act
R&D	Research & Development
RG	risk group
RSV	rouse sarcoma virus

Living sources is the origin of biologic drugs. Biologics are usually more complex in nature than other medications as they are derived from diverse sources. For example, they involve more efforts to purify, process, and produce. These drugs are highly sensitive to light and temperature. A wide range of diseases from autoimmune disorders to cancer is cured with minimum side effects. Copying complex biologic drugs to make more affordable versions, known as biosimilars, is indeed a more complex process compared with how generics are usually made.

Techniques for Downstream process for Biologic Drugs and Vaccines. https://doi.org/10.1016/B978-0-443-19157-2.00007-8

Biologics are different from most of the medications stocked on pharmacy shelves. In fact, they even go through a separate approval process with the FDA, and they do not have generics. They have highly similar and less expensive versions called biosimilars.

9.1 Towards biologics

Most of the biologics are based on the recombinant proteins conjugated with glycans, unnatural amino acids, and other biological or nonbiological moieties, which help in increasing the half-life of the drugs with improved in vivo pharmacokinetics properties. Currently, plasminogen activator and erythropoietin help in dissolving blood clots, and help in maintaining amount of red blood cells, respectively. Since the successful global market of human tissue plasminogen activator in 1987 and erythropoietin in 1988, the global biologic drugs market has gone up. In 2020, the global demand for biologics was about USD 447.92 billion and it is expected to reach USD 833.34 billion by 2027, growing at a CAGR of 7.02% during the forecast period 2020–27 (Newark, NJ, Sept10, 2020; BLOBE NEWAWRE).

Hence, the world's top biotechnology-based companies (Table 9.1) are actively busy in designing and manufacturing biopharmaceuticals, focusing on therapeutic proteins.

The current living style and adverse climatic changes have increased the incidence of the number of life threatening diseases like Crohn's diseases, ulcerative colitis, rheumatoid arthritis and other autoimmune diseases. The biotechnology-based drug discovery faces several challenges in meeting the demand for biologic drugs in the global market. The global protein therapeutic market grew at a CAGR of about 7% during 2014–19. It is expected that the therapeutic proteins market will be continuing to grow moderately within the next 5 years [1].

Unlike small molecule drugs, therapeutic proteins cannot be produced via a sequence of chemical reactions. Natural sources of a variety of host cells belonging to plants, animals, mammals, and microbes have been in use after suitable modification by advanced genetic engineering and protein engineering techniques. Some such techniques are presented in detail in the following sections.

Million U.S. dollars, 2015

Company	Million U.S. dollars, 2015
Johnson & Johnson	74,331
Pfizer	49,605
Novartis	49,551
Roche	47,749
Merck	42,237
Sanofi	41,244
GlaxoSmithKline	35,835
Gilead Sciences	24,890
AstraZeneca	24,696
Abbott Laboratories	20,247

Table 9.1 Top 10 biopharmaceutical companies of the world.

9.1.1 Development of fermentation technology

Fermentation is a process in which microbes are used as biocatalysts to convert carbohydrate to alcohol or organic acids in anaerobic conditions. The fermentation process is of two types, based on product development: alcohol fermentation and lactic acid fermentation. In alcoholic fermentation or ethanol fermentation, pyruvate (from glucose metabolism) is broken down into carbon dioxide and ethanol by using bacteria or yeasts as biocatalysts. The commercial production of beer, bread, and wine are based on alcohol fermentation. Pyruvate molecules from glucose glycolysis may be further fermented to lactic acid. In this process, lactose is converted to lactic acid.

With the gradual changes in the conceptual approaches, the process of classical definition has been changed to "*the conversion of organic materials into relatively simple substances by micro-organisms-essentially efficient, flexible bio-factories.*" Some of the most used microorganisms in the biopharmaceutical industry are:

(i) Prokaryotes (e.g., *Escherichia coli*, *Staphylococcus aureus*),
(ii) Streptomycetes (e.g., *Streptomyces* spp., *Actinomyces* spp.),
(iii) Filamentous Fungi (e.g., *Nigrospora* spp., *Aspergillus* spp.),
(iv) Some yeasts (e.g., *Saccharomyces cereviciae*, *Pichia pastoris*)

Most of the biopharmaceutical industries use short peptides and low molecular weight organic molecules, larger molecules like proteins, nucleic acids, and macromolecules such as lipids and carbohydrate polymers, and complex form of biomolecules such as lipopolysaccharides, lipopeptides, and peptidoglycan for biologics. These molecules are in use as active pharmaceutical ingredient (API).

9.1.2 Therapeutic proteins manufacturing process

The magnificent growth and development of the biopharmaceutical industry (life sciences industry) could be possible because of the magical development of digitalized fully automatic fermentation equipment for the mass culture of host cells derived from plants, animals, or insects, and genetically upgraded microbes. The latest developed fermentation technology contributes a variety of anticancer cytotoxic drugs, antiinfectious disease antibiotics and vaccines. The sophisticated microchips-based manufacturing process has not only reduced the burden of heavy manpower involvement but also helped in the reduction of expenses for overall maintenance. Biopharmaceutical manufacturing technologies are innovative from the upstream process to downstream process involved in the therapeutic proteins production (Fig. 9.1).

Besides, manufacturing in-house R&D-based products, the multiple and complex upstream process can be modified on end-user need basis by using the strain improvement technique, either by following classical enrich culture methods or through developing genetically engineered host cells by recombinant DNA technology.

By using modern genetically engineered techniques, it is relatively easy to add pieces of foreign DNA molecule (a vector). The most critical factor in incorporating foreign gene in a bacterial (or host cells) genome is its stability during fermentation process. By adding foreign DNA, we can generate the desired host cells (may be from plants, animals, mammals, or microbes) to produce commercialized biologics. Recovering purified products through the downstream process is more cost-oriented than

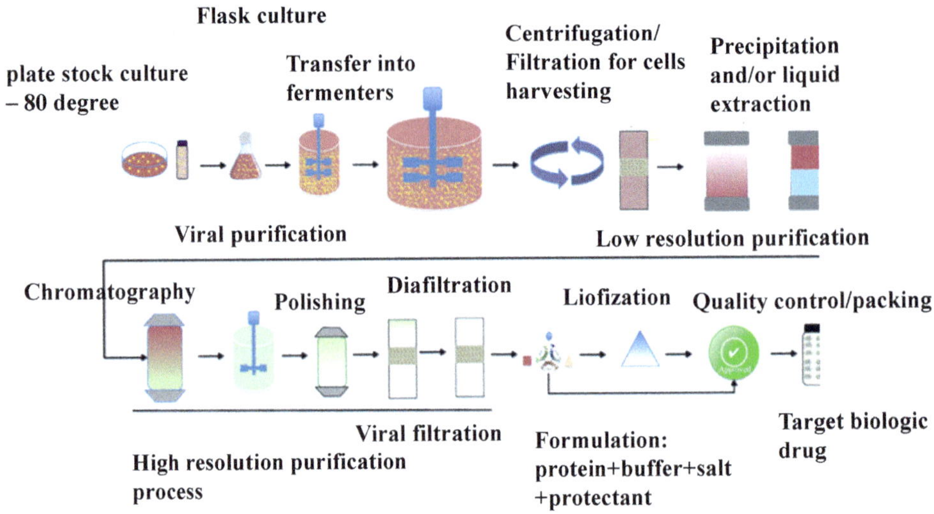

FIG. 9.1

Upstream & downstream process for therapeutic protein production.

the upstream process. The crystallization and polishing of biologics before packing is a more responsible task due to the short life of biologics.

Running and maintaining turnkey scale fermented for commercial production need intense supervision on quality by a well-trained quality assurance team. In addition, necessary bio-safety measures are taken, especially when Risk Group 2 (RG-2) pathogens are used. RG-2 pathogen can cause human or animal diseases but not to laboratory workers, livestock or community. Biosafety Level 2 Large Scale (BSL2-LS) containment facility is designed for special operational protocol of this process. As these products can be toxic and hazardous, their recovery and purification require adequate chemicals in biochemical facilities and equipment, including isolators for handling high-potent APIs (HPAPIs). Under cGMP fermentation procedures, quality is built into the entire process ensuring that regulatory agencies requirements are met in terms of safety, product identity, quality, and purity.

9.1.3 Advantages of microbial fermentation

Fermentation technology is a highly acclaimed method of production of APIs, where the bacteria or any other host cell simultaneously synthesis a targeted therapeutic molecule and catalyze the process. Bacteria and fungi as well as eukaryotic cells like CHO cells and insect cells are mostly used for manufacturing therapeutic proteins. The overall fermentation process depends on biological (concentration of microbe/host cells), physical (temperature, pH, O_2 concentration, starring process), and chemical factors (nutrients, essential elements, vitamins, carbohydrate). Most of the commercially produced enzymes, such as lipase, invertase, and rennet are produced by using genetically modified strains. Besides, the production of therapeutical biomolecules and commercialized enzymes, biomass production such as yeast biomass for bakery industry and starter culture for cheese making is also commercially manufactured. The fermentation process is mainly applicable for the production of biomass (viable cellular materials), production of extracellular materials, production of intracellular components, and transformation of substrate.

Sometimes, the biologics obtained through the fermentation process contain additional nontarget group(s) which may be toxic in nature. In order to make the product free from such toxic group(s), the impure therapeutic molecules are subjected to chemical modification, which nullifies the toxicity and increases potency. Therapeutic proteins produced through fermentation are of high molecular weight. The structural stability of such large molecular weight proteins depends on their secondary and tertiary structures, and posttranslational modifications (PTMs), mainly glycosylation. In this context, bacteria are simple organisms having a high capacity of adaptation with variable surroundings and diversity of survival strategies. Bacterial system lacks PTMs of protein, although glycosylation and phosphorylation have a significant role in the functioning of hemostatic proteins and are critical in the setting of disease diagnosis. Besides this, secondary level changes to hemostatic proteins have a wide range of effects on their ability to interact with other proteins. The therapeutically active protein produced by bacterial fermentation does not require PTMs as microbial systems, such as *E. coli*. Interestingly, there are glycoproteins which do not require glycans to exert their function. It was discovered that the recombinant human interleukin-2 (IL-2) produced in *E. coli* is having the same biological activity as the glycosylated IL-2 version isolated from the cultured mammalian cells. In 2006, Chiron Corporation (part of Novartis since April 2006) marketed nonglycosylated IL-2 to treat patients with metastatic melanoma. The main disadvantage with this biologic is its partial solubility in the blood stream, after delivery. This limited solubility of the nonglycosylated IL-2 was a serious problem not only because the dose of the therapeutic protein has to be increased to compensate for the lost activity, but because subsequent aggregation may cause immunogenicity. However, presently, recombinant therapeutic proteins have captured the global market due to their body friendly nature. The United States (US) tops the therapeutic market at the global level.

Industrialists mostly prefer therapeutic protein production by using recombinant *E. coli* as it is simple with short culture time, easy genetic manipulation, and low-cost media. Most of the fusion proteins are produced from a fusion gene, which is created by joining parts of two different genes. Fusion genes may occur naturally in the body by the transfer of DNA between chromosomes. For instance, the BCR-ABL gene found in some types of leukemia is a fusion gene that makes the BCR-ABL fusion protein. Fusion genes and proteins can also be made in the laboratory by combining genes or parts of genes from the same or different organisms. Insulin consists of two chains, the A- and B-chain, which are connected by two disulfide bridges. To produce a novel kind of Ec-fusion protein, the A-chain as well as the B-chain are fused to Ec-fragments containing either "knob" or "hole" mutation.

In 1970s, the discovery of recombination DNA technology for gene cloning and expression in bacteria had brought in a revolutionary milestone in the therapeutic biomolecule development process. Within the span of a few years, i.e., in 1978–79, the successful expression of two mammalian hormones, first Somatostatin and then Humalin by using bacteria mesmerized the scientific world [2] (Fig. 9.2).

In the beginning stage, the yield of insulin was low. But joint efforts of Genentech and then at Eli Lilly increased yield and led to the commercial production of human insulin by Eli Lilly. In 1982, FDA was given approval to insulin with the brand name "Humulin," resulted from the recombinant DNA technology for human use (Fig. 9.3).

While making insulin, there was no idea about neither the messenger RNA (mRNA) sequence nor the gene sequence of human insulin. The then knowledge was limited to the protein sequence

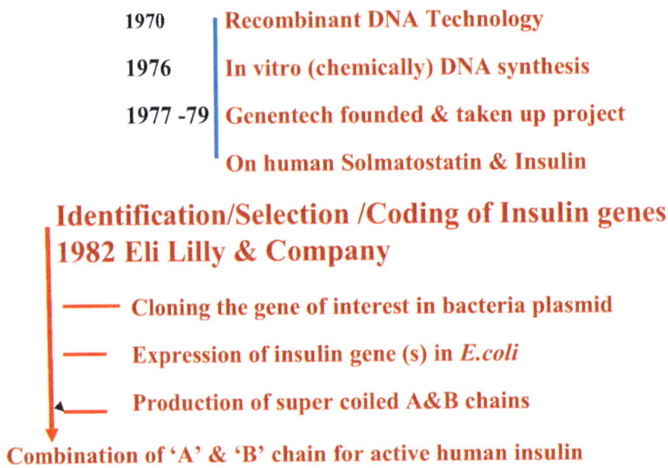

FIG. 9.2

Towards from recombinant DNA technology to insulin production.

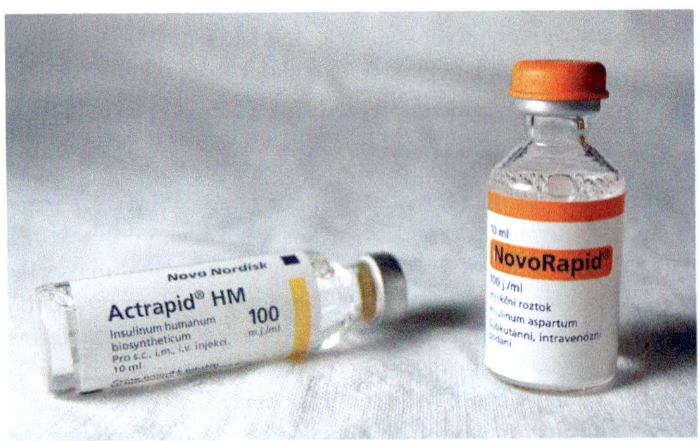

FIG. 9.3

Commercially available insulin.

and structure of insulin. The proceeding for Humalin was initiated with in vitro chemical synthesis of A and B chains of insulin by using the data-based knowledge on the sequence of amino acids for the insulin A and B peptide chain and then the genetic code, with a selection of codons preferred by *E. coli* [3]. Ultimately, the genes for the A and B chains of human insulin were ready for further processing linked to the human insulin production. It was wonderful to notice that the genes could function well in *E. coli*.

9.2 Biologic production process

Following are the complex sequential processes on how insulin coding DNA fragment is inserted into a plasmid and ultimately used for the expression of products through suitable host cell.

9.2.1 Plasmid expression system

Intensification of knowledge on recombinant technology and genetic engineering could make it possible to knock-out the desired genes and overexpression of target material by dismissing other vital steps in microbial systems or other host cells in a cost-effective way. The entire process of over expression of target biomolecules has become easier due to the development of synthetic biology and availability of data bank on life sciences. Following are the sequential steps on overexpression of desired biologic drugs from genetically engineered microbes or host cells belonging to plant tissues, animals, mammalians, and insects.

Genes of interest to be incorporated are optimized for cordon and iRNA usages, GC content, ribosome-binding sequences, and removal of unnecessary secondary structure, which improves the recombinant protein expression. It is advisable to integrate into transcriptionally active chromosomal region.

Generally, the promoter vector is derived from viruses, mice, and human. Commonly, human cytomegalo virus (HCMV) or murine is used as strong promoter vector for gene expression whereas low potential simian virus 40(SV_{40}) or rouse sarcoma virus (RSV) can be used for the expression of selection marker genes. In the CHO system, CHEF1-α promoter is more suitable as compared to the SV40 promoter alone [4,5]. Simian virus 40 (SV40) is also a strong promoter for the production of therapeutic protein in mammalian cells with lower expression yield, but higher stability than the EF1 and CMV promoters [6]. Mouse phosphoglycerate kinase 1 (PGK) promoter is also used to drive transgene expression in different cell line [7].

Synthetic promoters and hybrid promoters are also in use for long-term expression of a transgene [8,9]. The most common hybrid CAG promoters are designed by the fusion of cytomegalovirus (CMV) and chicken beta-actin promoter, and used for transgene expression in different cell lines [10,11]. The recombinant transcriptional activity in CHO cells can be monitored by integrating synthetic promoters in the gene. Currently, a number of synthetic promoters are available for enhancing transgene expression [12].

In order to select the expressed cells for desire protein, the plasmid should also contain selectable markers such as the *neo* gene generating aminoglycoside 3′-phophotransferase (APH 3′II) for G418 geneticin resistance, the *hyp* phosphotransferase gene for hygromycin resistance, the *Sh ble* gene for zeocin resistance, or puromycin N-acetyl-transferase encoded by the *pac* gene for puromycin resistance. Mostly, industries use DHFR or GS, which is present in plasmids when using DHFR-negative or GS-negative host cell lines, such as CHO-DG_{44} and NSo-GS/$CHOK_1$SVGS Nock out, respectively.

In order to generate transcriptionally active genomic environments in plasmid structure, scaffold or matrix-attachment regions are sequences in the DNA of eukaryotic chromosome [13,14].

The CHO host cells can also be engineered with attB recombination sites and plasmids with attP sequences [15]. The artificial chromosome expression (ACE) system consisting of a mammalian-based artificial chromosome known as platform ACE, an ACE targeting vector (ATV), and a mutant λ integrase (ACE integrase) is also used for target recombination [16].

ACE systems are a type of clone vectors, containing the DNA sequence elements, which can replicate and are stable in the host cell, and are expressed in daughter cells upon cell division. On need

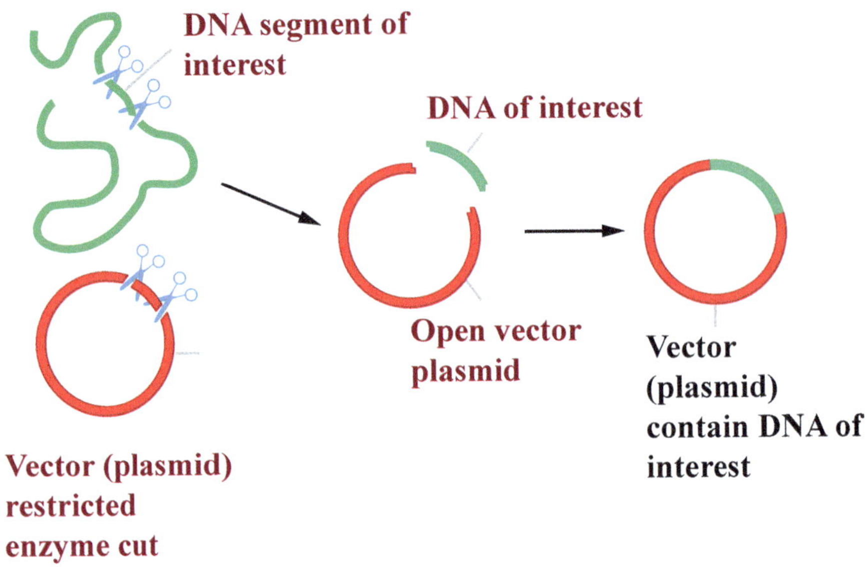

FIG. 9.4

Incorporation of foreign gene into a circular plasmid.

base, ACE for bacteria [17], mammalian cell, and human [18] yeast has been available as the ACE platform for the expression of desired proteins.

The ACE systems are used in the biological engineering of eukaryotic cells, including mammalian cells, with application in biopharmaceutical production. In this connection, insertion of synthetic gene(s) coded with insulin production is the easiest to understand the process.

Plasmids are essentially tiny, self-contained genomes that replicate independently, and genetically do not have any adverse effects on the host cells. Hence, they are unique biological tools for manipulating genes to obtain the desired recombinant therapeutic proteins. Insulin (biologic drugs) directly derived from animals are not friendly with the human system and may cause allergic symptoms. So, most commonly, the gene of interest from the human system is artificially synthesized in "in vitro" conditions or is isolated from the human pancreas cells and implanted in plasmid (Fig. 9.4).

The circular plasmid containing the gene of interest can be further inserted into the suitable bacterial strain to obtain pure super coiled plasmid DNA (Fig. 9.5).

Genes of interest to be incorporated are optimized for codon and iRNA usages, GC content, ribosome-binding sequences, and removal of unnecessary secondary structure, which improves the recombinant protein expression. It is advisable to integrate into transcriptionally active chromosomal region.

9.2.2 Foreign DNA transfer into host cells

Transfection is a process by which a desired fragment of foreign nucleic acid can be introduced into host cells or plasmid to obtain targeted biomolecules. The gene transfer vehicle is known as a vector. The vector may be a virus or plasmid. The foreign nucleic acids commonly used are plasmid DNA, RNA, SiRNA, and oligonucleotides. Ideal vectors transfer the precise amount of genetic material into

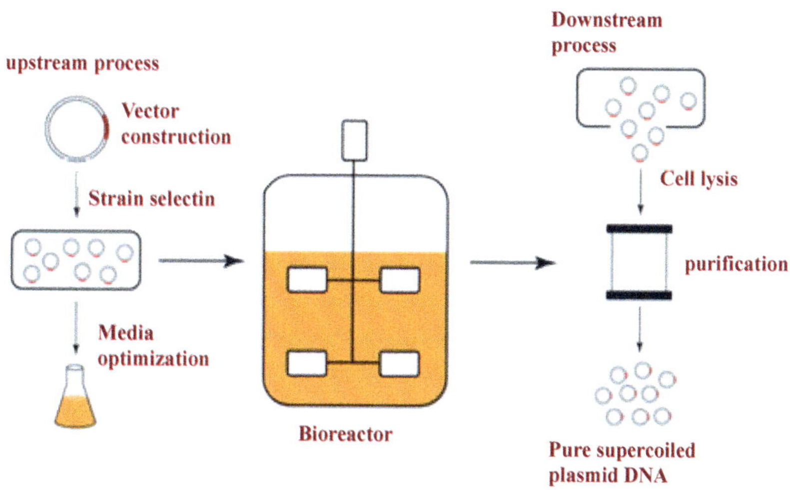

FIG. 9.5

Production of pure super coiled plasmid DNA.

the host cells, thereby allowing for the expression of the gene product without causing toxicity. Gene transfer via the viral vectors is called transduction while transfer via the nonviral vectors is called transfection.

The transfection process can also be possible through physical or chemical methods [19,20]. The process of transfection can be broadly divided into two categories: (i) existence of inserted foreign DNA fragment for a short period in the host cells and (ii) the inserted foreign DNA fragment persists in the host cells for a long-term and is passed to the progeny of the transfected cell.

The chemical method involves various cationic reagents such as calcium phosphate, lipids, and polymers. $CaPO_4$ precipitation is the earliest method, but has been surpassed by the more convenient and consistent lipid-based reagents such as Lipofectamine, Fugene, and Transfectin. Polyethyleneimine (PEI) is a low-cost alternative delivery method. The physical methods for DNA delivering includes direct injection by micro-needle, magnetofection by magnetic nanoparticles, and electroporation [20–22].

9.2.3 Amplification systems

Gene amplification processes play a vital role in designing and establishment of overexpression of cell lines with desired proteins. But, long-term stability of such overexpressed cell lines in the manufacturing process is highly challenging. Currently, Chinese hamster ovary cells are being used for the production of therapeutic proteins, especially monoclonal antibody (mAb) [23,24].

Although, several methods are presently available for gene amplification, the use of methotrexate (MTX) to induce gene amplification in the CHO DG44 system is the most successful. This method is noticed to be more time consuming. Methionine sulfoxine (MSX), which is glutamine synthetase (GS) inhibitor method has been widely used for the amplification process in recombinant antibody and protein production in mammalian cell culture. This method is an improved version of MTX with lesser time consumed with some drawbacks at various MSK concentrations [25,26].

Plasmid encoding method is also in use for gene amplification. In this process, a plasmid encoding a mammalian replication initiation region (IR) and a matrix attachment region (MAR) is used for increasing the copy number of the gene of interest in animal cells. This method is also known as the IR/MAR method. This method was initially used in basic cell biology research and at a later phase, it was also used for recombination protein production [27].

The overall responses to gene amplification processes vary with the type of systems to be used. For example, the IR/MAR is having low gene amplification in CHOK1 cells which is lower than that in CHO DG44 and COLD 320 cells [28–30].

The production of recombinant protein is higher in CHOK1 cells than in CHO DG44 cells [31]. So, IR/MAR is supposed to be a powerful method for the production of recombinant proteins production.

The following is a simple way of understanding plasmid containing DNA insertion into the host cell and large-scale insulin production.

9.2.4 Commercial production of insulin

The genetically modified strain of bacteria (*E. coli*) or yeast (*Saccharomyces cerevisi*) [32,33] (Fig. 9.6) is used for the production of insulin. The main reasons behind this technique are to make it easier to work with the DNA sequence. Once inside bacteria, a stretched DNA can readily be copied and its sequence determined. In addition, it induces the production of foreign protein within bacteria or a suitable host cell.

To clone a stretch of DNA (such as a gene) into a vector, restriction enzymes are used to cut out the DNA of interest and to open up the vector. The DNA is added to the vector by mixing the two together in the presence of the enzyme DNA ligase. This process is called molecular cloning or gene cloning. In case of nonavailability of adequate DNA for successful cutting or no suitable restriction enzyme recognition sites around the gene, more copies of DNA is made by polymerase chain reaction (PCR). By designing their PCR primers carefully, they can introduce new restriction sites on either sides of the copied DNA sequence.

The same restricted enzyme can be used to turn the vector into a linear molecule and make it ready to accept the new piece of DNA. The final step in cloning is to incorporate the DNA of interest into the vector. This is managed by mixing the gene and the opened vector together with a bacterium enzyme called DNA ligase. The ligase sticks DNA ends together to form a single circular molecule that includes both the vector and the gene.

Once a vector that contains foreign DNA has been constructed in the laboratory, it is introduced into bacterial cells. The insert of genetically engineered plasmid is done by creating tiny holes (pores) within the bacterial cell membrane, by sudden heating of the bacterial culture by several degrees or by passing an electric shock through the culture. The vector enters the bacteria while the pores are open. Once bacteria have recovered from the process of introducing DNA (called transformation), they can be cultured in the laboratory. Because the vector has an origin of replication, it is copied and passed to daughter cells in the same way as the bacterium's own DNA. Bacteria that contain foreign DNA are considered to be new, genetically modified organisms (GMOs). For this, conditions of their use are restricted. Most GM (genetically modified) bacteria are produced to be laboratory tools for making copies of DNA, producing protein and other therapeutic molecules like insulin in turnkey level (Fig. 9.7).

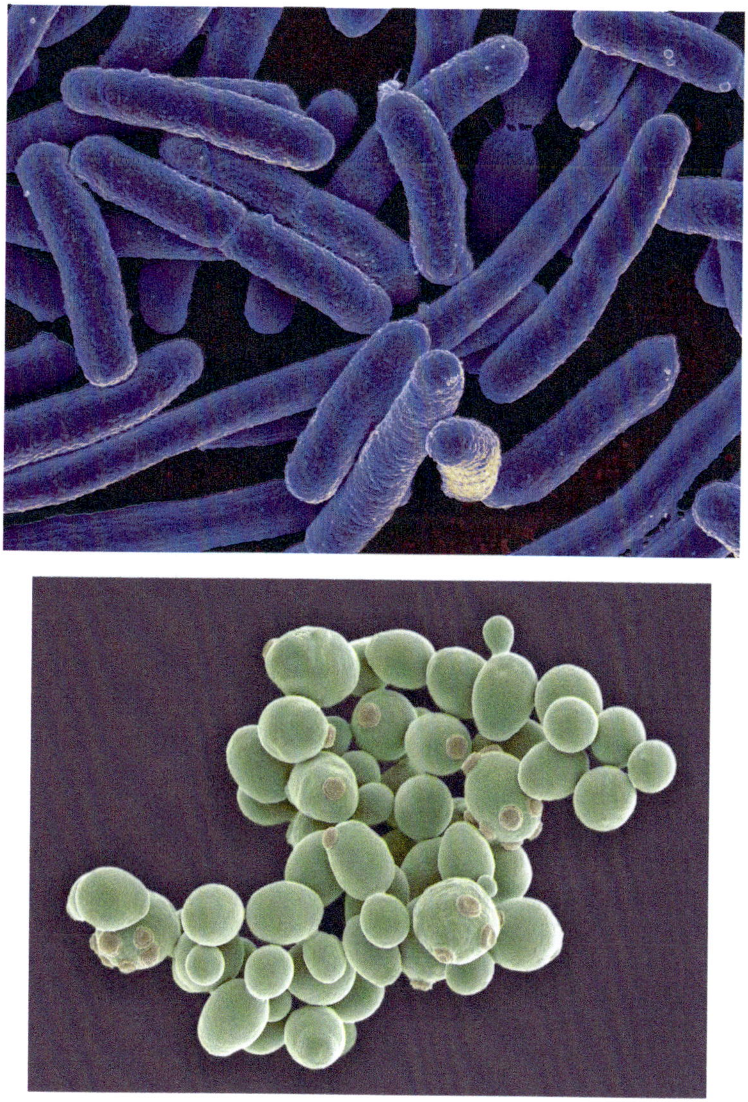

FIG. 9.6

Microbes used for genetically modified bacteria *E. coli* (top) and yeast *Saccharomyces cerevisi* (bottom).

9.3 What makes biologics unique?

Since a decade, biologics are emerged as new products in the market. As discussed earlier, biologics are originated from living sources and composed of large complex molecules, most of which are proteins or polypeptides. Chemically synthesized drugs can be better characterized by the current analytical tools and techniques. In contrast, biologics are difficult to be characterized due to their nature of origin from

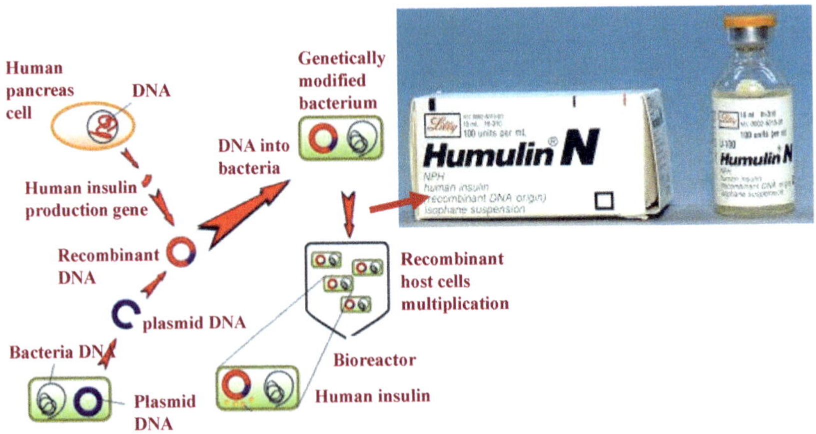

FIG. 9.7

Commercial production of insulin from GM bacteria through recombinant DNA technology.

a cell system in which it is produced, the fermentation media, or operating conditions. Hence, the most unique in biologics is genetical manipulation of living organisms. It is a challenging task to scale up biologics from laboratory quantities used for early analysis and preclinical testing to turnkey batches and producing quality product in consistency.

9.3.1 How is biologic different from generic drug?

Commonly, small molecule drugs (generic medicine) are used for medications. Most of them are chemically synthesized and used in a variety of dosage forms, including capsules, liquid, and topical creams. In contrast, biologics are made from living sources. They are mostly available as injections and infusions. But they can sometimes come as an inhaled medication. This is because oral biological drugs have been more challenging to make. Biologic drugs are structurally (Fig. 9.8) and functionally (Table 9.2) different from generic medicines.

9.3.2 Are biologics are expensive

With the advancement in tools and techniques on recombinant DNA technology, biologic drugs are gaining momentum in the global market. These drugs are most expensive and difficult to meet with pocket money. For instance, the biologic drug Humaria (generic name adalimumab), which is used to treat autoimmune diseases like rheumatoid arthritis and Crohn's disease, costs over $6000 for 1 carton (2 pens) without insurance. The cost of other biologic drugs like Opdivo, Enbrel, Keytruda, Herceptin, Rituxan, and Avastin can also run into thousands of dollars. The life-saving biologic medicine Zolgensma was in the news recently as it costs $2.1 million for a one-time treatment of spinal muscular atrophy, a rare childhood disorder.

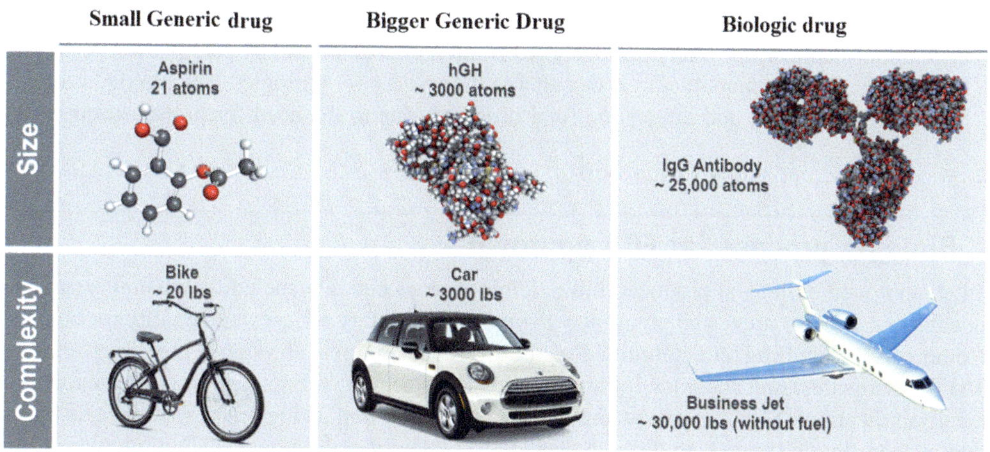

FIG. 9.8

Size difference generic drug vs. biologic drugs.

Table 9.2 Generic vs. biologic vs. biosimilar.

Monitoring	Generic	Biologic	Biosimilar
Development cost	2–3 million $	>800 million $	100–500$
Time for launching, years	2–3 years	8–10 years	7–8 years
Clinical studies	Bioequivalence studies in healthy volunteers	Phase I–III studies/efficacy	Phase pharmacokinetic comparison studies in phase III
Patients	20–50	800–1000	100–500
Postauthorization activities	Pharmacovigilance	Phase IV, risk management plan including Pharmacovigilance	Phase IV, risk management plan including Pharmacovigilance

Following are a few important reasons:

(i) Biologic drugs are derived from natural sources from humans, animals, and other living organisms. Biologics are composed of proteins, sugars, DNA, or even living cells. The manufacturing processes of biologics are highly complex as compared to chemically synthesized traditional pharmaceuticals. Hence, biologics production process is highly expensive.

(ii) The overall budget for establishing the development of biologics is substantially higher than most drugs. This is mainly due to the use of genetic engineering methodologies for a target biologic.

(iii) Due to the sophisticated and expensive manufacturing process, there is not much competition for biologic drug commercialization. Besides, some biologics are developed for treating rare diseases. As a result, no similar drugs from other pharmaceutical companies are available. Due to the structural complexity, it is difficult to replicate. Cheaper generic versions are available for most brand-name drugs. Many biologics are brand names protected by patent laws. As a result of this monopoly, drug manufacturers can charge high prices for biologics. The biologics are highly temperature-sensitive and affects the final drug cost due to the need for proper storage.

9.4 Biologic process for FDA approval

All FDA-approved biological products (biologic/biopharmaceutical), including biosimilars and interchangeable biosimilars, are tested vigorously to ensure their safety efficacy and quality for human use.

Commonly, small drugs are submitted to the FDA for approval through a new drug application (NDA) for their safety and accuracy for human use. Their generic versions go through an abbreviated new drug application (ANDA) to show they will work just as well as the brand versions. As biologics originated from living sources go through a special FDA-approval pathway called a biologic license application (BLA). Since biologic are more complex, the BLA process focuses more on where and how the medication is made, compared to the NDA's process. Biosimilar version goes through an abbreviated BLA process to show they are highly similar to their FDA-approved biologic counterpart.

Biosimilars are types of biologic similar to biologics (reference product) from safety, purity, potency, and effectiveness points of views. Biosimilars are evaluated for approval, based on all the evidence presented by the manufacturer. FDA approves biosimilars an abbreviated pathway. The target for biosimilar development program is to demonstrate the biosimilarity between the proposed biosimilar and its reference product, not to independently establish the safety and effectiveness of the proposed biosimilar. Biosimilar manufacturers do not need to conduct as many expensive and lengthy clinical trials. FDA evaluates each proposed biosimilar individually and advises manufacturers on the scope and extent of testing needed to show biosimilarity. FDA has the discretion to determine that an element is unnecessary in a proposed biosimilar application. FDA evaluates all the evidence, based on comparisons between the biosimilar and the reference product, in the context of the Agency's previous finding that the reference product is safe and effective.

On March 04, 2022-US Food and Drug administration brought in a few major amendments to ensure further safety and efficacy of new drugs, therapies, and treatments. In addition, the FDA wanted to quicken the navigating process by filling space of approval. On average, it takes at least 10 years for a new biologic to complete the journey from the initial discovery to the marketplace, with clinical trials alone taking almost 5–6 years (Fig. 9.9).

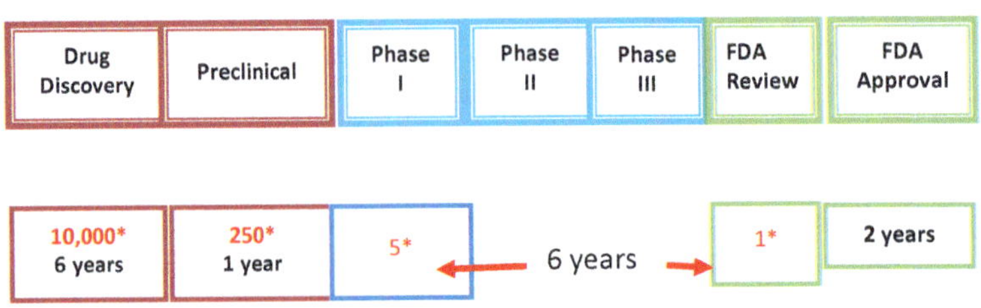

FIG. 9.9

Time taken for biologic drug discovery. *Components of drug substances.

The average cost to research and develop each successful drug is estimated to be $ 2.6 billion. This number incorporates the cost of failures of the thousands and sometimes million of compounds that may be screened and assessed early in the R&D process, only a few of which will ultimately receive approval. The overall probability of clinical success (the likelihood that a drug entering clinical testing will eventually be approved) is estimated to be less than 12%. While these numbers are daunting, a deeper understanding of the rigorous R&D process can explain why so many compounds do not make it and why it takes such a large, lengthy effort to get a new medicine to patients. So, to discover a therapeutic bio-molecule successfully needs a team of highly skilled scientists, provision of sophisticated equipment and tools, flexible manufacturing process, and highly efficient project management team.

9.4.1 FDA application types

The FDA has five common application types: NDA, ANDA, over-the-counter drug (OTC), BLA, and investigational new drug (IND):

(i) Investigational new drug (IND)

The current federal law requires that a drug has to conquer approval of marketing application, before bringing a drug into supply chain. This is mainly due to that a sponsor will probably want to ship the investigational drug to clinical investigators in many states; it must seek an exemption from the legal requirement. The IND is the means through which the sponsor technically obtains this exemption from the FDA.

(ii) New drug application (NDA)

After obtaining sufficient evidence on the drug's safety and effectiveness from FDA for marketing approval, the sponsor submits to FDA a NDA. The application must possess data on chemistry, pharmacology, medical, biopharmaceutics, and statistics. After the NDA approval, the product may be marketed in the United States.

(iii) Abbreviated new drug application (ANDA)

Commonly, for getting approval for generic drug from FDA, the applications are called "abbreviated" because they are generally not required to undergo any preclinical (animal) trial or clinical (human) trials. Instead, a generic applicant must scientifically demonstrate that its product is bioequivalent (i.e., performs in the same manner as the innovator drug). Once approved, an applicant may manufacture and market the generic drug product to provide a safe, effective, low-cost alternative to the American public.

(iv) Over-the-counter drugs (OTCs)

In the USA, some drugs are available to patients without prescription. These are known as "over-the-counter drugs (OTCs)", ranging from acne drug products to weight control drug products. As with prescription drugs, the Centre for Drug Evaluation and Research (CDER) oversees OTC drugs to ensure that they are properly labeled and that their benefits outweigh their risks.

(v) Biologics license application (BLA)

A BLA is a submission that contains specific information on the manufacturing process, chemistry, pharmacology, clinical pharmacology, and medical effects of the biological product. If the information provided meets the FDA requirements, the application is approved and a license is issued, allowing the firm to market the product.

9.4.2 Name-brand approval process

By the name-brand approval process of identifying, a viable name for a company, product, or service is given which includes research, brainstorming, and trademark vetting to ensure the name is available from a legal standpoint. There are many things that go into an effective brand name. Make it sure that the proposed name should meet the following criteria.

- Aligned with brand positioning;
- Embodies brand personality;
- Embodies (or more) brand benefits;
- Avoids negative or stigmatized concepts;
- Has an available trademark;
- Has an available URL.

In addition, there are many other factors that affect brand names like:

Highlighting the founder or inventor's name (like Eli Lilly & Company)
Describe the nature of product (Ubrelvy)
Describe an experience or image (Energy Drink)
Take a word out of context (Apple)
Make up a word (Google)

Distinctiveness: A unique brand name should be symbolically highly expressive within the competitive landscape in which it will live.

Brevity: A short brand name is highly acceptable to the customers. It should be easy to spell, say, and lucid to hear. It should also be compatible with the URL and email length.

Depth: A multilevel operational brand name is always welcome by the public. It should carry certain profound meaning that is related to the brand's story and/or positioning.

Energy: Effective brand names jump off the page. They work their way into your head and bounce around with a kinetic energy that other names lack.

Sound: The well acceptable name with crisp and bright, rolling off the tongue without much difficulty and falling on the ear with a pleasing resonance.

Discovery: Irrespective of company naming or product naming, the discovery process is vital. The naming process should be based on data analysis of brand research. It is well necessary to select brand name keeping audience to target and their acceptability. It helps to do a competitive brand audit, to get a sense of the types of names that exist in the market landscape. For getting the brand name granted, it is necessary to coordinate with all key stakeholders' opinion. This would play a significant role while finalizing the brand name.

9.4.3 Generic drug approval process

FDA approval is also mandatory for generic drugs. However, for generic drug approval, there is no need of preclinical (animal trial) or clinical trial (human trial) to establish safety the way brand-name products do for NDAs. Because of this, a different application, called the ANDA, is required for the FDA approval of generic drugs. It is the responsibility for a drug maker to provide proof that the generic product ingredient is effective against the condition/illness being treated and is the same as that of the name-brand product. It must show that the same amount of product enters the bloodstream in the same amount of time as the name-brand drug.

9.4.3.1 Over-the-counter drug (OTC) approval process

OTC approval process refers to a process which is carried through an NDA or an OTC monograph. As defined by FDA, OTC monograph is a "rule book" for each therapeutic category establishing conditions, such as active ingredients, uses (indications), does, labeling, and testing, under which an OTC drug is generally recognized as safe and effective (GRASSE) and can be marketed without an NDA and FDA premarket approval. Nonprescription drug products marketed under the OTC Drug Review are referred to as OTC monograph drug.

9.4.3.2 Medical device approval process

The overall process for medical device approval has, also, to follow the FDA approval. Both the preclinical testing clinical testing is performed with the medical devices for the FDA approval.

On the basis of risk, the approvals of medical devices are classified as follows:

Class I devices are low-risk and nonsurgical invasive devices such as bandages, handheld surgical instruments, and nonelectric wheelchair.

Class II devices are moderate-to-high risk devices such as syringes, catheters, and contact lenses.

Class III devices are high-risk devices. These are used for sustaining or supporting life or preventing impairment of human health such as pacemakers, fetal blood sampling monitors, and implants.

9.4.4 Patents

The life of a new patent in the US lasts for 20 years after the application filling date. Before a patent expires, the generic drug company is federally allowed to synthesize a generic version of the patented product.

While a generic drug applicant files for FDA approval should bring the notice of the original patent holder to make sure that it does not infringe on the name-brand patent or the patent is invalid.

After the name-brand company has been notified, it takes 45 days to take legal action against the generic drug maker. If there are grounds for a legal argument, the approval will be delayed for 30 months unless the patent reaches expiration or is determined to be invalid or not infringed upon. If the generic drug maker wins in court, the approval process will resume.

9.4.4.1 Reasons for approval, delay, and denial

The main reasons for delay in drug approval are missing or inadequate clinical or nonclinical studies that are usually included in NDAs submitted to US Food and Drug Administration (FDA). In several applications, study populations were either too small or inadequate to characterize risks anticipated during clinical use. Resubmission of failed applications is costly, delaying marketing approval and the availability of new drugs to patients.

9.4.4.2 Pay-for-delay agreements

Pay-for-delay agreements legally allows a name-brand drug maker to financially compensate a generic drug maker for delaying or even cancelling the generic product release. Pay-for-delay patent settlements essentially restrict generic drug market competition and keep name-brand drug price high.

9.4.4.3 Rapid approval classifications

In 1992, as per Prescription Drug User Fee Act (PDUFA), the FDA amended two types of time-based approval processes: standard review and priority review. Normally, the FDA is allotted 10 months to review new drugs, whereas, under priority review, that time is reduced to 6 months. The PDUFA permits the FDA to collect fees from drug manufacturers, and these user fees play an integral role in the expedition of the approval process.

PDUFA approval process is the first track process than anywhere in the world. Since the PDUFA was passed, more than 1000 drugs and biologics have come to market (i.e., HIV/AIDS antiretroviral drugs). The FDA granted Merck & Co, accelerated approval in 2021 for its drug Keytruda (pembrolizumab), which is used to target HER2-positive gastric cancer.

Generally, a breakthrough therapy is aimed to accelerate the development and review of drugs used to treat severe, life-threatening conditions. Breakthrough therapy must have preliminary clinical data suggesting that the drug has significant treatment advantages over the existing treatments on the market. In 2019, psilocybin-assisted therapy gained FDA designation as a breakthrough therapy for major depressive disorders and treatment-resistant depression. Hence, drugs that treat severe diseases and fulfil an unmet medical need qualify for fast-track approval. As requested by Pfizer and BioNTech, the mRNA COVID-19 vaccine was fast-tracked in July 2020 to prevent death and serious illness during the global pandemic.

At present, the FDA is piloting artificial intelligence (AI) and machine learning (ML) technology to automate the early stages of review. Using algorithms and statistical models in AI will allow for outcome productions with little to explicit programming.

References

[1] ResearchAndMarkets.com, Protein Therapeutics Market: Global Industry Trends, Share, Size, Growth, Opportunity and Forecast, 2020, pp. 2020–2025.

[2] A.S. Jones, J.V. O'Connor, Chemical characterization of methionyl human growth hormone, in: Hormone Drugs, Proceedings of the FDA—USP Workshop on Drug and Reference Standard for Insulin, Somatotropins and Thyroid Axis Hormones, U.S. Pharmacopeial Convention, Inc., Rockville, Maryland, 1981.

[3] R. Crea, A. Kraszewski, T. Hirose, K. Itakura, Chemical synthesis of genes for human insulin, Proc. Natl. Acad. Sci. U. S. A. 75 (12) (1978) 5765–5769.

[4] J.R. Deer, D.S. Allison, High-level expression of proteins in mammalian cells using transcription regulatory sequences from the Chinese hamster EF-1 alpha gene, Biotechnol. Prog. 20 (2004) 880–889.

[5] K.K. Chan, S.M. Wu, P.M. Nissom, S.K. Oh, A.B. Choo, Generation of high-level stable transgene expressing human embryonic stem cell lines using Chinese hamster elongation factor-1 alpha promoter system, Stem Cells Dev. 17 (2008) 825–836.

[6] S.C. Ho, Mariati, H.M.Y. Yeo, S.G. Fang, Y. Yang, Impact of using different promoters and matrix attachment regions on recombinant protein expression level and stability in stably transfected CHO cells, Mol. Biotechnol. 57 (2015) 138–144.

[7] N.S. Yew, M. Przybylska, R.J. Ziegler, D. Liu, S.H. Cheng, High and sustained transgene expression in vivo from plasmid vectors containing a hybrid ubiquitin promoter, Mol. Ther. 4 (2001) 75–82.

[8] H.L. Sladitschek, P.A. Neveu, Bidirectional promoter engineering for single cell MicroRNA sensors in embryonic stem cells, PLoS One 11 (2016), e0155177.

[9] H.W. Seo, et al., Evaluation of combinatorial cis-regulatory elements for stable gene expression in chicken cells, BMC Biotechnol. 10 (2010) 69.

[10] M.R. Schlabach, J.K. Hu, M. Li, S.J. Elledge, Synthetic design of strong promoters, Proc. Natl. Acad. Sci. U. S. A. 107 (2010) 2538–2543.

[11] J.M. Kim, J.S. Kim, D.H. Park, H.S. Kang, J. Yoon, K. Baek, Y. Yoon, Improved recombinant gene expression in CHO cells using matrix attachment regions, J. Biotechnol. 107 (2) (2004) 95–105.

[12] M. Zahn-Zabal, M. Kobr, P.A. Girod, M. Imhof, P. Chatellard, M. de Jesus, F. Wurm, N. Mermod, Development of stable cell lines for production or regulated expression using matrix attachment regions, J. Biotechnol. 87 (1) (2001) 29–42.

[13] M. Campbell, S. Corisdeo, C. McGee, D. Kraichely, Utilization of site-specific recombination for generating therapeutic protein producing cell lines, Mol. Biotechnol. 45 (3) (2010) 199–202.

[14] M.L. Kennard, D.L. Goosney, D. Monteith, L. Zhang, M. Moffat, D. Fischer, J. Mott, The generation of stable, high MAb expressing CHO cell lines based on the artificial chromosome expression (ACE) technology, Biotechnol. Bioeng. 104 (2009) 540–553.

[15] H. Shizuya, B. Birren, U.J. Kim, V. Mancino, T. Slepak, Y. Tachiiri, M. Simon, Cloning and stable maintenance of 300-kilobase-pair fragments of human DNA in Escherichia coli using an F-factor-based vector, Proc. Natl. Acad. Sci. U. S. A. 89 (1992) 8794–8797.

[16] A.P. Monaco, Z. Larin, YACs, BACs, PACs and MACs: artificial chromosomes as research tools, Trends Biotechnol. 12 (1994) 280–286.

[17] Y. Kazuki, M. Oshimura, Human artificial chromosomes for gene delivery and the development of animal models, Mol. Ther. 19 (2011) 1591–1601.

[18] P.A. Ioannou, C.T. Amemiya, J. Garnes, P.M. Kroisel, H. Shizuya, C. Chen, M.A. Batzer, P.J. de Jong, A new bacteriophage P1-derived vector for the propagation of large huan DNA fragments, Nat. Genet. 6 (1994) 84–89.

[19] A. Kichler, Gene transfer with modified polyethylenimines, J. Gene Med. 6 (Suppl. 1) (2004) S3–S10.

[20] S. Mehier-Humbert, R.H. Guy, Physical methods for gene transfer: improving the kinetics of gene delivery into cells, Adv. Drug Deliv. Rev. 57 (5) (2005) 733–753.

[21] T.K. Wong, E. Neumann, Electric field mediated gene transfer, Biochem. Biophys. Res. Commun. 107 (2) (1982) 584–587.

[22] R. Kunert, D. Reinhart, Advances in recombinant antibody manufacturing, Appl. Microbiol. Biotechnol. 100 (2016) 3451–3461.

[23] C.C. Kuo, et al., The emerging role of systems biology for engineering protein production in CHO cells, Curr. Opin. Biotechnol. 51 (2018) 64–69.

[24] M.I. Cockett, C.R. Bebbington, G.T. Yarranton, High level expression of tissue inhibitor of metalloproteinases in Chinese hamster ovary cells using glutamine synthetase gene amplification, Biotechnology (NY) 8 (1990) 662–667.

[25] S.C. Jun, M.S. Kim, H.J. Hong, G.M. Lee, Limitations to the development of humanized antibody producing Chinese hamster ovary cells using glutamine synthetase-mediated gene amplification, Biotechnol. Prog. 22 (2006) 770–780.

[26] S.S. Tanaka, S.H. Mitsuda, N. Shimizu, How a replication origin and matrix attachment region accelerate gene amplification under replication stress in mammalian cells, PLoS One 9 (2014) 29, https://doi.org/10.1371/journal.pone.0103439.

[27] Y. Araki, T. Hamafuji, C. Noguchi, N. Shimizu, Efficient recombinant production in mammalian cells using a novel IR/MAR gene amplification method, PLoS One 7 (2012), e41787, https://doi.org/10.1371/journal.pone.0041787.

[28] C. Noguchi, Y. Araki, D. Miki, N. Shimizu, Fusion of the Dhfr/Mtx and IR/MAR gene amplification methods produces a rapid and efficient method for stable recombinant protein production, PLoS One 7 (2012), e52990, https://doi.org/10.1371/journal.pone.0052990.

[29] H. Yoshimura, et al., High levels of human recombinant cyclooxygenase-1 expression in mammalian cells using a novel gene amplification method, Protein Expr. Purif. 80 (2011) 41–46.

[30] N. Shimizu, N. Hanada, K. Utani, N. Sekiguchi, Interconversion of intra- and extra-chromosomal sites of gene amplification by modulation of gene expression and DNA methylation, J. Cell. Biochem. 102 (2007) 515–529.

[31] M. L., et al., Auditioning of CHO host cell lines using the artificial chromosome expression (ACE) technology, Biotechnol. Bioeng. 104 (2009) 526–539.

[32] British National Formulary: BNF 69, sixty-ninth ed., British Medical Association, 2015, pp. 464–472, ISBN: 978-0-85711-156-2.

[33] N.A. Baeshen, M.N. Baeshen, A. Sheikh, R.S. Bora, M.M. Ahmed, H.A. Ramadan, et al., Cell factories for insulin production, Microb. Cell Factories 13 (1) (October 2014) 141, https://doi.org/10.1186/s12934-014-0141-0. 4203937 25270715.

CHAPTER

Biologics market size, shape and trend

10

Abbreviations

APSI	Applied Prevention Science International
CCSA	Canadian Centre on Substance Use and Addiction
CMO	Contract Manufacturing Organization
EMDDA	European Medical Device Distributors Agency
EUSPR	European Society for Prevention Research
FDA	Food and Drug Administration
FTC	Federal Trade Commission
MAbs	monoclonal antibodies
NCDs	noncommunicable diseases
PD	pharmacodynamic
PK	pharmacokinetics
PMx	pharmacometrics
QbD	quality by design
SPR	Society for Prevention Research
UNDCP	United Nations International Drug Control Programme
USFDA	United States Food and Drug Administration
WHO	World Health Organization

10.1 Biologic drugs (biopharmaceuticals) market

The biologic drugs market is referred to the sales of biopharmaceuticals and related services by drug agencies or traders at the national and international markets. The biologics market consists of companies manufacturing biopharmaceuticals derived from genetically modified microbes, insect cells, animal cells, plants, etc. A wide range of biologic products such as recombinant therapeutic proteins, gene therapy tissues, somatic cells, vaccines, and allergenics are available in the market. Since decades, the biologic drugs market in developed countries such as the United States is having high demand (Fig. 10.1).

As the biologics are derived from living systems, they have to undergo stringent rules and regulations prescribed by international drug control and regulating agencies. These include the European Medical Device Distributors Agency (EMDDA), the Canadian Centre on Substance Use and Addiction (CCSA), the Society for Prevention Research (SPR), the European Society for Prevention Research (EUSPR), World Health Organization (WHO), Applied Prevention Science International (APSI),

Techniques for Downstream process for Biologic Drugs and Vaccines. https://doi.org/10.1016/B978-0-443-19157-2.00010-8

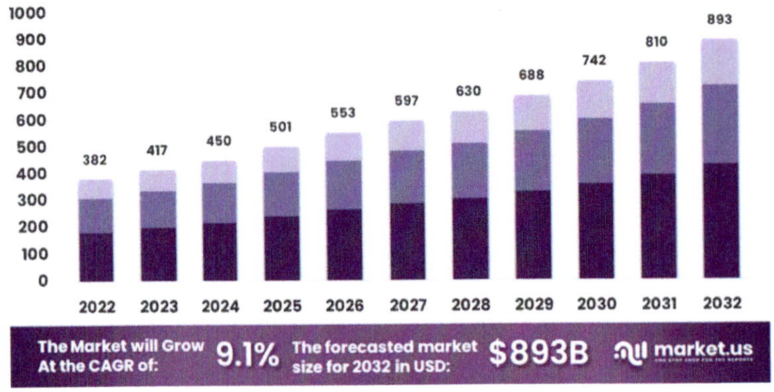

FIG. 10.1

Global biologics market (size by source 2023–32) (USD billion).

United Nations International Drug Control Programme (UNDCP), Codex Alimentarius Commission of Food and Agriculture Organization (CACFAO), and United States Food and Drug Administration (USFDA).

The demand for biologic drugs in the global market is increasing due to significance decline in productivity of small-molecule drug R&D. Patients accept these novel drugs more because of their increased efficacy in curing Crohn's disease and rheumatoid diseases. Biopharmaceutical production will increase with the development of manufacturing capabilities. R&D work around the world is in progress to find new species and expression methods. Revenue generation of biologics market is expected to be stimulated by the development of cell lines and reagents. A wide range of factors influence the growth and development of biopharmaceutical industries. Some of these factors include:

10.1.1 Research and development

A strong skilled research and development team is mainly responsible for new and innovative biologic product development. Continued research and development is necessary to manage consistency in product quality, which is responsible for generating the market's revenue growth.

10.1.2 Technological advancements

The technological advances in biotechnology are aiding the introduction of better quality products, thus boosting the financial growth of the biologics market.

10.1.3 Government/public funding initiatives

Biologic drugs originate from genetically modified expressing host cells from various living systems. The entire process of biologics is highly complex, risky and high-cost oriented. It has been a tough problem for the biopharmaceutical industry of a developing country to develop the infrastructure without government subsidy or financial assistance.

Public and philanthropic funding for drug development and clinical trials is a novel practice, particularly if directed toward nonprofits. Patent-led consortia reduces the need for a high rate of return for the commercial firms that ultimately manufacture and market a new drug. Such funding potentially could attract more industry investment in these therapies.

10.1.4 Rising prevalence of chronic diseases

Rising incidents of chronic diseases such as cancer, diabetes, and heart diseases are creating the necessity for a better option for remediation where generic drug fails. Biologics are highly targeted, efficacious, and represent an increasingly important part of the physicians' better option for the prevention or complete eradication of a chronic disease from the root cause of suffering.

10.2 Realization of necessity of biologic

In 1982, the first recombinant biologic, a human insulin produced in *Escherichia coli*, was introduced on the European Union (EU) market under the brand name Humulin [1]. Since then, more than 100 recombinant biologics have been developed and marketed. With the outbreak of COVID-19, people started realizing the urgency of biologic vaccine for the prevention/eradication of the corona pandemic.

In 2021, a handful of pharmaceutical companies created a world record on selling COVID-19 vaccines and treatments. In 2021, Pfizer alone generated $42.63 billion sales in vaccine (for COVID-19) jointly with BioNTech. Continued demand for the Pfizer-BioNtech vaccine is likely to keep Pfizer as the biggest firm of 2022. After an FDA's emergency use authorization in December 2020, the company's BioNtech-partnered COVID-19 became the most prominent in the world pharmaceutical market. The company received supply order from governments around the globe. Together with its partners, Pfizer produced 3 billion COVID-19 vaccine doses in the end of 2021 and reached about 4 billion at the end of 2022.

The second-best-selling pharmaceutical of 2021 was AbbVie's injectable biologic Humira, which generated $20.7 billion in sales. Moderna's COVID-19 captured the third position in selling in the international market which drove $17.7 billion in revenue. Merck's mega blockbuster Keytruda rounded out the top four with $17.2 billion in revenue. Although, Novartis and Roche desperately struggled to commercialize COVID-19 therapies, but Roche could develop a series of SARS-CoV-2 diagnostics. Subsequently, many pharmaceutical giants started manufacturing a variety of biologic drugs on the basis of market needs (Table 10.1).

10.3 Top best-selling biopharmaceuticals for the year 2021

10.3.1 Comirnaty COVID-19 vaccine

Comirnaty is a monovalent COVID-19 vaccine that is approved by FDA on August 23, 2021. This was developed, jointly, by Pfizer and BioNTech. The vaccine has been known as the Pfizer-BioNtech COVID-19 Vaccine, and the approved vaccine is marketed as Comirnaty, for the prevention of COVID-19 in individuals 12 years of age and older. Since the great successful global market of human tissue plasminogen activator, antiblood clots, in 1987 and erythropoietin, red blood cells enhancer, in

Table 10.1 Fifty best selling pharmaceuticals in 2021.

Drug name	Manufacturer(s)	2021 Sales	Indication(s)
Comirnaty COVID-19 vaccine	Pfizer/BioNTech	$59,108,942,000.00	Reduce the risk of COVID-19 infection
Humira (adalimumab)	AbbVie	$20,694,000,000.00	Rheumatoid and psoriatic arthritis, ankylosing spondylitis, Crohn's disease, ulcerative colitis
Spikevax COVID-19 vaccine	Moderna	$17,675,000,000.00	Reduce the risk of COVID-19 infection
Keytruda (pembrolizumab)	Merck	$17,186,000,000.00	Various cancers
Eliquis (apixaban)	Bristol Myers Squibb and Pfizer	$16,732,000,000.00	Blood clots
Revlimid (lenalidomide)	Bristol Myers Squibb (Celgene)	$12,821,000,000.00	Myelodysplastic syndrome, multiple myeloma, and mantle cell lymphoma
Imbruvica (ibrutinib)	Pharmacyclics (AbbVie) and Janssen (Johnson & Johnson)	$9,777,000,000.00	Chronic lymphocytic leukemia/small lymphocytic lymphoma with 17p deletion, Waldenström's macroglobulinemia
Eylea (aflibercept)	Regeneron Pharmaceuticals, Bayer	$9,244,294,000.00	Age-related macular degeneration, macular edema and diabetic retinopathy
Stelara (ustekinumab)	Janssen (Johnson & Johnson)	$9,134,000,000.00	Plaque psoriasis, psoriatic arthritis, ulcerative colitis and Crohn's disease
Biktarvy (bictegravir, emtricitabine, and tenofovir alafenamide)	Gilead Sciences	$8,624,000,000.00	HIV
Xarelto (rivaroxaban)	Janssen (Johnson & Johnson)/Bayer	$8,039,505,000.00	Reducing the risk of stroke in nonvalvular atrial fibrillation, deep vein thrombosis, pulmonary embolism, DVT prophylaxis following knee or hip replacement surgery
Opdivo (nivolumab)	Bristol Myers Squibb	$7,523,000,000.00	Various forms of cancer
Trulicity (dulaglutide)	Eli Lilly	$6,471,900,000.00	Type 2 diabetes
Dupixent (dupilumab)	Sanofi Genzyme, Regeneron Pharmaceuticals	$6,209,567,000.00	Atopic dermatitis, asthma, chronic rhinosinitus with nasal polyps.
Darzalex (daratumumab)	(Janssen) Johnson & Johnson	$6,023,000,000.00	Multiple myeloma
REGEN-COV	Regeneron Pharmaceuticals, Bayer	$5,828,000,000.00	Reduce the risk of COVID-19 infection
Trikafta/Kaftrio	Vertex Pharmaceuticals	$5,697,200,000.00	Cystic fibrosis

Table 10.1 Fifty best selling pharmaceuticals in 2021 ***Continued***

Drug name	Manufacturer(s)	2021 Sales	Indication(s)
Gardasil/Gardasil 9 (human papillomavirus 9-valent vaccine)	Merck	$5,673,000,000.00	Various cancers caused by human papillomavirus
Veklury (remdesivir)	Gilead Sciences	$5,565,000,000.00	COVID-19 requiring hospitalization
Ibrance (palbociclib)	Pfizer	$5,437,000,000.00	HR-positive and HER2-negative breast cancer
Januvia/Janumet (sitagliptin)	Merck & Co.	$5,288,000,000.00	Type 2 diabetes
Prevnar family (pneumococcal vaccine)	Pfizer	$5,272,000,000.00	Pneumococcal vaccine
Tagrisso (osimertinib)	AstraZeneca	$5,015,000,000.00	Non-small-cell lung carcinomas
Cosentyx (secukinumab)	Novartis	$4,718,000,000.00	Plaque psoriasis, psoriatic arthritis, ankylosing spondylitis
Ocrevus (ocrelizumab)	Roche	$4,622,292,000.00	Relapsing or primary progressive multiple sclerosis
Enbrel (etanercept)	Amgen	$4,465,000,000.00	Plaque psoriasis, rheumatoid arthritis, psoriatic arthritis, juvenile idiopathic arthritis and ankylosing spondylitis
Xtandi	Astellas	$4,173,232,862.57	Treatment of prostate cancer
Invega Sustenna, Xeplion, Invega Trinza/ Trevicta	Janssen (Johnson & Johnson)	$4,022,000,000.00	Schizophrenia, schizoaffective disorder
Vaxzevria	AstraZeneca	$3,917,000,000.00	Reduce risk of COVID-19 infection
Entyvio	Takeda	$3,908,309,048.65	Ulcerative colitis, Crohn's disease
Lynparza	AstraZeneca/ Merck & Co.	$3,737,000,000.00	Ovarian cancer, breast cancer
Perjeta (pertuzumab)	Roche	$3,616,452,000.00	HER2-positive breast cancer
Gilenya (fingolimod)	Novartis	$3,548,000,000.00	Multiple sclerosis
Pomalyst/Imnovid (pomalidomide)	Bristol-Myers Squibb	$3,332,000,000.00	Multiple myeloma
Orencia (abatacept)	Bristol-Myers Squibb	$3,306,000,000.00	Rheumatoid arthritis, juvenile idiopathic arthritis
Actemra/RoActemra (tocilizumab)	Roche	$3,257,092,800.00	Rheumatoid arthritis, forms of juvenile idiopathic arthritis and giant cell arteritis as well as CAR T cell-induced severe or life-threatening cytokine release syndrome
Prolia (denosumab)	Amgen	$3,248,000,000.00	Osteoporosis, low bone mass
Remicade (infliximab)	Janssen (Johnson & Johnson)	$3,190,000,000.00	Crohn's disease, ulcerative colitis, rheumatoid arthritis, ankylosing spondylitis, psoriatic arthritis, plaque psoriasis

Continued

Table 10.1 Fifty best selling pharmaceuticals in 2021 ***Continued***

Drug name	Manufacturer(s)	2021 Sales	Indication(s)
Tecentriq (atezolizumab)	Roche	$3,031,236,000.00	Urothelial carcinoma, non-small cell lung cancer and triple-negative breast cancer
Farxiga	AstraZeneca	$3,000,000,000.00	Type 2 diabetes mellitus, heart failure
Lantus (insulin glargine)	Sanofi-Aventis	$2,950,402,000.00	Diabetes
Skyrizi	AbbVie	$2,939,000,000.00	Plaque psoriasis, psoriatic arthritis
Genvoya (elvitegravir, cobicistat, emtricitabine and tenofovir alafenamide)	Gilead Sciences	$2,879,000,000.00	HIV-1 infection
Avastin (bevacizumab)	Roche	$2,794,406,400.00	Colorectal, lung, glioblastoma, kidney, cervical and ovarian cancer
Entresto (sacubitril/ valsartan)	Novartis	$2,787,000,000.00	Chronic heart failure and reduced ejection fraction
Hemlibra	Roche	$2,763,316,800.00	Hemophilia A
Symbicort (budesonide formoterol)	AstraZeneca	$2,728,000,000.00	Asthma, airflow obstruction in patients with COPD
Triumeq (dolutegravir, abacavir and lamivudine)	GlaxoSmithKline	$2,590,384,800.00	HIV
Herceptin (trastuzumab)	Genentech (Roche)	$2,463,393,600.00	Breast, stomach and esophageal cancer
Xeljanz (tofacitinib)	Pfizer	$2,455,000,000.00	Rheumatoid arthritis, psoriatric arthritis and ulcerative colitis

1988, the global biologic drugs market has gone up. In the beginning of 2022, BioNtech joined with Matina Biopharma in developing Lipid Nanocrystal (LNC) delivery platform technology to advance novel formulation for mRNA vaccines, including a potential formulation for oral vaccines. Pfizer is also betting on the future potential of mRNA-based vaccines. Although mRNA is not the only solution for game-changing impact on global health; hence, Pfizer with the partnership with BioNTech and Matinas BioPharma wanted to develop a robust mRNA strategy and are aggressively involved in the further application of mRNA-based vaccine development in other forms.

10.3.2 Humira (adalimumab)

Adalimumab (brand name—Humira) is a monoclonal antibody (MAb) used to treat moderate to severe rheumatoid arthritis (RA), such as joint swelling, pain, and fatigue.

Amjevita is a biosimilar to Humira (adalimumab), an anti-TNF-α MAb. The active ingredient of Amjevita is having the same amino acid sequence as Humira. Humira was approved for medical use in the United States in 2002. In 2020, it was the 205th most commonly prescribed medication in the United States.

It was manufactured at BASF Bioresearch Corporation, developed by BASF Knoll (BASF Pharma), and ultimately manufactured and marketed by Abbott Laboratories after Abbott's acquisition of BASF

Pharma. In 2013, Abbott split into two companies, one retaining the Abbott name and the other named AbbVie who has taken over the development and marketing of Humira.

10.3.3 Moderna

The Moderna COVID-19 vaccine (INN: elasomeran), is in the market under the brand name Spikevax.

10.3.3.1 Spikevax COVID-19

This was developed by the American company Moderna, the United States National Institute of Allergy and Infectious Diseases (NIAID), and the Biomedical Advanced Research and Development Authority (BARDA).

It is an mRNA vaccine composed of nucleoside-modified mRNA (modRNA) encoding a spike protein of SARS-CoV-2, which is encapsulated in lipid nanoparticles [2].

In 2022, the bivalent version of the vaccine (Moderna COVID-19 Vaccine, Bivalent) containing elasomeran/elasomeran 0-ommicron (Spikevax Bivalent Zero/Omicron) [3] was authorized for the use as booster doses in individuals aged 18 or older in the United Kingdom, Switzerland, Australia, Canada, the European Union, and the United States [4–12]. The second component of the version of the bivalent vaccine version used in other countries (mRNA is based on the Omicron BA1 variant).

10.3.4 Keytruda (pembrolizumab)

Pembrolizumab is sold under the brand name Keytruda, a humanized antibody used in cancer immunotherapy that treats melanoma, lung cancer, head and neck cancer, Hodgkin's lymphoma, stomach cancer, cervical cancer, and certain types of breast cancer. It is given by slow injection into a vein.

It was first approved in 1914 and marketed by Merck & Co. It stimulates the body's immune system to fight cancer cells. Pembrolizumab targets and blocks a protein called PD-1 on the surface of certain immune cells called T-cells. Blocking PD-1 triggers the T-cells to find and kill cancer cells.

10.3.5 Eliquis (apixaban)

Apixaban was developed jointly by Bristol Myers Squibb and Pfizer. This is sold under the brand name Eliquis. It is an anticoagulant medication used to treat and prevent blood clots and to prevent stroke in people with nonvalvular atrial fibrillation through the directly inhibiting factor, Xa. Specifically, it is used to prevent blood clots following hip or knee replacement and in those with a history of prior clots. It is taken orally.

10.4 Biologic drug market analysis

The biologics market can be well analyzed on the basis of the diversity in products which include MAbs, vaccines, recombinant hormones/proteins, cellular-based biologics, gene-based biologics and other products. A wide range of life-threatening communicable and noncommunicable diseases (NCDs) (cancer, infectious diseases like corona, autoimmune diseases, and other applications) which are not curable by generic medicines can be prevented or eradicated by biologic drugs.

10.4.1 Present status of biologics market size

The Market Research Reports says "The global biological market reached US$321.0 Billion in 2022. Looking forward, IMARC Group expects the market to reach US$535.5 Billion by 2028, exhibiting a growth rate (CAGR) of 8.1% during 2023-2028" [13].

There are many internationally reputed biologic (biopharmaceuticals) survey agencies around the world predicting the global market of biopharmaceutical products. The overall survey report says that the demand for biologic drugs is in the ascending order in the international market [14–19].

The biopharmaceutical market is majorly driven by the increase in capital investment from key players, increase in magnitude of chronic diseases, loss of patent rights of the leading biologic drugs, and growing demand of biopharmaceuticals due to their lesser side effects and target oriented nature. The following are the top 10 biotechnology and pharmaceutical companies ranked by revenue in USD billion.

1 Pfizer

With the manufacture of mRNA COVID-19 vaccine, Pfizer stands the world's biggest manufacturer of biologics with the sales of USD 72 billion, 37 billion of which was generated by COVID Comirnaty-developed in partnership with BioNTech-alone. In R&D, the company spent USD 13.829 billion. This makes Comirnaty the best-selling blockbuster medicines in a single year of all times by a huge margin.

2 AbbVie

Adalimumab is sold under the brand name Humira. It is a fully human MAb used to treat rheumatoid arthritis, psoriatic arthritis, ankylosing spondylitis, Crohn's disease, ulcerate colitis, and many other diseases. It is administered by subcutaneous injection.

In 2022, AbbVie Inc. a pharmaceutical company, headquartered in Chicago got USD 11 billion boost in sales, primarily driven by its flagship Humira drug. In 2022, the company's sales revenue was USD 55.041 billion. But, Humira's sales have plummeted in Europe due to the entry of biosimilar competition into the market.

Adalimumab was approved by the US Food and Drug Administration (FDA). It was derived from phage display. The medicine is a result of the collaboration between BASF Bioresearch Corporation and Cambridge Antibody Technology, United Kingdom.

3 Novartis

Novartis is one of the largest pharmaceutical companies in the world. It is a Swiss-American multinational pharmaceutical corporation. It is based in Basel, Switzerland and Cambridge, Massachusetts, United States. Novartis is consistently ranked among the global top five and the fourth largest by revenue in 2022 with a sales value of USD 51.128 billion.

4 Johnson & Jonson (J&J)

Johnson and Jonson (J&J) is a U.S. based multinational pharmaceutical company engaged in developing medical devices, pharmaceuticals, and consumer packaged goods. During COVID-19 outbreak, J&J committed over $1 billion for manufacturing COVID-19 vaccine on "not for profit" based. This viral vector vaccine requirement is single dose for prevention corona vaccine.

During COVID-19 pandemic, the Janssen COVID-19 vaccine, sold under the brand name Jcovden, is a COVID-19 vaccine that was developed by Janssen Vaccines in Leiden, The Netherlands and its Belgian parent company Janssen Pharmaceuticals, a subsidiary of American company Johnson & Johnson.

It is a viral vector vaccine based on a human adenovirus that has been modified to contain the gene for making the spike protein of the SARS-CoV-2 virus that causes COVID-19. The vaccine requires only one dose and does not need to be stored frozen.

5 Roche

F. Hoffmann-La Roche AG, also known as Roche, is a Swiss multinational healthcare company that operates under two divisions, Pharmaceuticals and Diagnostics. The company headquarters is located in Basel. Roche is the fifth largest pharmaceutical company in the world by revenue, and the leading provider of cancer treatments globally. Roche has grown into one of the world's largest biotech companies and a leading provider of in vitro diagnostics across major disease areas.

6 Bristol Myers Squibb

Bristol Myers Squibb (BMS) is an American multinational pharmaceutical company located in New York City. BMS manufactures prescription pharmaceuticals and biologics in several therapeutic areas, including cancer, HIV/AIDS, cardiovascular diseases, diabetes, hepatitis, rheumatoid arthritis, and psychiatric disorders. In 1989, Bristol-Myers and Squibb merged and became Bristol-Myers Squibb.

Among the best achievements, in 2014, BMS received FDA approval for the use of the PD-1 inhibitor nivolumab (Opdivo) in treating patients whose skin cancer cannot be cured or have not responded to previous drug therapies.

Presently, the company is known for biological and pharmaceutical research, and its blood thinners Plavix and Eliquis, as well as cancer drug Opdivo (nivolumab). Opdivo is a prescription medicine used in combination with Yervoy (ipilimumab) as a first treatment for adults with a type of advanced stage lung cancer (called nonsmall cell lung cancer).

7 Merck & Co., Inc.

Merck & Co., Inc. is a multinational pharmaceutical company based on New Jersey. It was founded in Germany in 1668. The company deals with medicines, vaccines, biologic therapies, and animal health products. In 2020, the company had eight blockbuster drugs or products, each with over $1 billion in revenue. At present, Merck is in process to develop a VSA-based COVID-19 vaccine in partnership with the International AIDs Vaccines Initiative (IAVI), a nonprofit that was originally set up to push forward HIV vaccines but has broadened its portfolio to include vaccines for a wide range of other diseases.

8 Sanofi

Sanofi S.A. is a French-based multinational pharmaceutical and healthcare company, headquartered in Paris. The company was established in 1973 and merged with Synthelabo in 1990 for Sanofi-Synthelabo. Again, in 2004, Sanofi-Synthelabo merged with Aventis and renamed to Sanofi-Aventis. The Sanofi-GSK COVID-19 vaccine sold under the brand name VidPrevtyn Beta, is a COVID-19 vaccine developed by Sanofi Pasteur and GSK. The Sanofi-GSK COVID-19 vaccine was approved for medical use in the Europe Union in November 2022. VidPrevtyn Beta is a monovalent, recombinant-protein COVID-19 vaccine developed by Sanofi, modified on the Beta variant spike antigen and includes GSK's pandemic adjuvant. The same recombinant-protein technology is used in Sanofi's approved seasonal flu vaccines.

9 AstraZeneca

AstraZeneca Plc is a British-Swedish multinational pharmaceutical and biotechnology company based in Cambridge Biomedical Campus in Cambridge, England. It has a portfolio of products for major diseases in areas including oncology, cardiovascular, gastrointestinal, neuroscience, respiratory and inflammation.

The Oxford-AstraZeneca COVID-19 vaccine is sold under the brand names Covishield and Vaxzevria. It is a viral vector vaccine for the prevention of COVID-19. Developed in the United Kingdom by Oxford University and British-Swedish company AstraZeneca, using the modified chimpanzee adenovirus ChAOx1 as a vector. The vaccine is given by intramuscular injection. In January 2021, India approved the use of the Oxford-AstraZeneca vaccine, paving the way for a mass immunization campaign in the world's second most populous country. It was announced that the Oxford-AstraZeneca shot would be made locally by the Serum Institute of India (SII) with a brand name COVISHIELD.

10 Glaxo Smith Kline Plc

Glaxo SmithKline plc (GSK plc) was established in 2000 by a merger of GlaxoWellcome and SmithKline Beecham. In 2022, *Fortune Global* placed GSK plc in the rank294 among world top 500 companies (Geneal); and became the tenth largest pharmaceutical company. The Company's product portfolio includes prescription medicines and vaccines. It develops and manufactures a range of vaccines, for the prevention of hepatitis A, hepatitis B, invasive disease caused by H, influenza, chickenpox, diphtheria, pertussis, tetanus, rotavirus, cervical cancer and others.

In July 2020, the UK government signed up for 60 million doses of a COVID-19 vaccine developed by GSK and Sanofi. It uses a recombinant protein-based technology from Sanofi and GSK's pandemic technology. In 2021, the companies had an agreement to produce 1 billion doses, subject to successful trials and regulatory approval. In the same year, the company also agreed to a $2.1 billion deal with the United States to produce 100 million doses of vaccine.

10.4.2 Future prospects of biologic drugs

One of the best survey reports on biologic drugs says: "it is faster-growing sector in the pharmaceutical industry; with total revenue reaching up to $163 billion with annual growth surpasses 8%, whereas conventional pharmaceuticals barely reach 4%" [20].

The main reason for such success in acceptability of biologic drugs by the people is its efficacy and safety with minimum side effects, as compared to the incapability of other generic medicines treatment, earlier. Since the development of the first recombinant DNA product "Humulin," developed by Genentech and marketed by Lilly in 1982, biologics have brought enormous trust among the people on the magical benefit of biologics to patients. Presently, more than 250 biologic drugs and vaccines in the global market have to meet the emergency on prevention/eradication of life-threatening epidemics/diseases, around the world. The last decade has witnessed the gradual shifting of investment dollars from small-molecule drugs to biologics [21].

Since 1995, the pattern of acceptability of biologic drugs has been in the order of 25% per year. Currently, there are about >200 approved drugs in the global market and over 1500 drugs are being evaluated in clinical trials, and many more drugs are in the pipeline [22].

Despite many successes, there are still a number of hurdles for developing the manufacture process of biologics. Due to expiry of patents of many biologic drugs and their high cost, it has been difficult for common patients to meet the expenditure from pocket. The average daily cost of a biologic drug in the United States is $45 compared with only $2 for chemical (small-molecule) drugs.

Hence, in order to overcome such a problem, the emergence of biosimilars that are developed to have active properties similar to a biologic drug that has already been approved, may resolve the issue related to the high cost of biosimilars and may increase the competitiveness in the market [23].

Biosimilars is "highly similar" to a branded drug, which is referred to as the "reference product."

Regenerative therapy and personalized medicines are two of the developing areas in biologic drugs discovery. Advanced personalized medicines based on gene therapies, immunotherapy, and antibody drug conjugates are likely to be introduced, offering greater promise for focused disease remissions and regenerative medicine.

Inadequate budgetary provision and stringent regulatory acts not only affect patient filling but also impact market access and put pharmaceutical companies under constraint in their way of operational system functioning. Currently, the biopharmaceutical companies are ready to risk sharing agreements and collaboration projects undertaking to launch biologic products in a sustainable pattern. For instance, in order to take prevention majors against SARS-CoV-2 outbreak in Wuhan City, China, in February 2020, under WHO's Coordination, a group of experts with diverse backgrounds started developing vaccines against COVID-19. In addition, the group made declaration that:

> We are scientists, physicians, funders and manufacturers who have come together as part of an international collaboration, coordinated by the World Health Organization (WHO), to help speed the availability of a vaccine against COVID-19. While a vaccine for general use takes time to develop, a vaccine may ultimately be instrumental in controlling this worldwide pandemic. In the interim, we applaud the implementation of community intervention measures that reduce spread of the virus and protect people, including vulnerable populations, and pledge to use the time gained by the widespread adoption of such measures to develop a vaccine as rapidly as possible. We will continue efforts to strengthen the unprecedented worldwide collaboration, cooperation and sharing of data already underway. We believe these efforts will help reduce inefficiencies and duplication of effort, and we will work tenaciously to increase the likelihood that one or more safe and effective vaccines will soon be made available to all [24].

Despite barriers to commercialize biologics, it has provided a safe solution in many areas of unmet medical needs and their growth is expected to continue. In order to smoothen the process of approval, the FDA outlined 4 key goals in tackling various issues such as streamlining the approval process, improving regulatory clarity, increasing educational efforts to improve understanding among stakeholders, and collaborating with the Federal Trade Commission (FTC) to address anticompetitive behaviors. Hence, in July 2018, the FDA published its Biosimilars Action Plan, acknowledging the lack of competition in the biologics space.

10.4.3 Market analysis of biologic drugs

The dynamic market of biologic drugs is mainly due to increase in the burden of chronic diseases, expiry of patent rights of highly demanding biologic drugs, and increase in acceptability of people due to less side effects and target-oriented drug delivery into the body.

According to the WHO report [25] NCDs cause 41 million deaths yearly, equivalent to 74% of the total death rate, globally. Each year, 17 million people die from an NCD before age 70; 86% of these premature deaths occur in low- and middle-income countries. Of all NCD deaths, 77% are in low- and middle-income countries. Cardiovascular diseases account for most NCD deaths, or 17.9 million people annually, followed by cancer (9.3 million), chronic respiratory diseases (4.1 million), and diabetes (2.0 million, including kidney disease deaths caused by diabetes). In addition, tobacco use, physical inactivity, harmful use of alcohol and unhealthy diets all increase the risk of dying from NCDs. The increasing cases of chronic diseases are expected to increase the demand for biologics further in treating chronic diseases.

Since the last decade, there has been rapid dropping in research and development on small therapeutic molecules, and biologics are projected to grow rapidly in the next few years. Biologic drugs are among the most sophisticated and elegant achievements of modern science. Due to their effective performance and less side effects, biologics have been getting high acceptability by the people, as compared to generic synthetic therapeutic molecules. Hence, top pharmaceutical companies are in the process to develop less expensive biologics in the forms of biobetter and biosimilars, and continuing research is yielding entirely new types of products. Radically new concepts are making it to the market, such as the cell therapy, proven successful in cancer treatment, and somewhat further out, gene therapies, which offer even more amazing promises of regenerative medicine or disease remission.

The world's top key players have been showing interest in spending money for R&D in order to meet the demand of biologic drugs in the global market. To maintain their market image and position, the key firms of pharmaceutical sector are focusing on the introduction and development of many biopharmaceutical medications. For instant, IQVIA reports (US health data company) total global spending on COVID-19 vaccines is supposed to reach $157 billion by 2025, driven by mass vaccination programs underway and "booster shot" expected every 2 years [26].

Looking at the rising demand for biologic drugs in the global market, the pharmaceutical players started showing interest in developing biologic drugs. For instance, the TXBio Pilot Grant Program is a research initiative led by the University of Texas at Austin College of Natural Sciences and Cockrell School of Engineering that focuses on the discovery, early development, and clinical translation of biological therapies developed across the university (UT Austin). TXBio (Texas Biologics) is funded by philanthropic gifts that are being used to improve the intellectual environment and infrastructure at UT Austin that is relevant to biologic discovery and translation.

Furthermore, product launches by the key market players are anticipated to drive the market's growth. For instance, in October 2021, Cadila Pharmaceuticals (Cadila) launched two similar biologics in the Indian market, namely NuPTH, a teriparatide biologic, and Cadalimab, a similar biologic of the auto-immune treatment adalimumab. Additionally, in 2021, Amgen reported that its partner AstraZeneca had submitted a Biologics License Application (BLA) to the US Food and Drug Administration (FDA) for tezepelumab, a potential first-in-class medicine for severe asthma.

However, there are operational and technical challenges. Consistency in reproducing large molecules at an industrial scale requires manufacturing capabilities of a previously unknown sophistication. For instance, a molecule of aspirin consists of 21 atoms. A biopharmaceutical molecule might contain anything from 2000 to 25,000 atoms. The biotechnology behind producing such big therapeutic and target-oriented biomolecules is highly sophisticated, and the genetically modified host cells (expressing system) must be preserved in the cell bank, where proper frozen conditions are maintained to keep consistency in the expression of host cells during the fermentation process for manufacturing products of interest. However, the stringent regulatory processes, high capital investment, and loss of patent exclusivity are likely to restrain the growth of the market.

Copying biopharmaceuticals is not a simple task. The budgetary provision for large-scale life science industry is high cost-oriented with the budget about $200 million to $500 million or even more, compared with similar-scale small-mole facilities that may cost just $30 million to $100 million with a time-bound period of 4 to 5 years. These facilities are costly to run, too, with long process duration, low yields, expensive raw materials, and, not the least, the need for a team of highly skilled experts to operate them. There are myriad reasons the rapid growth and increasing importance of the industry is producing new challenges and opportunities. To keep pace, biopharma players must revisit and fundamentally reassess many of the strategies, technologies, and operational approaches they currently use.

The key drivers of the biologics market are subjected to condition of adequate funding system to initiate the process. For instance, to meet the urgency of manufacturing vaccine for the COVID-19 pandemic needs large amounts of funding from the public and private organizations to enhance the process of drug development and supports research of new drugs. For instance, in 2020, MTF Biologics, which is a global nonprofit organization has awarded over $1 million in funding to 11 research studies across the US through its 2020 Extramural Research Grants Program. Welcome Trust, a UK-based research-charity and the Gates Foundation announced collaboration with the credit card company MasterCard in a $125 million push to speed up the development of drugs and vaccines for treating COVID-19 infection. Also, the Bill & Melinda Gates Foundation committed around $200 billion for the development of vaccines for different diseases. Hence increasing public-private funding for life science research will enhance the research and development which will drive growth of the biologics market.

10.4.4 Biologics market restraints

As discussed earlier, due to its origin from living sources, biologics production process is very stringent under well-prescribed rules and regulation from highly reputed regulatory authorities. The regulatory agencies are responsible to ensure the safety and efficacy of newly developed biologic drugs for human use. For this purpose, the regulatory bodies conform the results on pharmacokinetics (PK), pharmacodynamic (PD), and pharmacometrics (PMx) of the results taken from human trials. These studies are mainly for understanding the characteristic of drugs and how they behave in different patient population.

Biologic medicines present unique and complex challenges, from both technical and regulatory perspectives. While manufacturing biologic drugs, it is necessary to look after the quality of the raw materials, maintenance of contaminant free surrounding around manufacturing process, and deviations from process parameters. Minor differences in the manufacturing process can have a significant effect on product quality. Due to the complexity in structure, it requires special skill for managing the manufacturing process, involving living cells that are inherently more variable than the components involved in chemical synthesis. In addition, these medicines need critical quality attributes associated with product safety, potency, PKs, and overall quality.

Before releasing to market, it is mandatory for a new biologic drug to get approval from FDA's Current Good Manufacturing Practices (US), as well as similar guidelines from the WHO. This is mainly to control quality and safety of the biologic drugs, before human use. Through such practice, a manufacturer can establish a framework to identify, scientifically evaluate, and control potential risk to product quality. Hence, quality manufacturing by design (QbD) is a key to fostering a healthy, sustainable marketplace for all biologic medicines, including biosimilar and biobetter.

On the basis of origin from living resources and complexity, it is necessary for all manufacturers of biologic drugs to be deeply committed to quality. By adhering to rigorous global standards, the industry can help in developing public confidence and continue to earn trust among regulators, patients, and prescribers.

10.4.5 Biologics market trends

In order to properly streamline the biologic drugs manufacturing process, the US Food and Drug Administration (FDA) has brought many amendments in its regulatory acts to eliminate outdated biologic requirements, thus allowing drug manufacturers to employ the latest manufacturing technologies and testing capabilities. FDA has given alternate option to pharmaceutical industries to avail protocol from other reliable agencies other than FDA's Centre for Biologics Evaluation and Research (CBER) which would be helpful to ensure safety, purity and potency of biologics. The FDA has also removed a rule (section 610.21 of FDA code) which specifies minimal potency limits to be met for certain antibodies and antigens. The FDA has also amended section 610.53 regarding storage periods and storage condition for biologics. These amendments would be immensely helpful in the currently operational/manufacturing process of biologics in a much conventional way. As a result, the number of new biologics approved was 48 in 2019, and it has been increased to 53 in 2020.

10.4.6 Higher profit margin for biologics

It has been difficult for common people to afford high cost-oriented biologics from pocket money. However, increase in efficacy with lesser side effects has attracted the attention of people for biologic drugs [27–32]. Hence, the developed countries such as the US want to reduce the cost of biologics. In order to meet the demand of essential biologic drugs, it is high time for policy makers to find some alternatives to design biosimilar drugs to replace the expiry-patented biologics [33–37].

The top players of the global pharmaceutical industry regulate the biologic marketing division with a high margin [38]. In 2015, 25 largest pharmaceutical companies accounted for 73% profit value [39]. In response for such a high price rise in biologics, the pharmaceutical companies justify that this is mainly to show obligation to shareholders.

There have been extensive reports on the adverse consequences of high drug prices, but limited information on industry profits. Hence, in order to bring down the cost of biologics, it is necessary to understand how the price of a drug is fixed and the profit returns of an industry. According to US Food and Drug Administration (FDA), most traditional medications are made by mixing together chemical ingredients. Biologics, however, are made with genetically living organisms such as bacteria, yeasts, animal cells, and mammalian cells. These materials can be more sensitive and unwieldy to work with than chemicals in a "recipe." Christine Simm, the executive director of the Biosimilars Council and executive vice president of Policy and Strategic Alliances for the Association for Accessible Medicines in Washington, D.C. says "the rise in high cost is due to research, development and ultimately sophisticated manufacturing cost." Another important cost factor is the lack of competition in the global market. Making the "generic" version of a biologic (called a biosimilar medication) is costly and involves a rigorous approval process.

10.5 Alternate method for saving money on biologics

Despite the high cost of biologics, still, there are ways to save money on these medications. Biosimilar medications are very close in function and structure to the biologic medications as reference drug. Currently, there are more than 33 biosimilar medications approved by the FDA in the global market, and dozens more in the pipeline. Biosimilars are usually 15% to 35% cheaper than their biologic counterparts. Examples of biosimilars include:

- **Mvasi and Zirabev (biosimilar for Avastin)**
  Bevacizumab sold under the brand name Avastin is mainly used for the treatment of cancer and a specific eye disease. Mvasi is the biosimilar medication approved to treat cancer, followed by Zirabev.

- **Zarxio and Nivestym**
  Neupogen is one of the oldest biologics. Its biosimilar, Zarxio was the first biosimilar approved by the FDA, in 2015. Zarxio and Nivestym like Neupogen stimulate white blood cell production.
- **Semglee**
  Semglee is long-acting insulin that was first approved in June 2020. In July 2021, it became the first interchangeable biosimilar to Lantus. Lantus (insulin glargine) was first approved by FDA in 2000. And over 20 years later, Semglee (insulin glargine-yfgn) was approved as the first interchangeable biosimilar to Lantus. This means it is highly similar to Lantus.

10.6 Biologic market segmentation

10.6.1 Introduction

Biologic market segmentation refers to the process of grouping the mass market for better understanding and integration of different sectors to meet the demand of goods in quality and safety [39]. It is mainly for the development of the competitive market and discharge of functions in more effective manner [40,41].

10.6.2 Types of biologic market segmentation

For convincing and better understanding, the entire market size can be grouped into the following:

(i) By type
The biologics market is segmented into:
(a) Therapeutic proteins
The term therapeutic protein was first used to describe medicines that are genetically engineered versions of naturally occurring human proteins. The complex structure of proteins can be represented in several different ways to understand and correlate the structure with function. Hence, therapeutic proteins can be subgrouped into (i) metabolic disorders therapeutic proteins; (ii) cancer therapeutic proteins; (iii) cardiovascular therapeutic proteins; (vi) immunological therapeutic proteins; and (v) other therapeutic proteins.
Monoclonal antibodies
MAbs are produced from cell lines by cloning white blood cells. They bind to certain target body, such as antigens on the surface of cancer cells or any other foreign pathogens. There are varieties of MAbs, and each MAb is made so that it binds to only one antigen. MAbs are being used in the diagnosis and treatment of many diseases, including some type of cancer. They can be used alone or to carry drugs, toxins, or radioactive substances directly to cancer cells. MAbs can be well studied by grouping into: (i) anticancer MAbs; (ii) immunological MAbs; (iii) antiinfective MAbs; (iv) neuropharmacological MAbs; (v) cardiovascular and cerebrovascular MAbs; and (vi) other MAbs.
(b) Vaccines
A vaccine is a biological preparation that provides active acquired immunity to a particular infection or malignant disease. The safety and effectiveness of vaccines have been widely studied and verified. Vaccines can be subgrouped into: (i) antiinfective vaccines; (ii) autoimmunity vaccines; and (iii) other vaccines.

(ii) By route of administration
Based on route of administration, the market is segmented into injection and infusion. Oral administration is currently not possible for biologics. Barriers in the gastrointestinal tract severely limit the systematic absorption of biologics. Different research strategies have been explored to enable oral delivery of biologics.

(c) Drug classification.
The biologics market is segmented by drug classification into: (i) branded drug and (ii) Generic drugs. The branded drugs market is the largest of the biologics market segmented by drug classification, accounting for about 80% of the total in 2020. The global biologics market analysis report for 2022–25 says: "The global biologics market reached a value of nearly $253,414.8 million in 2020, having increased at a compound annual growth rate (CAGR) of 8.7% since 2015. The market is expected to grow from $253,414.8 million in 2020 to $420,551.5 million in 2025 at a rate of 10.7%. The market is then expected to grow at a CAGR of 9.6% from 2025 and reach $664,744.2 million in 2030" [42].

(d) By mode of purchase
The biologics market is segmented by mode of purchase into (i) prescription drugs and (ii) over-the-counter (OTC) drugs.

The prescription drugs market is the largest segment of the biologics market segmented by the mode of purchase, accounting for 78.2% of the total in 2020. By prescription drugs, OTC drugs segment is expected to be the fastest growing segment in the biologics market segmented by mode of purchase, at a CAGR of 10.9% during 2020–25.

(e) By distribution channel.
On the basis of distribution pattern, biologics can be grouped into (i) hospital pharmacies, (ii) retail pharmacies, and (iii) online pharmacies. The hospital pharmacies market.

The hospital pharmacies market was the largest segment of the biologics market segmented by hospital pharmacies, accounting for 50.1% of the total in 2020. Going forward, the online pharmacies segment is expected to be the fastest growing segment in the biologics market segmented by distribution channel, at a CAGR of 16.4% during 2020–25.

(f) By geography
The biologics market is segmented by geography into:

- North America (United States, Canada, Mexico)
- Western Europe (United Kingdom, Germany, France, Italy, Spain, Austria, Belgium, Denmark, Finland, Ireland, The Netherlands, Norway Portugal, Sweden, Switzerland)
- Asia Pacific (China, India, Japan Australia, Indonesia, South Korea, Hong Kong, Malaysia, New Zealand, Philippines, Singapore, Thailand, Vietnam)
- Eastern Europe (Russia, Czech Republic, Poland, Romania)
- South America (Brazil, Argentina, Chile, Colombia, Peru, Venezuela)
- Middle East (Israel, Saudi Arabia Turkey, UAE)
- Africa (Egypt, Nigeria, South Africa)

Research and Markets reports say: "North America was the largest region in the global biologics market, accounting for 53.2% of the total in 2020. It was followed by Western Europe, Asia-pacific and then other regions. Going forward, the faster-growing regions in the biologics market will be South America and Asia-pacific, where growth will be at CAGRs of 14.8% and 11.4% respectively, during 2020–2015" [43].

10.7 Contract manufacturing pharmaceuticals

Contact manufacturing is the process of outsourcing of the work of manufacturing or producing medication and other related materials. More specifically, contact manufacturing is the outsourcing of certain production activities that were previously performed by the manufacturer to a third-party. In this mutual and legalized understanding, a company may outsource the manufacturing of certain components of the product or outsource the assembly of the product. Nowadays, outsourcing companies have become specialists in a multitude of services for manufacturers including design, production, assembly, and distribution.

10.7.1 Benefit of contract manufacturing

The main purpose of contract manufacturing is to save money on labor management, materials storing, and other expenses related to production. Currently, the third-party contact manufacturers are commonly seen in developing countries with an abundant supply of cheap labor and minimum regulations. The contract manufacturing process is helpful for a company to lower its production costs, maintain the quality of its production, and increase its profit margins. Hence, contract manufacturing is gaining momentum at present owing to the increase in the investments by the pharmaceutical giants for the research and development of various biologics and biosimilars.

At present, North America has captured the global pharmaceutical market, having significant shareholders in the global pharmaceutical contract packaging market and is expected to grow at a CAGR of 9.6% during the forecast period [44].

The leading Biopharmaceuticals Contract Manufacturing companies such as Boehringer Ingelheim International GmbH, Lonza, Rentschler Biopharma SE, Binex Co., Ltd., Novartis AG, ProBioBioGen AG, TOYOBO CO. LTD., Samsung Biologics, ADMA Biologics, Inc., Baxter BioPharma Solutions (Baxter), iBio., Catalent, Inc., Camberbrex Corporation, Pfizer Inc., and others are currently working in the biopharmaceuticals contract manufacturing market.

Cipla Limited, a leading pharmaceutical player had an agreement for a joint venture project with Kemwell Biopharma Private Limited to develop, manufacture and commercialize biosimilars for global markets.

During COVID-19 outbreak, on August 28, 2020, iBio and Planet Biotechnology entered into an exclusive worldwide license agreement for the development of COVID-19 therapeutics.

10.7.2 Market dynamics of contract manufacturing

In order to avoid some of the costly expenditure such as the cost of specialized manufacturing equipment, the expertise to navigate regulatory concerns, and others issues, leading biopharmaceutical companies prefer to have contract manufacturing. For instance, on April 13, 2021, Lonza, a leading global Biopharmaceuticals Contract Manufacturing (CDMO/CMO) and Junshi Biosciences. An innovative Chinese pharmaceutical company collaborated to accelerate the development and manufacturing of biologics using Lonza GS Xceed Gene Expression System. Currently, various contract manufacturing organizations are involved in expending their activities to provide end-to-end manufacturing of biological drug substances and product supply. In November 2021, Bionova established USD 25 million single-use plant for expending its protein, cell line, and process development services into clinical and commercial manufacturing space. The high cost of establishing production facilities for biologic drugs and regulatory constrains for CMOs in gaining product approval are some of the factors likely to impede the Biopharmaceuticals Contract Manufacturing market growth.

A remarkable growth in biopharmaceutical companies has been observed, during the outbreak of unprecedented COVID-19 pandemic. On March 08, 2021, Baxter Bio Pharma Solutions, one of the largest contract manufacturing organizations (CMO) agreed with Moderma, Inc., a biotechnology company pioneering messenger RNA (mRNA) therapeutic and vaccines to provide fill/finish sterile manufacturing services and supply packaging for about 60–90 million doses of the Moderna COVID-19 vaccine in 2021.

References

[1] J. Sandow, W. Landgraf, R. Becker, G. Seipke, Equivalent recombinant human insulin preparations and their place in therapy, Eur Endocrinol. 11 (1) (2015) 10–16.
[2] U.S. Centers for Disease Control and Prevention (CDC), COVID-19 Vaccines for Moderately to Severely Immunocompromised People, 13 August, Centers for Disease Control and Prevention, 2021.
[3] M. Stevis-Gridneff, The E.U. drug regulator approves the Moderna vaccine, The New York Times (2021). 6 January.
[4] C. Hussey, Moderna Reports Second Quarter Fiscal Year 2021 Financial Results and Provides Business Updates, Moderna, 2021. (Press release) 5 August.
[5] C. Hussey, Moderna Announces Clinical Update on Bivalent COVID-19 Booster Platform, Moderna, 2022. (Press release). 19 April.
[6] First Bivalent COVID-19 Booster Vaccine Approved by UK Medicines Regulator, Medicines and Healthcare Products Regulatory Agency (MHRA), 2022. (Press release) 15 August.
[7] Regulatory Approval of Spikevax Bivalent Original/Omicron Booster Vaccine, Medicines and Healthcare products Regulatory Agency (MHRA), 2022. 15 August.
[8] Swissmedic Approves First Bivalent COVID-19 Booster Vaccine in Switzerland, Swissmedic, 2022. (Press release). 29 August.
[9] TGA Provisionally Approves Moderna Bivalent COVID-19 Vaccine for Use as a Booster Dose in Adults, Therapeutic Goods Administration, 2022. (Press release) 30 August.
[10] Regulatory Decision Summary—Spikevax Bivalent, Health Canada, 2022. 1 September.
[11] N. Boisvert, Health Canada approves updated Moderna vaccine for omicron variant, CBC News (2022). 1 September.
[12] A. Capobianco, Coronavirus (COVID-19) Update: FDA Authorizes Moderna, Pfizer-BioNTech Bivalent COVID-19 Vaccines for Use as a Booster Dose, U.S. Food and Drug Administration (FDA), 2022. (Press release) 31 August.
[13] https//www.imrogroup.com/healthcare.
[14] https://market.us/report/biologics-market/request-sample/.
[15] https://scoop.market.us/.
[16] https://media.market.us/.
[17] https://news.market.us/.
[18] https://medicalmarketreport.com/.
[19] https://chemicalmarketreports.com/.
[20] R. Otto, A. Santagonisto, U. Schrader, Rapid Growth in Biopharma: Challenges and Opportunities, McKinsey, 2014. Retrieved March 07, 2017, from http://www.mckinsey.com/industries/pharmaceuticals-and-medical-products/our-insights/rapid-growth-in-biopharma.

[21] The Global Use of Medicines: Outlook Through 2017 (Rep.), 2013. Retrieved March 7, 2017, from IMS Institute for Healthcare Informatics website: https://www.imshealth.com/files/web/IMSH%20Institute/Reports/The_Global_Use_of_Medicines_2017/global%20use%20of%20med%202017%20right6%20Biologics_Market.pdf.

[22] D.A. Emerton, Profitability in the biosimilars market: can you translate scientific excellence into a healthy commercial return? BioProcess Int. 11 (6 Suppl) (2013). 6–14, 23.

[23] A. Cassels, Why biosimilars should be interchangeable with biologics, Pharm. J. (2017). Retrieved 7 March 2017, from http://www.pharmaceutical-journal.com/opinion/insight/why-biosimilars-should-be-interchangeable-with-biologics.

[24] Public Statement for Collaboration on COVID-19 Vaccine Development, World Health Organization, 2020. https://www.who.int/news/item.

[25] Global Burden of Disease Collaborative Network, Global Burden of Disease Study 2019 (GBD 2019) Results, Institute for Health Metrics and Evaluation—IHME, 2020. https://vizhub.healthdata.org/gbd-results/.

[26] https://www.reuters.com/healthcare-pharmaceuticals.

[27] National Academies of Sciences, Engineering, and Medicine, Making Medicines Affordable: A National Imperative, National Academies Press, Washington, DC, 2018.

[28] Skyrocketing Drug Prices: Year One of the Trump Administration, Committee on Oversight and Government Reform, Washington, DC, 2018. (Press release) May 11.

[29] R. Sachs, Prescription Drug Policy: The Year in Review, and the Year Ahead, Health Affairs Blog website, 2019. https://www.healthaffairs.org/do/10.1377/hblog20190103.183538/full/. Published January 3.

[30] A.S. Kesselheim, J. Avorn, A. Sarpatwari, The high cost of prescription drugs in the United States: origins and prospects for reform, JAMA 316 (8) (2016) 858–871, https://doi.org/10.1001/jama.2016.11237.

[31] United States Government Accountability Office Drug Industry, Profits, Research and Development Spending, and Merger and Acquisition Deals, United States Government Accountability Office, Washington, DC, 2017. https://www.gao.gov/assets/690/688472.pdf.

[32] R.M. Califf, A. Slavitt, Lowering cost and increasing access to drugs without jeopardizing innovation, JAMA 321 (16) (2019) 1571–1573, https://doi.org/10.1001/jama.2019.3846.

[33] M. Angell, The pharmaceutical industry—to whom is it accountable? N. Engl. J. Med. 342 (25) (2000) 1902–1904, https://doi.org/10.1056/NEJM200006223422509.

[34] N. Sood, T. Shih, K. Van Nuys, D. Goldman, The Flow of Money through the Pharmaceutical Distribution System, Leonard D. Schaeffer Center for Health Policy & Economics, Los Angeles, CA, 2017. https://healthpolicy.usc.edu/wp-content/uploads/2017/06/USC_Flow-of-MoneyWhitePaper_Final_Spreads.pdf.

[35] L. Chen, The most profitable industries in 2016, Forbes (2015). December 21 https://www.forbes.com/sites/liyanchen/2015/12/21/the-most-profitable-industries-in-2016/.

[36] R. Anderson, Pharmaceutical industry gets high on fat profits, BBC News (November 6) (2014). https://www.bbc.com/news/business-28212223.

[37] A. Kirzinger, B. Wu, M. Brodie, Kaiser Health Tracking Poll (2018): Views on Prescription Drug Pricing and Medicare-for-All Proposals, The Henry J. Kaiser Family Foundation Website, 2018. https://www.kff.org/health-costs/poll-finding/kaiser-health-tracking-poll-march-2018-prescription-drug-pricing-medicare-for-all-proposals/.

[38] W. Lazonick, M. Hopkins, K. Jacobson, M.E. Sakinç, Ö. Tulum, US pharma's financialized business model, 2017. https://www.ineteconomics.org/uploads/papers/WP_60-Lazonick-et-al-US-Pharma-Business-Model.pdf. Institute for New Economic Thinking Working Paper No. 60.

[39] W. Pride, O.C. Ferrell, B.A. Lukas, S. Schembri, O. Niininen, R. Cassidy, Marketing Principles, third Asia-Pacific ed., Cengage, 2018, p. 200.

[40] S. Madhavaram, S.D. Hunt, The service-dominant logic and a hierarchy of operant resources: developing masterful operant resources and implications for marketing strategy, J. Acad. Mark. Sci. 36 (1) (2008) 67–82.

[41] P.R. Dickson, J.L. Ginter, Market segmentation, product differentiation, and marketing strategy, J. Mark. 51 (2) (1987) 1.
[42] The Business Research Company, Global Biologics Opportunities and Strategies Market, 2022. http://www.Thebusinessresearchcompany.com/report.
[43] Global Biologics Market Analysis Report 2022-2025 & 2030 Featuring Merck & Co, AbbVie, F. Hoffmann-La Roche, Johnson & Johnson, & Pfizer, 2022. http://www.prmewswire.com/news-releases/global.
[44] https://straitsresearch.com/paper&packaging.

Index

Note: Page numbers followed by *f* indicate figures and *t* indicate tables.

R

S

T

U

V

Y

CPI Antony Rowe
Eastbourne, UK
August 04, 2023